Advances in Extraterrestrial Drilling

Advances in Extraterrestrial Drilling

Ground, Ice, and Underwater

VOLUME II

Advances in Terrestrial and Extraterrestrial Drilling:
Ground, Ice, and Underwater

Edited by
Yoseph Bar-Cohen
Kris Zacny

CRC Press
Taylor & Francis Group
Boca Raton London New York

CRC Press is an imprint of the
Taylor & Francis Group, an **informa** business

First edition published 2021
by CRC Press
6000 Broken Sound Parkway NW, Suite 300, Boca Raton, FL 33487-2742

and by CRC Press
2 Park Square, Milton Park, Abingdon, Oxon, OX14 4RN

CRC Press is an imprint of Taylor & Francis Group, LLC

Library of Congress Cataloging-in-Publication Data
Names: Bar-Cohen, Yoseph, editor. | Zacny, Kris, editor.
Title: Advances in terrestrial and extraterrestrial drilling : ground, ice,
and underwater / Yoseph Bar-Cohen, Kris Zacny.
Description: Boca Raton, FL : CRC Press/Taylor & Francis Group, LLC, 2021.
| Includes bibliographical references and index. | Summary: "This 2
volume set will include the latest principles behind the processes of
drilling and excavation on Earth and other planets. It will cover the
categories of drills, the history of drilling and excavation, various
drilling techniques and associated issues, including rock coring as well
as unconsolidated soil drilling and borehole stability"-- Provided by
publisher.
Identifiers: LCCN 2020037780 (print) | LCCN 2020037781 (ebook) | ISBN
9780367653460 (v. 1 ; hardback) | ISBN 9780367653477 (v. 2 ; hardback)
| ISBN 9780367674861 (v. 1 ; ebook) | ISBN 9781003131519 (v. 2 ; ebook)
Subjects: LCSH: Boring. | Space probes--Equipment and supplies. | Drilling
and boring machinery.
Classification: LCC TN281 .A234 2021 (print) | LCC TN281 (ebook) | DDC
622/.3381--dc23
LC record available at https://lccn.loc.gov/2020037780
LC ebook record available at https://lccn.loc.gov/2020037781

ISBN: 978-0-367-65347-7 (hbk)
ISBN: 978-1-003-13151-9 (ebk)

Typeset in Times
by SPi Global, India

Contents

Preface

Drilling can simply be defined as the action of penetrating into solid media. For terrestrial applications, drilling technology is well-established with a range of commercial tools that are readily available. There are also significant research and development efforts that enhance the drilling capability for much harsher conditions while reducing operational cost and increasing profit margins. Drilling or coring on other planets requires a highly detailed understanding of drilling processes and development of new capabilities for operation under extreme temperature, pressure, as well as low gravity. Some of these conditions are also applicable on Earth and will be key in reaching extreme depths to tap into new oil reservoirs, drilling geothermal wells in very hot rocks, and/or exploring ice layers in the Antarctic. This book has been compiled to cover the latest scope of knowledge in drilling as provided by leading scientists and engineers around the world.

We focused on drilling with emphasis on penetration of various subsurface materials, including rocks, permafrost, ice, soil, and regolith, to name a few. We have covered a range of mechanical and other drilling techniques as well as related issues including cuttings transportation and their disposal, borehole stability, the current and future levels of drilling autonomy. We have also covered the need for sample acquisition, caching and transport and restoration of *in situ* conditions necessary for eventual integrated science instruments and data interpretation. We describe the drilling process from basic science and the associated process of breaking and penetrating various media, and the required hardware and the process of excavation and analysis of the sampled media.

Chapter 1 covers Extraterrestrial Drilling. Chapter 2 covers Environmental Drilling/Sampling and Offshore Modeling Systems. Chapter 3 covers Scientific rationale for planetary drilling and Chapter 4 covers Biological Contamination Control and Planetary Protection Measures as Applied to Sample Acquisition. The latter is particularly important as we expand our *in-situ* exploration of other bodies in the Solar System and trying to avoid cross contamination and the related consequences.

Yoseph Bar-Cohen
Pasadena, CA

Kris Zacny
Pasadena, CA
July 2020

Acknowledgments

The editors would like to acknowledge and express their appreciation to the following individuals who took the time to review various book chapters. Their contributions are greatly appreciated and helped making this book of significantly greater value to readers. The individuals who served as reviewers of chapters in this book are as follows:

Chapter 1
Rohit Bhartia, Photon Systems Inc., Covina, CA
Dara Sabahi from Honeybee Robotics, Altadena, CA
Yoseph Bar-Cohen, Jet Propulsion Laboratory/California Institute of Technology, Pasadena, CA
Stewart Sherrit, Jet Propulsion Laboratory/California Institute of Technology, Pasadena, CA

Chapter 2
Rohit Bhartia, Photon Systems Inc., Covina, CA
Mircea Badescu, Jet Propulsion Laboratory/California Institute of Technology, Pasadena, CA
Dara Sabahi, Honeybee Robotics, Altadena, CA

Chapter 3
Alfred William (Bill) Eustes III, Petroleum Engineering Department at the Colorado School of Mines, Golden, CO
Yang Liu, Jet Propulsion Laboratory/California Institute of Technology, Pasadena, CA

Chapter 4
Andy Spry, NASA Planetary Protection Officer, NASA Headquarters, Washington, DC
Elaine Seasley, Deputy Planetary Protection Officer NASA Headquarters, Washington, DC

Editors

Dr. Yoseph Bar-Cohen is the Supervisor of the Electroactive Technologies Group (http://ndeaa. jpl.nasa.gov/) and a Senior Research Scientist at the Jet Propulsion Lab/Caltech, Pasadena, CA. In 1979, he received his PhD in Physics from the Hebrew University, Jerusalem, Israel. His research is focused on electro-mechanics including planetary sample handling mechanisms, novel actuators that are driven by such materials as piezoelectric and EAP (also known as artificial muscles), and biomimetics.

Dr. Kris Zacny is a Senior Scientist and Vice President of Exploration Systems at Honeybee Robotics, Altadena, CA. His expertise includes space mining, sample handling, soil and rock mechanics, extraterrestrial drilling, and In Situ Resource Utilization (ISRU). Dr. Zacny received his PhD from UC Berkeley (2005) in Geotechnical Engineering with focus on planetary drilling and space mining, ME from UC Berkeley (2001) in Petroleum Engineering, and BSc cum laude from University of Cape Town (1997) in Mechanical Engineering. He spent several years working in South African mines and tested space drills in Antarctica, Arctic, Greenland, and the Atacama.

Contributors

Grayson Adams
Honeybee Robotics Spacecraft Mechanisms
    Corporation
Altadena, California

Leslie Alarid
Honeybee Robotics Spacecraft Mechanisms
    Corporation
Altadena, California

Michael Amato
NASA Goddard Space Flight Center
Greenbelt, Maryland

Colin Andrew
Honeybee Robotics Spacecraft Mechanisms
    Corporation
Altadena, California

Shigeru Aoki
Shimizu Corporation
Chuo City, Tokyo, Japan

Mircea Badescu
Jet Propulsion Laboratory
California Institute of Technology (Caltech)
Pasadena, California

Pietro Baglioni
European Space Agency, ESA-ESTEC
Noordwijk, The Netherlands

Jameil Bailey
Honeybee Robotics Spacecraft Mechanisms
    Corporation
Altadena, California

Marek Banaszkiewicz
Space Research Centre PAS
Warsaw, Poland

Xiaoqi Bao
Jet Propulsion Laboratory
California Institute of Technology (Caltech)
Pasadena, California

Yoseph Bar-Cohen
Jet Propulsion Laboratory
California Institute of Technology (Caltech)
Pasadena, California

Phil Beard
Honeybee Robotics Spacecraft Mechanisms
    Corporation
Altadena, California

L. W. Beegle
Jet Propulsion Laboratory
California Institute of Technology (Caltech)
Pasadena, California

James N. Benardini
Jet Propulsion Laboratory
California Institute of Technology (Caltech)
Pasadena, California

Dean Bergman
Honeybee Robotics Spacecraft Mechanisms
    Corporation
Altadena, California

Jocelyn Bergman
Honeybee Robotics Spacecraft Mechanisms
    Corporation
Altadena, California

Rohit Bhartia
Optical Informatics, LLC.
Altadena, California

Andrew Bocklund
Honeybee Robotics Spacecraft Mechanisms
    Corporation
Altadena, California

Natasha Bouey
Honeybee Robotics Spacecraft Mechanisms
    Corporation
Altadena, California

Ben Bradley
Honeybee Robotics Spacecraft Mechanisms
    Corporation
Altadena, California

Michael Buchbinder
Honeybee Robotics Spacecraft Mechanisms
    Corporation
Altadena, California

Lee Carlson
Honeybee Robotics Spacecraft Mechanisms
    Corporation
Altadena, California

Conner Castle
Honeybee Robotics Spacecraft Mechanisms
    Corporation
Altadena, California

C. M. Caudill
Jet Propulsion Laboratory
California Institute of Technology (Caltech)
Pasadena, California

Colin Chen
Honeybee Robotics Spacecraft Mechanisms
    Corporation
Altadena, California

Paul Chow
Honeybee Robotics Spacecraft Mechanisms
    Corporation
Altadena, California

Phil Chu
Honeybee Robotics Spacecraft Mechanisms
    Corporation
Altadena, California

Evan Cloninger
Honeybee Robotics Spacecraft Mechanisms
    Corporation
Altadena, California

E. Cloutis
Jet Propulsion Laboratory
California Institute of Technology (Caltech)
Pasadena, California

Patrick Corrigan
Honeybee Robotics Spacecraft Mechanisms
    Corporation
Altadena, California

Tighe Costa
Honeybee Robotics Spacecraft Mechanisms
    Corporation
Altadena, California

Paul Creekmore
Honeybee Robotics Spacecraft Mechanisms
    Corporation
Altadena, California

Tom Cwik
Jet Propulsion Laboratory
California Institute of Technology (Caltech)
Pasadena, California

Kiel Davis
Honeybee Robotics Spacecraft Mechanisms
    Corporation
Altadena, California

J. Dickson
Jet Propulsion Laboratory
California Institute of Technology (Caltech)
Pasadena, California

Stephen Durrant
European Space Agency, ESA-ESTEC
Noordwijk, The Netherlands

Jack Emery
Honeybee Robotics Spacecraft Mechanisms
    Corporation
Altadena, California

Richard Fisackerly
European Space Agency, ESA-ESTEC
Noordwijk, The Netherlands

Zak Fitzgerald
Honeybee Robotics Spacecraft Mechanisms
    Corporation
Altadena, California

Chris Flesher
Stone Aerospace
Del Valle, Texas

Jean-Pierre Fleurial
Jet Propulsion Laboratory
California Institute of Technology (Caltech)
Pasadena, California

Steve Ford
Honeybee Robotics Spacecraft Mechanisms
   Corporation
Altadena, California

Masaki Fujimoto
Japan Aerospace Exploration Agency (JAXA)
Chuo-ku, Tokyo, Japan

Akihiro Fujiwara
Faculty of Science and Engineering
Chuo University
Hachiōji, Tokyo, Japan

Sam Goldman
Honeybee Robotics Spacecraft Mechanisms
   Corporation
Altadena, California

Stephen Gorevan
Honeybee Robotics Spacecraft Mechanisms
   Corporation
Altadena, California

Amelia Grossman
Honeybee Robotics Spacecraft Mechanisms
   Corporation
Altadena, California

Matthias Grott
German Aerospace Center (DLR)
Institute of Planetary Research
Berlin, Germany

Andrej Grubisic
NASA Goddard Space Flight Center
Greenbelt, Maryland

Jerzy Grygorczuk
Astronika Sp. z o. o.
Warsaw, Poland

Corey Hackley
Stone Aerospace
Del Valle, Texas

Jeffery Hall
Jet Propulsion Laboratory
California Institute of Technology (Caltech)
Pasadena, California

Ashley Hames
Honeybee Robotics Spacecraft Mechanisms
   Corporation
Altadena, California

Kevin Hand
Jet Propulsion Laboratory
California Institute of Technology
   (Caltech)
Pasadena, California

Patrick Harkness
School of Engineering
University of Glasgow
Glasgow, UK

John Harman
Stone Aerospace
Del Valle, Texas

Matt Heltsley
Honeybee Robotics Spacecraft Mechanisms
   Corporation
Altadena, California

Jason Herman
Honeybee Robotics Spacecraft Mechanisms
   Corporation
Altadena, California

Joe Hernandez
Honeybee Robotics Spacecraft Mechanisms
   Corporation
Altadena, California

P. Hill
Jet Propulsion Laboratory
California Institute of Technology (Caltech)
Pasadena, California

Ben Hockman
Jet Propulsion Laboratory
California Institute of Technology (Caltech)
Pasadena, California

Bart Hogan
Stone Aerospace
Del Valle, Texas

Will Hovik
Honeybee Robotics Spacecraft Mechanisms
   Corporation
Altadena, California

Samuel M. Howell
Jet Propulsion Laboratory
California Institute of Technology (Caltech)
Pasadena, California

Robert Huddleston
Honeybee Robotics Spacecraft Mechanisms
   Corporation
Altadena, California

Troy Lee Hudson
Jet Propulsion Laboratory
California Institute of Technology (Caltech)
Pasadena, California

Kevin Humphrey
Honeybee Robotics Spacecraft Mechanisms
   Corporation
Altadena, California

Stephen Indyk
Honeybee Robotics Spacecraft Mechanisms
   Corporation
Altadena, California

Shannon Jackson
Jet Propulsion Laboratory
California Institute of Technology (Caltech)
Pasadena, California

Anchal Jain
Honeybee Robotics Spacecraft Mechanisms
   Corporation
Altadena, California

Nathan Jensen
Honeybee Robotics Spacecraft Mechanisms
   Corporation
Altadena, California

Helen Jung
Honeybee Robotics Spacecraft Mechanisms
   Corporation
Altadena, California

Hiroshi Kanamori
Shimizu Corporation
Chuo City, Tokyo, Japan

Robert Kancans
Honeybee Robotics Spacecraft Mechanisms
   Corporation
Altadena, California

Hiroki Kato
Japan Aerospace Exploration Agency (JAXA)
Chuo-ku, Tokyo, Japan

Hiroyuki Kawamoto
Department of Applied Mechanics and
   Aerospace Engineering
Waseda University
Shinjuku, Tokyo, Japan

Bartosz Kędziora
Astronika Sp. z o. o.
Warsaw, Poland

Cecily Keim
Honeybee Robotics Spacecraft Mechanisms
   Corporation
Altadena, California

Sarineh Keshish
Honeybee Robotics Spacecraft Mechanisms
   Corporation
Altadena, California

Isabel King
Honeybee Robotics Spacecraft Mechanisms
   Corporation
Altadena, California

Christian Krause
German Aerospace Center (DLR)
Institute of Planetary Research
Berlin, Germany

Takashi Kubota
Japan Aerospace Exploration Agency (JAXA)
Chuo-ku, Tokyo, Japan

Tomasz Kuciński
Astronika Sp. z o. o.
Warsaw, Poland

Sherman Lam
Honeybee Robotics Spacecraft Mechanisms
　Corporation
Altadena, California

Andrea Lamore
Honeybee Robotics Spacecraft Mechanisms
　Corporation
Altadena, California

Caleb Lang
Honeybee Robotics Spacecraft Mechanisms
　Corporation
Altadena, California

Hyeong Jae Lee
Jet Propulsion Laboratory
California Institute of Technology (Caltech)
Pasadena, California

Scott Lelievre
Stone Aerospace
Del Valle, Texas

Roy Lichtenheldt
German Aerospace Center (DLR)
Institute of Planetary Research
Berlin, Germany

Alberto Lopez
Stone Aerospace
Del Valle, Texas

Ralph Lorenz
The Johns Hopkins University Applied Physics
　Laboratory
Laurel, Maryland

Jacob Madden
Honeybee Robotics Spacecraft Mechanisms
　Corporation
Altadena, California

Jessica Maddin
Honeybee Robotics Spacecraft Mechanisms
　Corporation
Altadena, California

Tibor Makai
Honeybee Robotics Spacecraft Mechanisms
　Corporation
Altadena, California

Michael Malaska
Jet Propulsion Laboratory
California Institute of Technology (Caltech)
Pasadena, California

Zach Mank
Honeybee Robotics Spacecraft Mechanisms
　Corporation
Altadena, California

Richard Margulieux
Honeybee Robotics Spacecraft Mechanisms
　Corporation
Altadena, California

Sara Martinez
Honeybee Robotics Spacecraft Mechanisms
　Corporation
Altadena, California

Yuka Matsuyama
Honeybee Robotics Spacecraft Mechanisms
　Corporation
Altadena, California

Andrew Maurer
Honeybee Robotics Spacecraft Mechanisms
　Corporation
Altadena, California

Molly McCormick
Honeybee Robotics Spacecraft Mechanisms
　Corporation
Altadena, California

Boleslaw Mellerowicz
Honeybee Robotics Spacecraft Mechanisms
    Corporation
Altadena, California

Brandon Metz
Jet Propulsion Laboratory
California Institute of Technology (Caltech)
Pasadena, California

Jerry Moreland
Honeybee Robotics Spacecraft Mechanisms
    Corporation
Altadena, California

Scott Moreland
Jet Propulsion Laboratory
California Institute of Technology (Caltech)
Pasadena, California

Phil Morrison
Honeybee Robotics Spacecraft Mechanisms
    Corporation
Altadena, California

Jurgen Mueller
Jet Propulsion Laboratory
California Institute of Technology (Caltech)
Pasadena, California

Robert Mulvaney
British Antarctic Survey
Cambridge, UK

Erik Mumm
Honeybee Robotics Spacecraft Mechanisms
    Corporation
Altadena, California

Krista Myers
Stone Aerospace
Del Valle, Texas

Seiichi Nagihara
Department of Geosciences
Texas Tech University
Lubbock, Texas

Taro Nakamura
Faculty of Science and Engineering
Chuo University
Hachiōji, Tokyo, Japan

Adoni Netter
Honeybee Robotics Spacecraft Mechanisms
    Corporation
Altadena, California

Phil Ng
Honeybee Robotics Spacecraft Mechanisms
    Corporation
Altadena, California

Peter Ngo
Honeybee Robotics Spacecraft Mechanisms
    Corporation
Altadena, California

Huey Nguyen
Honeybee Robotics Spacecraft Mechanisms
    Corporation
Altadena, California

Tyler Okamoto
Jet Propulsion Laboratory
California Institute of Technology (Caltech)
Pasadena, California

Avi Okon
Jet Propulsion Laboratory
California Institute of Technology (Caltech)
Pasadena, California

Joey Palmowski
Honeybee Robotics Spacecraft Mechanisms
    Corporation
Altadena, California

Christian Panza
Leonardo S.p.A
Nerviano (MI), Italy

Aayush Parekh
Honeybee Robotics Spacecraft Mechanisms
    Corporation
Altadena, California

Gale Paulsen
Honeybee Robotics Spacecraft Mechanisms
    Corporation
Altadena, California

Marco Peruzzotti
Leonardo S.p.A
Nerviano (MI), Italy

Fredrik Rehnmark
Honeybee Robotics Spacecraft Mechanisms
    Corporation
Altadena, California

Dario Riccobono
Jet Propulsion Laboratory
California Institute of Technology (Caltech)
Pasadena, California

Kristof Richmond
Stone Aerospace
Del Valle, Texas

Hunter Rideout
Honeybee Robotics Spacecraft Mechanisms
    Corporation
Altadena, California

Albert Ridilla
Honeybee Robotics Spacecraft Mechanisms
    Corporation
Altadena, California

Julius Rix
British Antarctic Survey
Cambridge, UK

Andrea Rusconi
Leonardo S.p.A
Nerviano (MI), Italy

Alexandra Rzepiejewska
Honeybee Robotics Spacecraft Mechanisms
    Corporation
Altadena, California

Dara Sabahi
Honeybee Robotics Spacecraft Mechanisms
    Corporation
Altadena, California

Luke Sanasarian
Honeybee Robotics Spacecraft Mechanisms
    Corporation
Altadena, California

Vishnu Sanigepalli
Honeybee Robotics Spacecraft Mechanisms
    Corporation
Altadena, California

H. M. Sapers
Jet Propulsion Laboratory
California Institute of Technology (Caltech)
Pasadena, California

Yasutaka Satou
Japan Aerospace Exploration Agency (JAXA)
Chuo-ku, Tokyo, Japan

Hirotaka Sawada
Japan Aerospace Exploration Agency (JAXA)
Chuo-ku, Tokyo, Japan

Karol Seweryn
Space Research Centre PAS
Warsaw, Poland

Jeff Shasho
Honeybee Robotics Spacecraft Mechanisms
    Corporation
Altadena, California

Kris Sherrill
Jet Propulsion Laboratory
California Institute of Technology (Caltech)
Pasadena, California

Stewart Sherrit
Jet Propulsion Laboratory
California Institute of Technology (Caltech)
Pasadena, California

Vickie Siegel
Stone Aerospace
Del Valle, Texas

Miles Smith
Jet Propulsion Laboratory
California Institute of Technology (Caltech)
Pasadena, California

Walter F. Smith
NASA Goddard Space Flight Center
Greenbelt, Maryland

David Smyth
Honeybee Robotics Spacecraft Mechanisms
    Corporation
Altadena, California

Pablo Sobron
Impossible Sensing
St. Louis, Missouri

Nancy Sohm
Honeybee Robotics Spacecraft Mechanisms
    Corporation
Altadena, California

Jesus Sosa
Honeybee Robotics Spacecraft Mechanisms
    Corporation
Altadena, California

Joey Sparta
Honeybee Robotics Spacecraft Mechanisms
    Corporation
Altadena, California

Tilman Spohn
German Aerospace Center (DLR)
Institute of Planetary Research
Berlin, Germany

Justin Spring
Honeybee Robotics Spacecraft Mechanisms
    Corporation
Altadena, California

Leo Stolov
Honeybee Robotics Spacecraft Mechanisms
    Corporation
Altadena, California

William Stone
Stone Aerospace
Del Valle, Texas

Moogega Stricker
Jet Propulsion Laboratory
California Institute of Technology (Caltech)
Pasadena, California

Miranda Tanouye
Honeybee Robotics Spacecraft Mechanisms
    Corporation
Altadena, California

Lisa Thomas
Honeybee Robotics Spacecraft Mechanisms
    Corporation
Altadena, California

Thomas Thomas
Honeybee Robotics Spacecraft Mechanisms
    Corporation
Altadena, California

Luke Thompson
Honeybee Robotics Spacecraft Mechanisms
    Corporation
Altadena, California

Ryan Timoney
School of Engineering
University of Glasgow
Glasgow, UK

Nick Traeden
Honeybee Robotics Spacecraft Mechanisms
    Corporation
Altadena, California

Melissa Trainer
NASA Goddard Space Flight Center
Greenbelt, Maryland

Ethan Tram
Honeybee Robotics Spacecraft Mechanisms
    Corporation
Altadena, California

Roland Trautner
European Space Agency, ESA-ESTEC
Noordwijk, The Netherlands

Elizabeth Turtle
The Johns Hopkins University Applied Physics
    Laboratory
Laurel, Maryland

Crystal Ulloa
Honeybee Robotics Spacecraft Mechanisms
    Corporation
Altadena, California

Naohiro Uyama
Shimizu Corporation
Chuo City, Tokyo, Japan

Dylan Van-Dyne
Honeybee Robotics Spacecraft Mechanisms
    Corporation
Altadena, California

Vincent Vendiola
Honeybee Robotics Spacecraft Mechanisms
    Corporation
Altadena, California

Kasthuri J. Venkateswaran
Jet Propulsion Laboratory
California Institute of Technology (Caltech)
Pasadena, California

Alex Wang
Honeybee Robotics Spacecraft Mechanisms
    Corporation
Altadena, California

Lillian Ware
Honeybee Robotics Spacecraft Mechanisms
    Corporation
Altadena, California

Gordon Wasilewski
Astronika Sp. z o. o.
Warsaw, Poland

Don Wegel
NASA Goddard Space Flight Center
Greenbelt, Maryland

Bobby Wei
Honeybee Robotics Spacecraft Mechanisms
    Corporation
Altadena, California

Hunter Williams
Honeybee Robotics Spacecraft Mechanisms
    Corporation
Altadena, California

Jack Wilson
Honeybee Robotics Spacecraft Mechanisms
    Corporation
Altadena, California

Dale Winebrenner
University of Washington
Seattle, Washington

Torben Wippermann
German Aerospace Center (DLR)
Institute of Planetary Research
Berlin, Germany

Łukasz Wiśniewski
Astronika Sp. z o. o.
Warsaw, Poland

Kevin Worrall
School of Engineering
University of Glasgow
Glasgow, UK

Nathan Wright
Stone Aerospace
Del Valle, Texas

Yasuyuki Yamada
Department of Design Engineering and
    Technology
Tokyo Denki University
Adachi-ku, Tokyo, Japan

Bernice Yen
Honeybee Robotics Spacecraft Mechanisms
    Corporation
Altadena, California

Ben Younes
Honeybee Robotics Spacecraft Mechanisms
    Corporation
Altadena, California

David Yu
Honeybee Robotics Spacecraft Mechanisms
    Corporation
Altadena, California

Kris Zacny
Honeybee Robotics Spacecraft Mechanisms
    Corporation
Altadena, California

Mike Zasadzien
Honeybee Robotics Spacecraft Mechanisms
    Corporation
Altadena, California

Raymond Zheng
Honeybee Robotics Spacecraft Mechanisms
    Corporation
Altadena, California

Wayne Zimmerman
Jet Propulsion Laboratory
California Institute of Technology (Caltech)
Pasadena, California

Paul Backes
Jet Propulsion Laboratory
California Institute of Technology (Caltech)
Pasadena, CA

1 Extraterrestrial Drilling and Excavation

Kris Zacny, Gale Paulsen, Phil Chu, Boleslaw Mellerowicz, Stephen Indyk, Justin Spring, Alex Wang, Grayson Adams, Leslie Alarid, Colin Andrew, Jameil Bailey, Ron Barkie, Dean Bergman, Jocelyn Bergman, Phil Beard, Andrew Bocklund, Natasha Bouey, Ben Bradley, Michael Buchbinder, Kathryn Bywaters, Lee Carlson, Conner Castle, Mark Chapman, Colin Chen, Paul Chow, Evan Cloninger, Patrick Corrigan, Tighe Costa, Paul Creekmore, Kiel Davis, Stella Dearing, Jack Emery, Zak Fitzgerald, Steve Ford, Sam Goldman, Barry Goldstein, Stephen Gorevan, Amelia Grossman, Ashley Hames, Nathan Heidt, Ron Hayes, Matt Heltsley, Jason Herman, Joe Hernandez, Greg Hix, Will Hovik, Robert Huddleston, Kevin Humphrey, Anchal Jain, Nathan Jensen, Marnie Johnson, Helen Jung, Robert Kancans, Cecily Keim, Sarineh Keshish, Michael Killian, Caitlin King, Isabel King, Daniel Kim, Emily Kolenbrander, Sherman Lam, Andrea Lamore, Caleb Lang, Joseph Lee, Carolyn Lee, John Lorbiecki, Kathryn Luczek, Jacob Madden, Jessica Maddin, Tibor Makai, Mike Maksymuk, Zach Mank, Richard Margulieux, Sara Martinez, Yuka Matsuyama, Andrew Maurer, Molly McCormick, Jerry Moreland, Phil Morrison, Erik Mumm, Adoni Netter, Jeff Neumeister, Tim Newbold, Joey Niehay, Phil Ng, Peter Ngo, Huey Nguyen, Tom O'Bannon, Sean O'Brien, Joey Palmowski, Aayush Parekh, Andrew Peekema, Fredrik Rehnmark, Hunter Rideout, Albert Ridilla, Alexandra Rzepiejewska, Dara Sabahi, Yoni Saltzman, Luke Sanasarian, Vishnu Sanigepalli, Emily Seto, Jeff Shasho, Sase Singh, David Smyth, Nancy Sohm, Jesus Sosa, Joey Sparta, Leo Stolov, Marta Stone, Andrew Tallaksen, Miranda Tanouye, Lisa Thomas, Thomas Thomas, Luke Thompson, Mary Tirrell, Nick Traeden, Ethan Tram, Sarah Tye, Crystal Ulloa, Dylan Van-Dyne, Robert Van Ness, Vincent Vendiola, Brian Vogel, Lillian Ware, Bobby Wei, Hunter Williams, Jack Wilson, Brian Yaggi, Bernice Yen, Sean Yoon, Ben Younes, David Yu,

Michael Yu, Mike Zasadzien, and Raymond Zheng
Honeybee Robotics Spacecraft Mechanisms Corp., Altadena, CA

*Yoseph Bar-Cohen, Paul Backes, Mircea Badescu,
Xiaoqi Bao, Tom Cwik, Jean-Pierre Fleurial, Jeffery Hall,
Kevin Hand, Ben Hockman, Samuel M. Howell,
Troy Lee Hudson, Shannon Jackson, Hyeong Jae Lee,
Michael Malaska, Brandon Metz, Scott Moreland,
Avi Okon, Tyler Okamoto, Dario Riccobono, Kris Sherrill,
Stewart Sherrit, Miles Smith, Jurgen Mueller, and
Wayne Zimmerman*
Jet Propulsion Laboratory (JPL)/California Institute of Technology
(Caltech), Pasadena, CA

*Michael Amato, Melissa Trainer, Don Wegel, Andrej
Grubisic, and Walter F. Smith*
NASA Goddard Space Flight Center,
Greenbelt, MD

Ralph Lorenz, and Elizabeth Turtle
The Johns Hopkins University Applied Physics Laboratory,
Laurel, MD

*Hirotaka Sawada, Hiroki Kato, Yasutaka Satou,
Takashi Kubota, and Masaki Fujimoto*
Japan Aerospace Exploration Agency (JAXA),
Sagamihara, Japan

*Pietro Baglioni, Stephen Durrant,
Richard Fisackerly, and Roland Trautner*
European Space Agency, Noordwijk, the Netherlands

Marek Banaszkiewicz, and Karol Seweryn
Space Research Centre PAS (CBK PAN),
Warsaw, Poland

Akihiro Fujiwara, and Taro Nakamura
Faculty of Science and Engineering, Chuo University,
Tokyo, Japan

Matthias Grott
German Aerospace Center (DLR), Institute of Planetary Research,
Berlin, Germany

*Jerzy Grygorczuk, Bartosz Kędziora, Łukasz Wiśniewski,
Tomasz Kuciński, and Gordon Wasilewski*
Astronika Sp. z o. o., Warsaw, Poland

Seiichi Nagihara
Department of Geosciences, Texas Tech University, Lubbock, TX

Rohit Bhartia
Optical Informatics, LLC, Altadena, CA

Hiroyuki Kawamoto
Dept. of Applied Mechanics and Aerospace Engineering,
Waseda University, Tokyo, Japan

Julius Rix, and Robert Mulvaney
British Antarctic Survey, High Cross, Cambridge, UK

Andrea Rusconi, Christian Panza, and Marco Peruzzotti
Leonardo S.p.A., Viale Europa s.n.c, Nerviano (MI), Italy

Pablo Sobron
Impossible Sensing, St. Louis, MO

Ryan Timoney, Kevin Worrall, and Patrick Harkness
School of Engineering, University of Glasgow,
Glasgow, Scotland

Naohiro Uyama, Hiroshi Kanamori, and Shigeru Aoki
Shimizu Corporation,
Japan, Tokyo

Dale Winebrenner
University of Washington,
Seattle, WA

Yasuyuki Yamada
Department of Design Engineering and Technology,
Tokyo Denki University, Japan

Tilman Spohn
International Space Science Institute (ISSI), Bern, Switzerland

Christian Krause
German Aerospace Center (DLR), Microgravity User Support Center
(MUSC), Space Operations and Astronaut Training,
Cologne, Germany

Torben Wippermann
German Aerospace Center (DLR), Institute of Space Systems,
Bremen, Germany

Roy Lichtenheldt
Institute of System Dynamics and Control, Münchener Straße 20, 82234,
Weßling, Germany

CONTENTS

1.1 WHY SUBSURFACE EXPLORATION?

The "easy" science related to in situ exploration can be claimed as been done already. The scientific questions posed by upcoming missions are becoming increasingly more difficult. Whether it's search for life (Beegle et al. 2007; McKay et al. 2013), trying to answer questions on planetary climate (Smith et al. 2020), geologic evolution and innerworkings of planetary bodies, or extract resources for In Situ Resource Utilization (ISRU) (Sanders et al. 2011). These challenges are becoming more difficult to address and requiring exploration in 3D (3rd dimension being depth). The 3D exploration in fact leads to the 4th dimension—time. Going below the surface means going back in time. Just like on Earth, ice cores from Antarctica provide a glimpse into climate history on Earth, ice cores on Mars would do the same.

There are two aspects to subsurface exploration of extraterrestrial bodies. The one is to bring subsurface materials to the surface for scientific analysis or practical use. The other is to install (or deploy) scientific instruments to the subsurface. With respect to NASA goals to "search for life," samples from greater depths are required since those samples would more likely be protected from deadly radiation. Penetrating subglacial oceans and lakes on Europa would significantly increase the probability of finding extraterrestrial life. Understanding the climate history of Mars requires penetrating tens of meters into the polar layer deposits. Mining water for In Situ Resource Utilization requires digging at least a meter on the Moon, and deeper into the surface of Mars in mid-latitudes (locations best suited for human settlement).

Over the past few decades there have been numerous sample acquisition systems developed for space exploration. Apart from the few instances where the spacecrafts and missions were the same (e.g. twin Mars Exploration Rovers, twin Viking landers, spacecrafts in the Venera program and so on), every sampling system has been a very different design. The main reason for the one-of-a-kind development was because the science requirements, mission requirements, and environmental constraints have been different in each of these missions.

This chapter presents numerous sampling systems developed to date, ranging from scoops, to near surface drills, all the way to ultra-deep drills. The presented examples are by no means all-inclusive but rather represent a selection of planetary sampling systems developed to date.

1.2 SCOOPS

Scoops are highly versatile sample acquisition systems. They are typically deployed from robotic arms and are used to capture surface granular regolith. Scoops are required to deliver samples to various instruments and as such, those instruments need to be placed within robotic arm's reach. That requirement sometimes significantly constrains placement of a robotic arm and instruments.

Scoops are not normally used for metering out samples hence intermediate portioning hardware may need to be incorporated to do this. For example, a scoop could deliver a sample to a cup and fill the cup completely before the cup is moved to an instrument for analysis.

Scoops unfortunately cannot penetrate competent material such as ice, icy soil, and rocks. In addition, if the soil is crusty (dried clayey soil) a scoop may have a hard time penetrating it as well. In these instances, a percussive scoop could be used to cut into material; however, the resultant broken-up sample will come in chunks as opposed to fine powder. Hence sample post processing may be required to reduce particle sizes to levels that are manageable by the instrument.

Scoops were deployed on the Lunar Surveyor, Mars Viking, Mars Phoenix, and Mars Curiosity missions, to name a few. In the following sections we provide additional information about selected missions.

1.2.1 VIKING 1 AND 2

The first excavators deployed on Mars were the scoops on the Viking 1 and 2 landers (Holmberg et al. 1980). The purpose of the scoop, called the Viking Surface Sampler Assembly (SSA) and shown in Figure 1.1, was to acquire, process, and distribute samples to various instruments

FIGURE 1.1 Viking lander with the Viking Surface Sampler Assembly (SSA). SSA was used to dig into Martian surface, capture and deliver material into instruments. *Source:* Courtesy of NASA.

(Seger and Gillespie 1974). The sampler consisted of a 3-m-long rolled-up tubular boom with a collector head at its end. The extendable/retractable boom was unlike traditional robotic arms with joints—it was akin to a tape measure. This significantly reduced volume and mass, as well as power. The boom combined with the integrated azimuth/elevation gimbal allowed the collector head to be placed at any location within the articulation limits of the boom. The collector head, with its Solenoid-operated lid, backhoe, and 108° rotation capability, was designed to acquire samples from a variety of potential surface materials and to deliver original granular material or 2000-µm sieved soil to the deck-mounted instruments.

Additional soil processing mechanisms were integrated within the individual instruments. For the Gas Chromatograph Mass Spectrometer (GCMS), the sample was sieved through a 2000-micron sieve, crushed to less than 600 microns, sieved again through a 300-micron sieve before metering out into a 1 cm³ volume. For the biology instrument, the sample was sieved through a 1500-micron sieve and then metered out into a 7 cm³ volume.

1.2.2 Mars Phoenix

The 2007 Phoenix mission was the first mission to land at the Northern Polar Regions of Mars. The lander carried two science instruments that needed Martian regolith samples: the Thermal Evolved Gas Analyzer (TEGA) and the Microscopy, Electrochemistry, and Conductivity Analyzer (MECA). To provide needed samples, the lander included a 4-degrees-of-freedom (DOF) robotic arm with a scoop mounted at the end (Figure 1.2). It was expected that the material at this latitude would include solid ice, and as such the scoop, called the Icy Soil Acquisition Device (ISAD), included a tungsten carbide drill called the Rasp (Chu et al. 2008). The purpose of the Rasp was to drill into ice and icy regolith and create chips.

During the nominal mission operations, the scoop was used to remove the layer of loose regolith and expose ice-bearing material underneath it. The flat-edged blade, attached below the scoop, was then used to scrape away icy regolith and level the surface. Once the surface was prepared, the robotic arm preloaded the scoop against the surface with up to 40 N of force and allowed the Rasp to cut into ice. The Rasp drilling operation would normally take 30–60 s, and during this time, icy chips would ballistically fall to the back of the scoop. The scoop would then rotate several times to transfer the ice chips from the back to the front.

Delivery of dry material was relatively easy. However, during delivery of icy soil, an off-nominal operation was observed. The nominal operational sequence was relatively straightforward: excavate several holes with the Rasp and capture icy chips, position the scoop over the instrument inlet port, and finally rotate and dump the material into the inlet port. The first sample was from the overburden

FIGURE 1.2 Components of the Mars Phoenix 2007 Icy Soil Acquisition Device. *Source:* Courtesy of NASA.

(ice-free) layer. It was successfully delivered to the Thermal and Evolved Gas Analyzer (TEGA) port with a screen mesh on top. The sample, however, got stuck on top of the mesh for a few Sols (Martian days) even when extensive vibration of the screen was employed. During the subsequent acquisition and delivery of an icy soil sample from the Snow-White trench, some of the samples that were successfully collected in the scoop would not fall out (Bonitz et al. 2009). Given the low temperature (−100°C) and a low-pressure environment (7 Torr), the expectation was that any ice would remain stable. However, the faint Sun was able to warm some of the ice inside the scoop. The resulting increase in material temperature, coupled to the unexpected antifreeze salts in the soil, was sufficient to thaw the ice. When the Sun angle dropped, the material refroze and adhered to the scoop, a phenomenon we are very familiar with on Earth. Fortunately, for Phoenix, the ISAD had a "percussive mode"; the Rasp created rattling which helped dislodge the adhered material from the scoop. Moving forward, the team resolved this problem operationally by sampling early in the morning and as quickly and efficiently as possible to limit solar heating of the scoop.

1.2.3 VIBRATORY AND PERCUSSIVE SCOOPS

The purpose of vibratory and percussive scoops is to reduce excavation forces during material acquisition and to allow easier discharge of material during material gravity transfer—something that was witnessed during operation of the Mars Phoenix scoop (Zacny et al. 2009). Vibration (as opposed to percussion) occurs when the motion is in a particular direction without any impacts. This is akin to a sonic toothbrush. Percussion occurs when there is an impact onto the blade—e.g. jack-hammers that are used to break up concrete. Using percussive scoops is not new; some commercial hammer drills include a scoop as an attachment. To use these scoops, the hammer drill is switched to hammer-only mode.

Figure 1.3 shows experimental setup for testing of a vibratory scoop (Zacny and Spring 2013). It includes a 4-DOF robotic arm with a vibratory scoop at the end. The scoop used a simple brushless DC motor to power an offset mass by way of a pair of helical gears. Two slightly different designs were used, whereby the plane of vibration of the offset mass was oriented differently relative to the direction of scooping. Figure 1.4 shows a model of the two scoop designs and test data. The scoop used for these excavation tests is similar to the scoop used by the Lunar Surveyor mission. Using similar geometry facilitates more meaningful comparison with the Surveyor data, as well as with

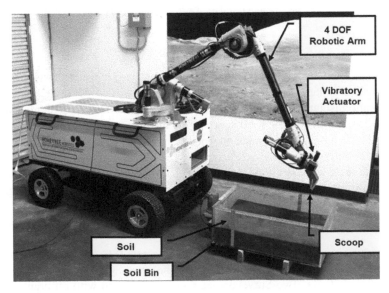

FIGURE 1.3 A Surveyor-style vibratory scoop mounted to a robotic arm on Honeybee's mobile test platform.

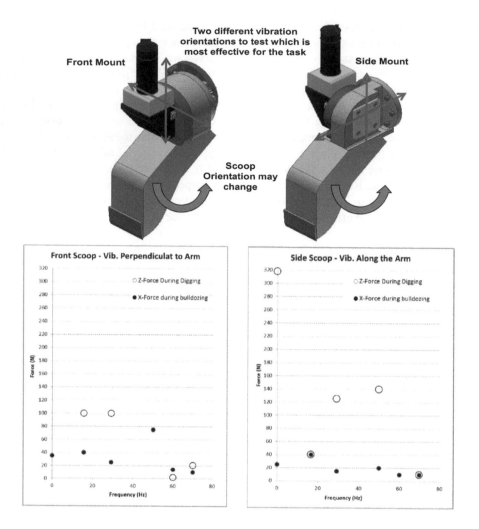

FIGURE 1.4 Vibratory scoop designs and test data. *Source:* Courtesy Honeybee Robotics.

tests performed by others using the same or similar geometry. Excavation tests were performed in JSC-1, a lunar mare simulant.

A simple vibration isolation system allowed the scoop to vibrate during digging while minimizing the vibration transmitted to the robotic arm. This mechanism used an off-the-shelf linear ball bearing slide and a set of springs to isolate the vibration. The springs were changed out to allow for different stiffnesses during testing.

The excavation tests were straightforward in their procedure. The scoop was pushed into the soil bin, and the robotic arm was used to drag the scoop across the soil in a straight move. This dragging motion produced negligible vertical and side motion. The 6-axis force torque sensor mounted between the arm and the scoop was used to collect force and torque loads at the tool/arm interface.

A number of tests were performed in vibrator compacted JSC-1a. The penetration depth was 30 mm and penetration rate (along the horizontal plane) was 20 mm/s. Vibration frequency was varied between 0 and 70 Hz. Two Surveyor-size scoops were used with different orientations: the Front Scoop (because the motor was mounted in front) produces vibrations orthogonal to the plane of the robotic arm. Side Scoop (because the motor is mounted on a side of the scoop) produces vibrations in the plane of the robotic arm. In each case, six tests were performed, and test data were plotted in the two side-by-side graphs in Figure 1.4. For ease of comparing the two orientations, the two graphs have the same Y scale. The graphs show forces (N) as a function of frequency (Hz). The two forces are Z forces experienced when the scoop penetrated the soil (backhoe-style) and X-force when the scoop was dragged backwards towards the rover (bulldozer-style).

It can be seen that even low frequency vibration reduces forces. However, the greatest reduction occurs above 60 Hz. For the X-forces (bulldozer), the forces are similar between 0–50 Hz, but drastically drop again above 60 Hz. Hence, this data suggests that a minimum frequency required to achieve significant force reduction is 60 Hz.

The two vibratory orientations did not result in significantly different excavation forces. However, side-mounted vibration (alongside the scoop as opposed to side-to-side) was shown to be slightly better.

As mentioned earlier, the underlying principle in percussive scoop is an impact energy that's delivered to the front edge of the scoop. If this impact energy is high enough, it could cut into competent material. For this reason, percussive scoops are significantly more powerful than vibratory scoops which are not optimal for crusty soils or more competent materials.

The photograph in Figure 1.5 shows an experimental setup that included a Surveyor-sized scoop mounted to a hammering system (Green et al. 2012). The hammering system was able to provide up to 5 J of impact energy at frequencies of up to 30 Hz. The hammering system was mounted onto a Z-stage that allowed the scoop to penetrate into the material. To measure excavation forces the hammering system was mounted onto 6-DOF force torque sensor. The excavation setup includes a soil bin with a vibrator mounted onto an X-stage (horizontal stage). Prior to testing, the vibrator compactor was turned on to compact the soil to desired density. To perform the test, the soil bin's X-stage was commanded to move at a desired rate from left to right (i.e. towards the scoop), while the scoop's- Z-stage was commanded to move down to a target depth. As such, penetration depth and excavation rate could be varied between the tests. The hammer was used in various frequencies and impact modes.

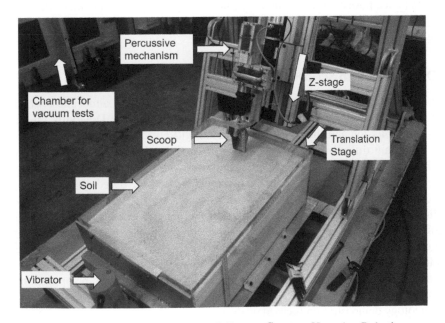

FIGURE 1.5 Percussive scoop experimental testbed. *Source:* Courtesy Honeybee Robotics.

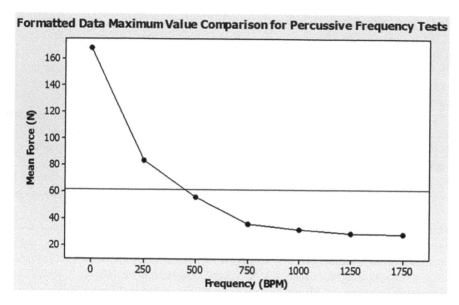

FIGURE 1.6 Percussive scoop test data. Excavation force vs. frequency (BPM = Blows Per Minute). *Source:* Courtesy Honeybee Robotics.

The curve in Figure 1.6 shows typical test data: force vs. frequency (Green et al. 2013). It can be seen that the excavation forces significantly drop as the frequency (Blows Per Minute) increased. In fact, as frequency goes from 0 to 750 BPM, the excavation force drops from over 160 to 30 N or so. This means that the excavation force provided by the arm would be at least five times less if a percussive system is employed. Since all the forces need to be reacted by the lander, the mass of the lander can be five times lower as well. It should be noted that increasing frequency above 750 BPM does not significantly affect excavation forces. For example, doubling the frequency from 750 to 1500 BPM would literally have no effect on the excavation forces, but the hammering power would double (since power is directly proportional to frequency). This illustrates the importance of optimizing percussive systems and trading excavation force vs. percussive power to optimally meet the requirements.

1.2.4 DRONE MOUNTED SCOOPS

A number of planetary bodies, including Venus, Titan, and Mars have atmosphere which allow for the deployment of a drone or a helicopter-based mission. In fact, the Dragonfly mission to Titan is a drone-based mission and the Mars 2020 mission will also feature the Mars helicopter called Ingenuity. The main difference between the two missions is that the Dragonfly is a stand-alone mission while Ingenuity is a technology demonstration subsystem of the Mars 2020 rover mission.

Free-flying systems enable far superior analyses relative to rover-based systems or humans, since valuable resources may be in areas that are too hard to access for a rover, or too hazardous for humans. In 2017 a free-flying robot equipped with a sampling tool was assembled and field-tested, successfully proving access to remote areas, airborne compositional surveys, and autonomous sample return using a touch-and-go sample acquisition strategy. These advances demonstrate capability increases in space exploration such (1) enabling access to sites that are too hard or dangerous for rovers to access, (2) handling of high-hazard materials, (3) identifying, mapping, and sampling available planetary resources in situ.

This technology demonstration focused on developing an Autonomous Airborne Sample Tracing And Recovery (AA*STAR) system for mineral exploration applications (Figure 1.7). AA*STAR

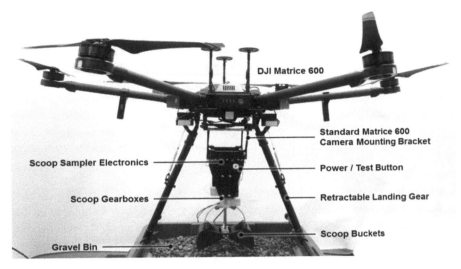

FIGURE 1.7 Scoop Sampler mounted on DJI Matrice 600 UAV. *Source:* Courtesy Honeybee Robotics.

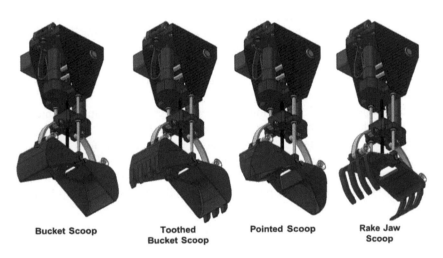

FIGURE 1.8 Various scoop designs for drones. *Source:* Courtesy Honeybee Robotics.

utilizes leading planetary exploration technologies in miniature high-performance sample acquisition, fault-tolerant processing, mature UAV autonomous navigation and control, as well as expertise within in situ resource exploration and utilization, to realize a UAV-based sample tracing and acquisition system for remote, autonomous resource recovery.

Four different scoop end effectors were designed and built to expand the functionality of the scooping mechanism in various sample types. These scoop end effectors include a Bucket Scoop, Toothed Bucket Scoop, Pointed Scoop, and a Rake Jaw Scoop used to capture rocks. These end effector designs are shown in Figure 1.8, and are described as follows:

- **Bucket Scoop**: This is a generic scoop bucket end-effector for sampling in soil and gravel. The edge of the scoop is flat, enabling a large amount of sample to be acquired.

- **Toothed Bucket Scoop**: This is a modification of the generic bucket scoop, with the addition of "teeth" to the edge of the bucket. This end-effector was modeled after construction equipment excavators, where these teeth can be used to help loosen large rocks from soil.
- **Pointed Scoop**: This is a modification of the generic bucket end-effector where the scoop is tapered so that the leading edge of the scoop is significantly thinner than the rest of the scoop. This scoop is designed to maximize the contact force at the leading edge of the scoop, allowing it to penetrate through harder materials like crusty soil, and soil embedded with gravel.
- **Rake Jaw Scoop**: This end-effector is designed specifically to acquire larger rocks and other objects. The slots on both sides of the scoop jaws are designed to pass by one another, allowing the scoop to pick up pebbles as well as small rocks.

The Scoop Sampler attaches to the base of the Standard Matrice 600 Camera Mounting Bracket. Not shown are simple clamp linkages which preload the top of the Scoop Sampler to the base of the Camera Mounting Bracket, providing a stiff connection point without the need for tooling. Fully integrated electronics, located near the top of the Scoop Sampler, control all of the functionality and logic required to control the Scoop Sampler's two DC brushed motors. The electronics board also includes a wireless receiver capable of communicating with a base unit as far as 2 km away (assuming clear visibility).

After integrating the custom landing gear and Scoop Sampler onto the DJI Matrice 600, we performed a field test of the system in the Mojave Desert. Figure 1.9 shows the completely integrated system in the field.

As shown, the custom tripod landing gear provides a stable platform for scoop sampling operations, and a remote camera mounted to a 2-degree-of-freedom gimbal system can be used both for navigation and viewing the scoop during sampling operations. The scoop has its own dedicated electronics and battery, which are independent of the Matrice 600. The scoop is initiated via wireless signal using a dedicated controller.

During our field test we met all of our minimum requirements and completed objectives successfully. We also completed a number of bonus success criteria. One of the major objectives accomplished was an end-to-end flight test and sampling using only the First-Person View (FPV), i.e. only seeing what the UAV camera sees, and not looking at the actual position of the UAV. This operation mimics a "real world" application, where the user might not always have direct line of site to the UAV. The Matrice 600 was flown to a remote location and landed. The Scoop Sampler was then

FIGURE 1.9   Scoop Sampler at field test site in Mojave Desert. *Source:* Courtesy Honeybee Robotics.

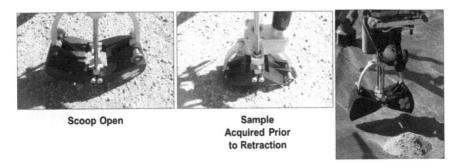

FIGURE 1.10   Pointed Scoop in dry soil. *Source:* Courtesy Honeybee Robotics.

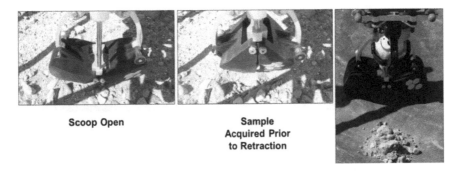

FIGURE 1.11   Pointed Scoop in dry soil with pebbles and gravel. *Source:* Courtesy Honeybee Robotics.

initiated, acquiring a sample, and the UAV was flown back to the home position, where the sample was dropped off onto a tarp laid on the ground.

After completing two end-to-end tests, a number of other tests were performed where the UAV was manually positioned onto specific sample sites, including loose soil, crusty soil, gravel, and gravel with large pebbles and rocks. For each of these different soil types, different scoop end-effectors were used to test their effectiveness in that type of soil. Figures 1.10 through 1.13 show a subset of the tests performed during this field test. Of the three bucket-style scoops tested, the Pointed Scoop (Figures 1.10 and 1.11) seemed to be the most adaptable to different soil types, and was able to acquire larger pebbles as well as loose and crusty soil.

The Toothed Bucket Scoop (Figure 1.12) did not perform very well in soil with gravel, as had been expected, but it performed very well in hard crusty soil compared to the basic scoop. The teeth on the bucket helped to break up the upper crusty soil prior to the bucket passing through the soil. The Rake Jaw Scoop (Figure 1.13) was not able to dig into the soil and acquire a buried rock. However, it was able to pick up a large non-buried rock and grasp it securely for the return flight. These scoop jaws were rapidly prototyped out of polymer, and it is expected that aluminum versions will improve performance significantly, and potentially allow for excavation of larger buried rocks.

Following performance testing of AA*STAR in the Mojave Desert, a planetary mission-analog field demonstration was performed in the Socompa location in Northern Chile (Figure 1.14). AA*STAR was remotely piloted to a location, collected soil and gravel samples, and returned to the base of operations. This testing was performed at ~5100 m above sea level. A number of functions were successfully executed during this field test, including several scooping operations in three

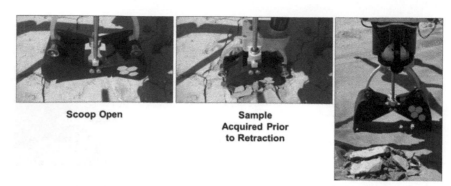

FIGURE 1.12   Toothed Bucket Scoop in crusty soil. *Source:* Courtesy Honeybee Robotics.

FIGURE 1.13   Acquiring large rock with Rake Jaws. *Source:* Courtesy Honeybee Robotics.

different soil deposits with two different scoop types, and a flight stress test of the Matrice 600 drone, which included flying up to approximately 80 m above ground level, and flying 5 km round trip without depleting the battery. Testing was also performed to demonstrate the ability to grab rocks with a claw end effector. Overall, this field test of a first-of-a-kind remotely operated Scoop Sampler was a success.

Follow-on developments are taking navigation and autonomous control several steps forward, by developing an innovative approach to human-in-the-loop robot control using a collaborative teleoperative approach. Our approach utilizes navigation-function-based controllers combined with human input to drive the flyer to a region of interest (ROI).

1.3   GRINDERS

Exposed rocks surfaces tend to be altered by radiation and oxidation. As such, looking at the rock surface does not reveal the true nature of the rock itself. For this reason, geologists typically carry rock hammers, to break or chip a rock and reveal its unexposed interior.

1.3.1   ROCK ABRASION TOOL (RAT)

The Rock Abrasion Tool (RAT) on the Mars Exploration Rovers (Gorevan et al. 2003) was the first grinder to be deployed on another planetary body (Figure 1.15).

The RAT is roughly the size of a soda can and operates with about 10 W of power. The RAT produces a hole approximately 45 mm in diameter and can grind to a depth of roughly 5 mm with less than 10 N of down-force on most targets. This is accomplished by spinning two diamond-embedded

FIGURE 1.14   AA*STAR in flight at field site and aerial view of field test site. *Source:* Courtesy Honeybee Robotics.

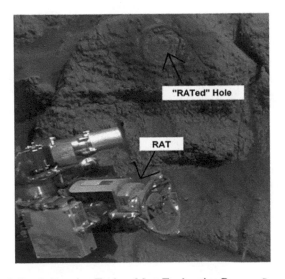

FIGURE 1.15   Honeybee's Rock Abrasion Tool on Mars Exploration Rovers. *Source:* Courtesy of NASA.

grinding tips at 3000 rpm. The RAT then uses two brushes to sweep dust from the boreholes. The main purpose of the RAT was to remove the weathered and thus altered outer surface of rocks and to reveal virgin rock structure for analysis. However, as the mission progressed, the RAT was also used as a brushing tool—during this operation it would position its cutting head to spin just above the rock surface but close enough for brushes to touch the rock.

Initial requirements for both RATs (on Spirit and Opportunity) were to grind at least three holes. The combined number of successful grinds performed on Mars was over 100. The cable shield of each RAT is made from aluminum recovered from the World Trade Center after the September 11 attacks. The RAT was a stand-alone system with all actuators packaged inside it.

1.3.2 Powder and Regolith Acquisition Bit (RABBit)

The Powder and Regolith Acquisition Bit (RABBit) is an alternative approach—it uses a drilling system to provide all the needed actuation—see Figure 1.16 (Zacny et al. 2012b). The RABBit attaches to a robotic drill just as if it were any other drill bit; however the internal mechanisms within the RABBit allow for controlled rock grinding, similar to the Rock Abrasion Tool on the Mars Exploration Rovers.

Tests were performed in both Travertine and Indiana Limestone under normal atmospheric conditions. After the Butterflies were preloaded against the grinding target, a Seek-Scan operation is issued which localizes the point at which to start grinding. Several successful tests were performed in various rock types to show this approach is viable for future missions.

It has also been demonstrated that the revolve brush can be used as a means of collecting the produced cuttings onto a miniature scoop and delivering them to a cup with the help of the robotic arm. A piece of brass shim was used as an improvised version of the scoop as this was a proof of concept. Some images from the test showing the sequence of events are shown in Figure 1.17. This simple add-on to the RABBit can add a large amount of value to any mission that uses an abrading tool. Having the capability to acquire abraded powder in this way could greatly simplify sample handling and sample processing mechanisms on future Mars or lunar missions.

The use of a drill for a powdered abrading tool is currently being implemented on the Mars 2020 mission. The mission will use a combination of rotation and percussion to create a ground hole. The

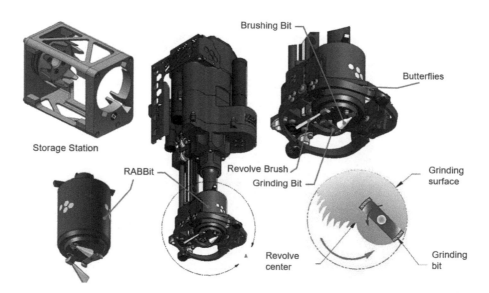

FIGURE 1.16   RABBit on RoPeC Drill. *Source:* Courtesy Honeybee Robotics.

FIGURE 1.17   A pile of cuttings is brushed onto a brass scoop using the slow moving revolve brush. An arm transfers the collected cuttings on the scoop into a cup. A few hammer blows from a drill help to deliver powder into a cup. *Source:* Courtesy Honeybee Robotics.

fine powder generated during the process will then be blown out using a stand-alone gas-based Dust Removal Tool (see next section).

1.4   DUST REMOVAL SYSTEMS

Rocks on planetary bodies are often covered by a layer of dust. This dust needs to be removed in order for in situ instruments to be able to analyze the pristine rock surface. There are several methods that can be employed to remove surface dust including brushes, compressed air, or high-speed propellers. The major difference between brush or air options is that brushes are in contact with a rock while approaches that use high-speed gas or atmosphere are not.

There are many pros and cons to each of these techniques. For example: brush wires wear out with time due to fatigue, can scratch the rock surfaces (if the rock is very weak), could potentially contaminate the surface with metal residue, propel dust where it may not be desired, and may have difficulty cleaning up dust from cracks. However, they have the advantage of being able to remove very cohesive dust since brushes apply significant mechanical force. High-speed gas or atmosphere options have the advantage of being at a standoff distance and can clean dust from deep cracks. However, compressed gas is a consumable while a propeller approach requires high-speed motor and arm movement (the cleaning is best directly underneath the tip of the spinning blades). Since the propeller approach uses atmospheric air, there is virtually no contamination potential while compressed gas may contaminate rock with the gas (if it's not sufficiently pure). A propeller can also be spun in reverse and suck a sample out of the cracks, if needed. Compressed gas, on the other hand, would be simple to implement since it's essentially a cold gas propulsion system.

1.4.1   ROCK ABRASION TOOL (RAT)

The first use of a brushing tool on a planetary mission was performed by the Rock Abrasion Tool (RAT) on the Mars Exploration Rovers (MER) as described in Section 1.3.1. The RAT was designed as an abrading tool but also had two sets of brushes to remove ground-up rock powder away from the abraded surface (Gorevan et al. 2003). These brushes were also used to remove dust without the abrading bit ever contacting the rock (vertical positioning was critical for this operation). The brushing technique was very useful for science observations and also helped preserve the life of the abrading bit—which is a consumable. At the conclusion of the mission, over 100 brushing operations have been done on MER Spirit and Opportunity.

FIGURE 1.18   MSL Dust Removal Tool (left) and cleaned area (right). *Source:* Courtesy of NASA.

1.4.2   DUST REMOVAL TOOL (DRT)

The next mission to employ a dust removal system was MSL Curiosity. The Dust Removal Tool (DRT) is a standalone brushing system which employs a set of compliant brushes. The DRT weighs 925 g and it fits within 100 mm diameter × 150 mm high volume (Davis et al. 2012). During the dust removal process, a set of brushes articulate to maintain surface contact as they rotate at high speed (500 rpm). DRT requires the arm to perform a sweeping motion (just as in conventional brushing). To date several brushing operations have been conducted successfully as shown in Figure 1.18. As of Sol 1575, the DRT has been used 66 times.

1.4.3   AIR DUST REMOVAL TOOL (AIRDRT)

The AirDRT is a propeller-based dust cleaning approach first demonstrated in October of 2014 (Zacny et al. 2017b). Numerous tests were performed in a Mars chamber with a Variable Pitch Propeller (VPP), off-the-shelf fixed propellers for the Earth ambient environment, ducted fans, and computer fans. The results satisfactorily showed it is possible to generate enough flow to blow dust off the surface at Mars pressure (Figure 1.19).

FIGURE 1.19   AirDRT tests of 16 mm and 5–10 mm deep M2020 ground holes. SB stands for Santa Barbara sandstone. *Source:* Courtesy Honeybee Robotics.

The second round of tests was conducted in March of 2016. Tests were performed with ducted fans at various angles and with various nozzles positioned above a previously RATed hole filled with fine rock dust. The main result was that it is possible to generate enough flow to blow the dust from a few mm deep RATed hole at Mars pressure (6–7 Torr).

The third round of tests was conducted in May of 2016. The test requirement was to demonstrate cleaning of dust from holes up to 16 mm deep with the AirDRT at least 10 mm above the rock surface (i.e. 26 mm above the bottom of the hole). The AirDRT's final design and an actual prototype weighs approximately 600 g (250 g AirDRT and 350-g actuator). It fits within a 100 mm diameter and 100-mm-long cylinder. The motor power is ~60 W.

This round of tests was pretty extensive with over 70 tests done with the VPP, fixed propellers, ducted fans, and centrifugal fans in the Mars chamber. It was found that almost any configuration (centrifugal fan, propeller, ducted fan) when spun at high enough rpm and with a certain shape prop can generate sufficient flow to clean the hole. Due to the volume restrictions and contamination issues, the "Desk Fan" design seems optimal. In this configuration, the propeller speed needs to be at least 30,000 rpm which requires 20–30 W at the output side. It was also found that the shape, size, number, and diameter of the props significantly affect cleaning. Some blades, in fact, did not clean at all. Optimized airfoils for Mars conditions are expected to perform at even greater efficiencies.

It was found that the propeller position with respect to the hole needs to be continuously adjusted for best cleaning. There are some sweet spots where the props generate optimum flow and in turn cleaning; these normally occur underneath the tips of the blades. Therefore, for the best cleaning, the center of AirDRT should follow the hole's rim. This requires a robotic arm to slowly move the AirDRT in a circular fashion. Other arm motions are also possible but won't result in the best cleaning. If the depth or distance requirement can be reduced, this will improve cleaning and possibly reduce the requirements for the robotic arm motion.

If a hole is only 5 mm deep (and the AirDRT 10 mm above the rock surface), 100% of the surface area can be cleaned when the rock is in the horizontal position (off horizontal position improves cleaning, hence horizontal position is the worst case). If a hole is 5–10 mm deep (while the AirDRT is 10 mm above the rock surface), >90% of the hole surface area can be cleaned when the rock is in the horizontal position. For a 16-mm deep hole and with the AirDRT 10 mm above the rock surface (test requirement), >70% of the surface area can be cleaned when the rock is in the horizontal position. The operation takes ~5 min.

1.4.4 GAS DUST REMOVAL TOOL (GDRT)

The Mars 2020 mission will use two arm-mounted instruments, PIXL and SHERLOC, to study the Mars rock smooth surfaces free of dust (Jens et al. 2017). To create smooth surfaces, Mars 2020 drill will use an abrading drill bit. Unlike the Rock Abrasion Tool which clears dust off abraded surfaces during the grinding process, the Mars 2020 abrading bit leaves residual dust behind. To clean the residual dust and particles from the surfaces, Mars 2020 will use a compressed gas system, called the Gas Dust Removal Tool (gDRT) as shown in Figure 1.20.

The gDRT is co-located with PIXL, SHERLOC, and the Corer on the Turret Assembly which is located at the end of the 5-degree-of-freedom robotic arm at the front of the Rover. The arm is used to position the Corer onto a rock surface and allow for abrading operation. After rock abrasion is complete, the Turret then rotates to position the gDRT above the abraded surface. The gDRT then performs three gas puffs to remove dust, each puff releasing 0.138 g GN2. After the abraded surface is clean, the Turret rotates again to position PIXL or SHERLOC above the surface. The gDRT consists of:

- A gas tank filled with 159 g of GN2.
- A plenum tank that is filled prior to each gDRT operation.
- Two redundant supply valves to transfer gas from the supply tank to the plenum tank.

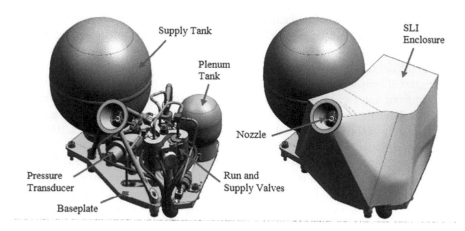

FIGURE 1.20   Gas Dust Removal Tool (gDRT) will be deployed on Mars 2020 Rover. *Source:* Courtesy of NASA.

- A run valve that releases the gas from the plenum tank.
- A nozzle that releases gas towards the surface.
- A fill-and-drain valve for loading and off-loading gas into the supply tank.
- A pressure transducer to monitor tank pressure.

In developing gDRT, over 400 tests were conducted at Martian pressure conditions to determine the effect of the gas mass flow rate, gas molecular weight, plenum size, ambient pressure, nozzle design, pulsed operation, and nozzle height above the surface on dust cleaning. These tests have shown that gDRT can effectively remove dust from 40 mm of a 50-mm diameter hole of up to 16 mm depth (Jens et al. 2018).

1.4.5   PUFFER-Oriented Compact Cleaning and Excavation Tool (POCCET)

There is a major drive to scale down the size of spacecrafts, whether these are landers, rovers, or their payloads. The goal is to try and achieve as much as with larger spacecraft but take advantage of new techniques, materials, and novel approaches. The Pop-Up Flat Folding Explorer Robot (PUFFER) is a next generation rover, the size of an iPhone, that can be stowed flat and can be assembled by activating a release mechanism (Karras et al. 2017).

One of the plausible excavation payloads for the PUFFER robot is a pneumatic excavation system. A gas-based system can be used not just to dig trenches in regolith, but also to clean dust rocks as well as solar panels. Honeybee Robotics developed a PUFFER-Oriented Compact Cleaning and Excavation Tool (POCCET), a 280-g gas-based excavation tool that can be easily mounted on and carried by the PUFFER. POCCET includes all the plumbing required for excavation—tanks, valves, pressure regulators, and nozzles. Depending on the required tasks, the pressure, flow rate, and nozzle type can be varied.

A POCCET prototype was built, integrated, and tested with one of JPL's latest PUFFER prototypes (Figure 1.21). All tests were performed in Mars vacuum conditions (~6 Torr) to demonstrate POCCET's capability to trench and clean rocks. On a horizontal surface, POCCET can clean off a layer of dust 10 mm deep and ~15 cm^2 in area in 6 s.

The POCCET prototype carries enough gas to support approximately 24 s of nominal cleaning operations. During this time, the volumetric flow rate is steady and cleaning performance is relatively consistent. Subsequently, the flow rate will begin to decrease as the tank is depleted. POCCET can continue to be used with this reduced flow, but the cleaning performance will eventually suffer.

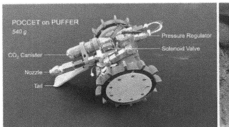

FIGURE 1.21 Honeybee's PUFFER-Oriented Compact Cleaning and Excavation Tool (POCCET) allows trenching and cleaning of rocks from small platform such as NASA JPL's PUFFER. *Source:* Courtesy Honeybee Robotics.

In testing, POCCET utilizes a nominal burst duration of 6 s, and can execute four full cleaning operations. However, this cleaning duration can be adjusted based on the expected amounts of dust or decreased to prolong the useful lifetime of the tool.

This iteration of POCCET was intended as a proof-of-concept prototype and uses commercial off-the-shelf hardware. The prototype weighs 280 g but can be further reduced by employing composite material. With POCCET mounted on top, PUFFER remains mobile, but trafficability on steeper slopes is noticeably diminished. A flight-like POCCET design would use custom hardware and has the potential to be significantly lighter and more compact. With this initial POCCET prototype, Honeybee Robotics successfully demonstrates the feasibility and usefulness of a compact cleaning and excavation tool for miniature planetary rovers.

1.5 SMALL BODIES SAMPLING AND EXCAVATION

Over the past few decades, a number of sampling technologies have been developed for various planetary bodies (Bar-Cohen and Zacny 2009; Ball et al. 2007; Yano et al. 2002; Zacny et al. 2013a). The vast majority of these developments focused on the exploration of Mars and the Moon, while the status of sampling technology for asteroids and comets is relatively low. The following sections describe promising sampling technologies for small bodies.

1.5.1 BiBlade Sampler

The BiBlade sampler is a touch-and-go (TAG) sample acquisition system deployed from a spacecraft using a robotic arm (Backes et al. 2017; Moreland et al. 2018) which was developed for a comet surface sample return application. The sampling process begins with the spacecraft several meters from the surface of a comet and the BiBlade sampler deployed by a robotic arm (Figure 1.22—arm not shown). The spacecraft continues its approach until the sampler contacts the surface and fires, with springs quickly driving two blades into the surface of the comet thereby acquiring and encapsulating the sample. The spacecraft would immediately thrust away from the surface with the sample in the BiBlade.

The primary components of the sampler are shown in Figure 1.23. To prepare the BiBlade for sampling, the actuator rotates the roller screw to drive the gripper (roller screw nut is part of the gripper) into contact with the shuttle. The fingers of the gripper passively lock onto the shuttle at the shuttle latching plate. The actuator then pulls the gripper back which pulls the shuttle back while compressing the sampling springs. The shuttle pulls the carriages and blades up the carriage rails using the pushrods. The gripper stops and is held in a retracted position just before the firing location. The gripper is then pulled back a few mm further which causes the back of the fingers to contact the rigid release plate and release the shuttle, which is then pushed down the shuttle rails

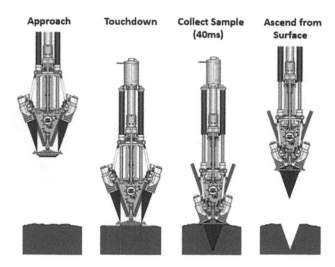

FIGURE 1.22   Touch-and-go sampling sequence with BiBlade sampler. *Source:* Courtesy of NASA JPL.

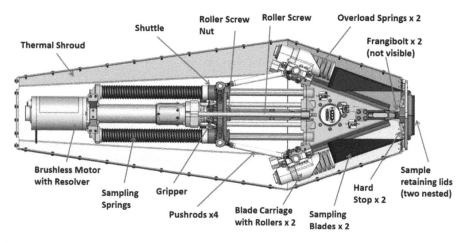

FIGURE 1.23   BiBlade design with sampling springs compressed and blades pulled back in touchdown (ready to sample) configuration. *Source:* Courtesy of NASA JPL.

by the sampling springs. The shuttle motion causes the blades to move down the carriage rails and penetrate the comet surface at a speed of approximately 10 m/s. The blade motion is stopped by hard stops and overload springs to absorb the residual energy that was not used by the blades to acquire the sample. The overload springs also prevent damage to the blades if impacting perfectly rigid surfaces.

The TRL 6 version of the BiBlade is shown in Figure 1.24 in the configuration where the sampling blades are extended after sampling. The BiBlade performance was validated in various test environments including thermal-vacuum conditions and full-system level end-to-end sampling system validation on a full-scale air-levitated 2500-kg spacecraft emulator, as shown in Figure 1.25. The testbed enabled testing of the integrated dynamics of the spacecraft and sampling system during the sampling process. Air bearings enabled 3-degree-of-freedom dynamic motion of the spacecraft

FIGURE 1.24 TRL 6 version of the BiBlade sampler in configuration with blades extended after sample capture. *Source:* Courtesy of NASA JPL.

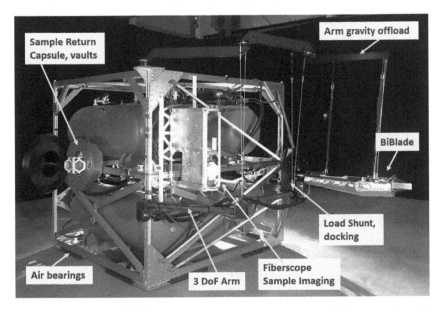

FIGURE 1.25 BiBlade on full-scale 2500-kg air-levitated robotically controlled spacecraft emulator for end-to-end sampling validation. *Source:* Courtesy of NASA JPL.

emulator and a passive mechanism enabled a gravity offloaded arm motion in the spacecraft plane of motion.

A load shunt was incorporated so that sampling forces would be transferred through the structure of the spacecraft rather than the robotic arm. Before a sampling event the robotic arm would dock the back of the BiBlade sampler to the load shunt structure. A wrist spring between the docking mechanism and BiBlade accommodates side loads and moments while the sampler is in contact with the surface.

A fiberscope imaging system was developed to directly measure the sample volume (Toda et al. 2016). It is shown on the spacecraft emulator in Figure 1.25. After the sampling event and spacecraft ascent from the surface, the robotic arm would insert the closed blades which encompass the sample into the sample measurement chamber. The blades would then be pulled back slightly, exposing the sample in the approximately 5-mm slits between them. Nine fiberscopes along the walls of the measurement chamber would passively transfer views of the surface of the sample to a common

FIGURE 1.26   TRL 6 BiBlade with 10 cm wide × 15 cm long sample collected from 5-Mpa cone penetration resistance strength comet simulant. *Source:* Courtesy of NASA JPL.

camera which would acquire one picture that includes images from all of the fiberscope locations. The picture would be analyzed to determine the quantity of sample.

A sample could be transferred to a vault in the Sample Return Capsule (SRC) or ejected. The SRC is a spacecraft element that is released upon return to Earth and brings the sample through the atmosphere and to the Earth's surface. For containerization the arm transfers the BiBlade to the SRC and inserts the blades into a sample vault. The blades are fully retracted and a lid is released from the BiBlade and remains attached to the sample vault to store the sample. A second lid on the BiBlade allows for containerization of a second sample.

Sampling was validated over a range of comet simulant strengths. The sample acquired from a 5-MPa comet simulant block with the BiBlade attached to the spacecraft emulator is shown in Figure 1.26.

1.5.2   Brush Wheel Sampler

The Touch-and-Go concept with brush wheel mechanisms was developed at NASA JPL and is shown in Figure 1.27 (Bonitz 2012). The main advantage of using brush wheels (as opposed to cutting wheels or other, more complex mechanisms) is that upon encountering soil harder than expected, the brushes could simply deflect and the motor(s) could continue to turn. The brush stiffness, and motor torque and speeds could be selected for greatest effectiveness in sampling soil of a specific anticipated degree of hardness.

1.5.3   Clamshell Sampler

The Clamshell Sampler developed at NASA JPL acquires a surface sample by driving two quarter-sphere buckets into the surface (Backes et al. 2014). A linear piston, or slider, drives a linkage that causes the two buckets to rotate about a common axis to close the buckets into each other (Figure 1.28). Once the sample is obtained, two torsion springs keep the buckets closed while the buckets are detached from the tool body in order to minimize the amount of volume that is stored in the return capsule. A benefit of the Clamshell Sampler is that one action with one actuator is used to acquire and retain the sample. Tests were performed with the system in both reactionless and non-reactionless configurations. Complete samples were acquired during tests that were performed on a range of simulants including 25- and 75-kPa floral foams and 300-kPa grill brick. The stronger grill brick required the highest sampling energy, 188 J.

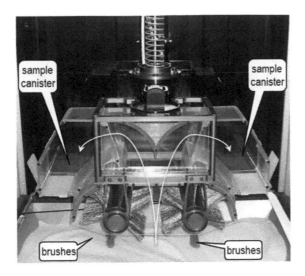

FIGURE 1.27 Brush Wheel Sampler (BWS). *Source:* Courtesy of NASA JPL.

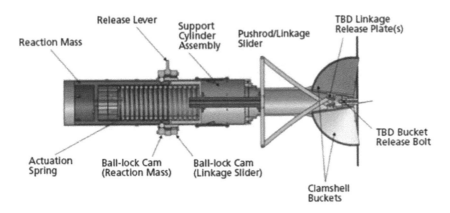

FIGURE 1.28 Clamshell Sampler. *Source:* Courtesy of NASA JPL.

1.5.4 Coring Sampler (C-SMP) for Martian Moons eXploration (MMX)

Martian Moons eXploration (MMX) is a sample return mission planned to launch in 2024 by the Japanese Aerospace Exploration Agency (JAXA) and will land in Phobos for detailed scientific observation and sampling acquisition (Kawakatsu et al. 2019). The sampling system uses the Coring Sampler (C-SMP) to collect regolith at depths greater than 2 cm from the surface and the Pneumatic Sampler (P-SMP or P-Sampler) from the surface. This section describes the C-SMP mechanism and the peripheral system that creates the MMX sampler system developed by JAXA (Figure 1.29), while the P-SMP is covered in Section 1.9.2.

While the coring mechanism is simple and widely used, the C-SMP mechanism is designed to rapidly perform subsurface sampling, where the mass of the sampled regolith will be more than 100 times greater than that of the Hayabusa 2 mission. It is equipped with a shooting actuator that employs a special shape memory alloy, SCSMA. When deformed by heat, the instantaneous volume

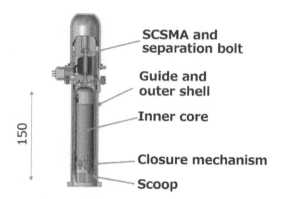

SCSMA and
separation bolt

Guide and
outer shell

Inner core

Closure mechanism

Scoop

150

FIGURE 1.29   Key components of the C-SMP. *Source:* Courtesy of JAXA.

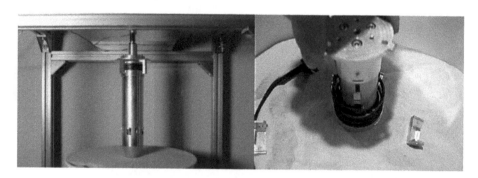

FIGURE 1.30   Testing of the C-SMP conceptual model. *Source:* Courtesy of JAXA.

change rate of 9% breaks the separation bolt and obtains an exceptionally large ejection energy for shooting the inner core with a scoop. The time to heat the actuator depends on the thermal conditions and is typically designed to operate in 30 s and up to a minute in flight. During instantaneous ejection, the regolith enters inside the inner core. After ejection, the closing mechanism is triggered to activate and prevent the obtained regolith from spilling out.

These functions were tested under 1G as well as microgravity in the drop towers (Figure 1.30). The SCSMA-based mechanism cannot be made to reliably eject while falling in the drop towers. Hence shooting energy of the actuator was measured in 1G tests, and the drop tower model, a spring-based shooting mechanism with associated shooting energy and acceleration profiles, was used in the drop tower. Silica sand with controlled particle sizes or the Phobos simulant created by the University of Tokyo (Miyamoto et al. 2019) are used for the experiment.

If a rock exists in the regolith, the C-SMP mechanism would bounce back, and fail to obtain regolith. For this, the MMX's sampler is equipped with a robotic arm that enables movement to the desired sampling locations within the landed site, avoiding rocks. Rocks on the surface can be observed using the equipped camera. For tactile inspection of subsurface rocks, a thin rod and force sensor are installed on the end effector of the robotic arm. After these inspections, the C-SMP mechanism is ejected (Kato et al. 2020). The robot arm has an end effector with three C-SMP mechanisms and they will be ejected at multiple landing sites.

The MMX sampler will collect regolith at the landing site with the entire sampling procedure being performed only in 2.5 hr because the spacecraft is not prepared for overnight operation. Both

C-SMP and P-SMP can collect samples quickly. The robotic arm will collect regolith from the ground by shooting the C-SMP mechanism. P-SMP that is installed nearby the footpad of the landing leg will be operated independently from the C-SMP. And after leaving the landing site, the robotic arm will transfer both C-SMP and P-SMP canister to the sample return capsule.

1.5.5 Dynamic Acquisition and Retrieval Tool (DART)

Dynamic Acquisition and Retrieval Tool (DART) developed at NASA JPL, is a stand-alone sampler that is ejected from the spacecraft towards the comet, impacts and penetrates into the comet's surface, and captures the sample (Figure 1.31). After collecting the sample, the sample canister is ejected towards the spacecraft and captured. DART has been designed for materials with 10–100-kPa shear strength in loose or consolidated form (Badescu et al. 2013).

The DART prototype was tested in consolidated simulant (rectangular floral foam blocks) and unconsolidated simulant (1–5-cm-sized crumbs of the floral foam). With the 100 J of impact energy the DART penetrated the predicted depth and collected 500 cc of sample in consolidated simulant. In the unconsolidated simulant, the DART uses a decelerator plate to prevent it from sinking into the simulant.

1.5.6 Harpoon Samplers

Harpoon samplers (Badescu et al. 2009; Purves and Nuth 2017; Smith 2018) can be fired into the surface of a small body, capture a sample during the course of penetration into the subsurface, and be reeled back into the spacecraft using a tether. All these operations could be accomplished at a relatively safe distance from the body to be sampled.

Honeybee Robotics developed a number of harpoon concepts that could be deployed in a variety of formations. Figure 1.32 shows some of the prototypes designed to be deployed from a balloon on Titan. These prototypes were tested with cryogenic ice at approximately −150°C. The tests were focused on demonstrating that the harpoon can impact ice surface at 45° and 0° angles and still capture and retain a sample. To perform the tests, each frozen sample container was placed at an

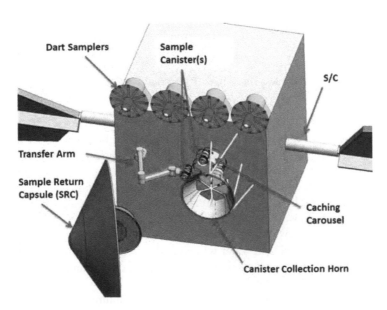

FIGURE 1.31 Dynamic Acquisition and Retrieval Tool (DART). *Source:* Courtesy of NASA JPL.

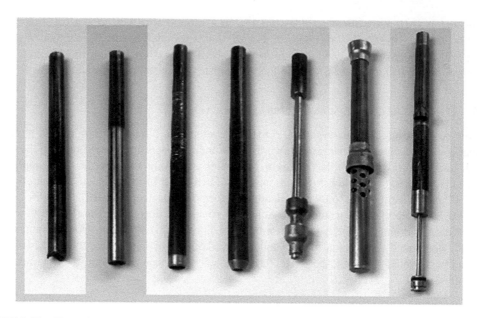

FIGURE 1.32 Honeybee harpoon sampler. *Source:* Courtesy Honeybee Robotics.

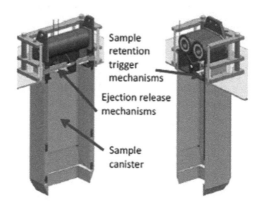

FIGURE 1.33 NASA JPL Reactionless Drive Tube (RDT). *Source:* Courtesy of NASA JPL.

appropriate underneath the harpoon. The tests have shown that a harpoon can acquire 1-g samples in cryogenic ice at orientations up to 45°. Additional tests were performed to determine if there was a minimum energy (i.e. harpoon velocity) required to capture a sample. It was determined that the sample collection starts to drop off at around 8 m/s. The harpoon was also tested in dry and wet sand. The results showed the harpoon sampler can collect samples greater than 1 g of dry or wet sand. In addition, the harpoon was plunged into room temperature water to simulate liquid methane sampling. A layer of sand approximately 3 in deep was placed in the bottom of a 5-gal bucket to absorb excess impact energy, and then water was poured into the bucket until the water depth was 10 in above the sand. Only two tests were performed and both of the samples that were taken show the device was able to retain more than 1 g of liquid.

NASA JPL developed a Reactionless Drive Tube (RDT) to capture samples to 10 cm depth while maintaining stratigraphy; see Figure 1.33 (Backes et al. 2014).

To reduce excavation forces, the system would eject a sacrificial mass. The sampler has an outer shell structure, an inner sample canister, a decelerator, sample retention mechanisms, and sample canister ejection mechanisms. The sampler is attached with sliders to a deployment rail. The energy source, e.g. a spring, drives the sampling tool down a rail and into the comet which drives the sample into the inner canister. The energy source accelerates the sampler toward the sampling media while accelerating a sacrificial mass in the opposite direction. After full penetration, a pull-guillotine door slices through the sample to cut and retain the sample in the inner canister. The inner canister with the enclosed sample is ejected from the outer structure. The RDT prototype worked as designed and successfully tested in a 75-kPa floral foam.

NASA Goddard Space Flight Center Sample Retrieval Projectile (SARP) is described in Section 1.5.7.

1.5.7 Sample Retrieval Projectile (SARP)

The harpoon-based Sample Acquisition System (SAS) developed by Goddard Space Flight Center delivers a Sample Retrieval Projectile (SARP) to its target with sufficient kinetic energy that will penetrate the surface regardless of regolith properties. A Triangular Rollable and Collapsible (TRAC) boom deploys with the projectile and provides a means to return the sample to the spacecraft. This allows the SARP to payout with a flexible "tether," impact the target, and then be retracted in its stiffer configuration that acts like a boom. Upon returning to the spacecraft, the sample cartridge is extracted by a robotic arm (RA) which in turn transfers the sample to a storage container inside of an Earth Return Vehicle (ERV). The storage container keeps the sample secure and cold while in transit back to earth.

By keeping the spacecraft 10 m above the surface, a low-risk profile to the spacecraft is maintained while permitting sample operations that otherwise would pose a much higher risk. By sending a projectile that is tethered, this strategy permits site selection that is most attractive for scientific reasons that would be inaccessible for other methods. This strategy minimizes risk to the mission while permitting sampling operations from most challenging, but scientifically compelling, locations by keeping the spacecraft a safe distance from the target. Unlike other methods, this strategy is capable of penetrating deep below the surface. Accessing deeper material at a lower risk creates an opportunity for a uniquely compelling sample return mission.

The SARP is a mechanically robust sampling projectile that tolerates the shock involved in sampling operations. The electronics on board are simple designs to trigger actuation at certain times during sampling. The SARP contains batteries that supply the actuators power. Figures 1.34 and 1.35 illustrates the major components in the SARP.

1.5.7.1 Sample Cartridge, Door and Release

The cartridge is a hollow shell with a square cross section that allows for a sample 10 cm deep. The SARP penetrates its target with the sample door open, allowing it to fill with sample material. Once the SARP comes to a stop, a mechanism closes the front door and captures the sample. The SARP door is a spring-loaded mechanism with a leading edge that behaves like a cutter. The door is released by a pin puller actuator that is triggered by onboard electronics. Design consideration is given to handle the possible cold temperatures of the sample and testing has verified the door shuts properly when cooled with liquid nitrogen. Specific design elements minimize the chances of the door getting stuck or hung up on components in the regolith.

1.5.7.2 Outer Sheath and Release

The outer sheath gives the SARP more momentum (via additional mass) to penetrate deeper while protecting the mechanisms inside. Once the cartridge door is closed, a second mechanism is actuated that releases the outer sheath from the SARP. The outer surface of the outer sheath is cylindrical with a tapered nose to funnel material into the cartridge. Several protrusions on the nose aid in

FIGURE 1.34   Sample Retrieving Projectile (SARP). *Source:* Courtesy of NASA GSFC.

fracturing the surface of the target to facilitate deeper penetration. The sheath release mechanism is in the rear of the SARP and is based on a fast-acting pin puller. Once released, the outer sheath stays in the target giving a clear exit path for the smaller diameter sample cartridge. This exit path has been cleared out by the incoming outer sheath and will aid in retracting the cartridge out of the target.

1.5.7.3   Inner Sheath, Hinge, Latch and Release

The cartridge returns to the spacecraft inside of the inner sheath which protects the assembly and mechanisms. The inner sheath also provides structure for a hinge that allows the cartridge to be flipped over. The hinge has a torsional spring to open the joint and a latch to hold it open. The hinge is held closed by a release that, once at the spacecraft, is triggered to allow the hinge to open. This action flips the cartridge over and exposes a set of features on the back side of the cartridge which have been shielded from contamination up to this point. These features allow the robotic arm to grip and manipulate the cartridge into the Earth Return Vehicle (ERV).

1.5.7.4   Boom Retraction and Deployment (BRAD)

The BRAD is an assembly that keeps the SARP connected to the spacecraft via the TRAC boom. The boom connects to the SARP on one end and to the spacecraft, via a drum, on the other (Figure 1.36). The BRAD system controls the payout and retraction of the TRAC boom allowing the SARP to successfully collect samples in a wide range of targets with varying compressive strength.

1.5.7.5   Triangular Rollable and Collapsible (TRAC) Boom

The TRAC boom allows the SARP to payout freely like a flexible tether and after impact, be retracted in its stiffer configuration, like a boom (Figure 1.37). Selecting the appropriate design for the tether required the help of the Air Force Research Lab (AFRL) in New Mexico. AFRL has been involved in the design and development of deployable booms for spacecraft instruments. A design developed at AFRL, designated as a TRAC boom (Banik and Murphey 2010), has similar application requirements needed for this application. By combining two leaves into an assembly, a TRAC boom can be wrapped around a drum yet become stiff when deployed. A baseline design, TRAC V5, was decided on as a compromise between stored packing envelope and deployed stiffness. This design choice effects the design of the BRAD drum attachment devices to the SARP and the launcher. The boom length was selected to ensure a minimum of 10-m deployment, taking into account variants such as the target surface topography and spacecraft height relative to the surface.

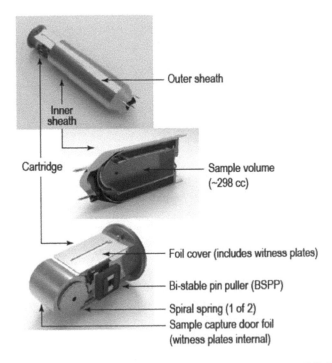

FIGURE 1.35 Outer sheath and inner cartridge of the SARP. *Source:* Courtesy of NASA GSFC.

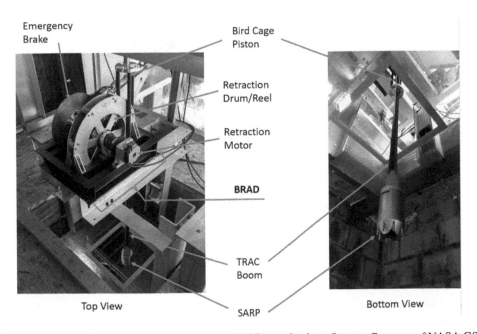

FIGURE 1.36 Boom Retraction and Deployment (BRAD) mechanism. *Source:* Courtesy of NASA GSFC.

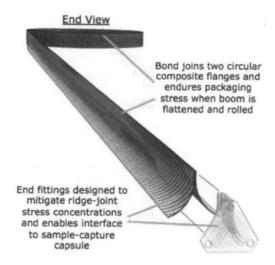

End View

Bond joins two circular
composite flanges and
endures packaging
stress when boom is
flattened and rolled

End fittings designed to
mitigate ridge-joint
stress concentrations
and enables interface
to sample-capture
capsule

FIGURE 1.37 TRAC boom. *Source:* Courtesy of NASA GSFC.

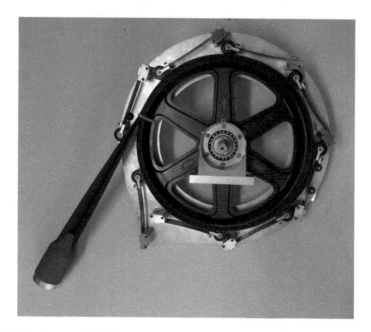

FIGURE 1.38 BRAD drum and TRAC boom. *Source:* Courtesy of NASA GSFC.

1.5.7.6 Drum

As a part of BRAD, the drum provides a way to stow the TRAC boom and manage it during deployment (Figure 1.38). The drum has to accommodate properties of the TRAC V5 design. Special consideration is given to how the drum attaches to the boom to react to all load cases. A mechanism attaches to the drum that arrests the residual momentum in the system after launch. The design of this arrestor took into account the range of expected conditions that would exist due to various interactions at the target's surface. The arrestor design also served to mitigate the shock to the spacecraft upon stopping the drum. After the sample is gathered, a retraction actuator on the drum reels the TRAC boom back to return the cartridge. This preload comes from the drum actuator continuing to apply torque to the drum and keeps the assembly stable. If the surface of the sampling target does not stop the SARP, the momentum arrestor in BRAD will absorb the energy in the system and prevent the TRAC boom from spooling off the drum. The drum design minimized inertia that would otherwise rob the projectile of energy. The design took care to provide sufficient structure for stowing the TRAC boom and handle the load associated with the launch process. The momentum arrestor acts through the structure of the drum so that load condition was the primary design driver for the structure.

1.5.7.7 Retraction System

Also, a part of BRAD, the retraction actuator is attached to the drum to reel the boom back to the spacecraft. The actuator is coupled to the drum through a clutch that allows free rotation at deployment and full engagement during retraction. Once the SARP is back at the spacecraft, this actuator applies a torque to the drum that preloads the assembly to mechanical ground. Upon return to the SC, the actuator has a spring-loaded brake which can maintain that tension with the power being removed. Holding the SARP tight to mechanical ground is critical to enable the robot actuator to extract the cartridge.

1.5.8 Hayabusa Sampler

Hayabusa (Figure 1.39) was the first mission to return material from another celestial body's surface other than the Moon. The Hayabusa sampler was designed to use a 5-g Tantalum pellet fired at 300 m/s into the surface to break up surface material and capture ejecta within a sampling horn connected to a sample container (Barnouin-Jha et al. 2004). This specific approach using a pellet and the horn was chosen because the surface properties were not known *a priori*. On Hayabusa1, a range of anomalies prevented the spacecraft from firing of a pellet. However, during the first of the two attempts, when the horn touched the surface, some surface particles made their way up the horn and into a sample chamber.

Hayabusa-2 used the sampler of almost the same design as that used in the Hayabusa1. However, small modifications were made to meet scientific requirements (Tachibana et al. 2013). A 5-g Tantalum bullet was also used to shoot onto the surface at the velocity of 300 m/s at the time of a

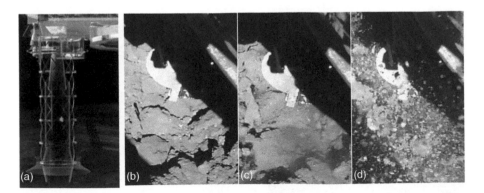

FIGURE 1.39 Hayabusa sampling system (a). Hayabusa-2's onboard camera captured shots of the spacecraft touch and go on asteroid Ryugu: 4 s before touchdown (b), touchdown (c), and 4 s after touchdown (d). *Source:* Courtesy of JAXA.

touchdown, and the ejecta was captured in the sample container. Hayabusa-2 sampler had also a backup sampling method: the tip of the sampler horn had upwards pointed teeth to lift the material towards the sample horn after touch and go. The captured material was propelled towards the sample canister as soon as the spacecraft decelerated. The sample container of the Hayabusa1 had two chambers while the container on Hayabusa-2 has three.

Hayabusa-2's first surface sampling event was performed on February 21, 2019. When the sampler horn touched the surface, a projectile (5-g Tantalum pellet) was fired at 300 m/s into the surface. The resulting ejecta particles were collected by a catcher at the top of the horn.

The requirement of Hayabusa-2 was to also capture a subsurface sample. The subsurface sample collection required an impactor to excavate a crater to eventually obtain material which has not been subjected to space weathering. On April 5, 2019, Hayabusa-2 deployed the Small Carry-on Impactor (SCI): a 2.5-kg copper projectile shot to the surface by an explosive charge. The copper impactor was shot to the surface from an altitude of about 500 m and it excavated a crater of about 10 m in diameter, exposing pristine material. The successful subsurface sampling took place on July 11, 2019 (Hasegawa 2019).

Hayabusa-2 plans to return samples in late 2020. The capsule will reenter the Earth's atmosphere at 12 km/s and will land at the Woomera Test Range in Australia.

1.5.9 OSIRIS-REx SAMPLER

OSIRIS-REx is a New Frontiers (NF-3) mission that targets the near-Earth carbonaceous asteroid called Bennu. The goal of the mission is to return at least 60 g of material (Figure 1.40). The sampling operation will be conducted using a Touch-and-Go Sample Acquisition Mechanism (TAGSAM).

Upon contacting the surface, an annular jet of nitrogen pointed at a surface will fluidize the regolith (Clark et al. 2016). This dusty gas would escape through a filter element within the round sampler. The filter would then capture regolith and let the nitrogen escape into space. During this time, the surface contact pads will also collect fine-grained material. Particle sizes will be limited to 2 cm. The surface operation is limited to 5 s to mitigate the chance of a collision with the asteroid. After 5 s, the back-away maneuver will initiate a safe departure from the asteroid surface. The spacecraft will use images and spinning maneuvers to verify the sample has been acquired and it will verify if the sample is in excess of the required 60 g. If the sampling event failed to capture a sample, another sampling event will occur. The TAGSAM allows for three attempts.

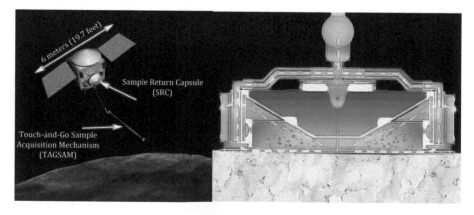

FIGURE 1.40 OSIRIS-REx asteroid sampler. (Left) Spacecraft. (Right) Close-up of TAGSAM sampler. The arrows show the flow of compressed gas. *Source:* Courtesy of NASA.

In addition to the bulk sampling mechanism, the contact pads made of steel Velcro on the end of the sampling head will passively collect dust grains smaller than 1 mm.

After the sampling attempt, the Sample-Return Capsule (SRC) lid will be opened to allow the sampler head to be stowed. The arm will then be retracted into its launch configuration, and the SRC lid will be closed and latched preparing to return to Earth. OSIRIS-REx return capsule would reenter Earth's atmosphere and land under a parachute at the Air Force's Utah Test and Training Range on September 24, 2023.

In December 2019, NASA announced that Nightingale had been selected as the primary sample site and Osprey was selected as the backup site (Figure 1.41). Both located within craters, Nightingale is near Bennu's North Pole while Osprey is near the equator. NASA's initial plans are to perform the first sampling in October 2020. In April 2020 NASA performed sample collection rehearsal that brought OSIRIS-REx 65 m from the surface.

1.5.10 PYRAMID COMET SAMPLER (PyCoS)

The PyCoS illustrates one family of sample capture systems: piercing blades. In fact, such a system could potentially include 2 blades (BiBlade), 3 blades (Tetrahedron), 4 blades (Pyramid) and more as shown in Figure 1.42. In addition, the blades could be of different shapes and in turn final captured volume could range from "conical" to multi-faced.

FIGURE 1.41 (Left) Nightingale had been selected as the primary sample site and Osprey was selected as the backup site. (Right) Sample collection rehearsal brought OSIRIS-REx 65 m from the surface. TAGSAM arm is fully extended and the Nightingale sample site is at the top of the frame. *Source:* Courtesy of NASA.

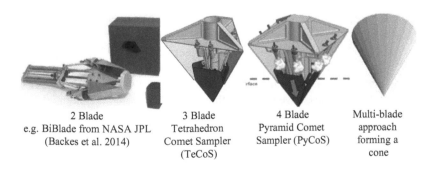

| 2 Blade e.g. BiBlade from NASA JPL (Backes et al. 2014) | 3 Blade Tetrahedron Comet Sampler (TeCoS) | 4 Blade Pyramid Comet Sampler (PyCoS) | Multi-blade approach forming a cone |

FIGURE 1.42 "Piercing blades" approach to sample capture. *Source:* Courtesy Honeybee Robotics.

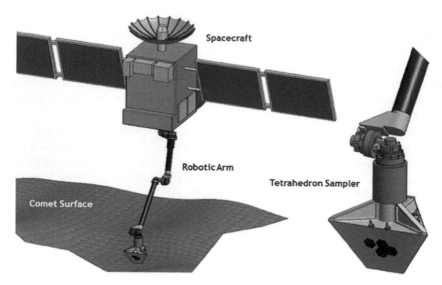

FIGURE 1.43 Initial Deployment of Pyramid Comet Sampler or PyCoS. *Source:* Courtesy Honeybee Robotics.

The final selection between 2, 3, 4 and more blade approach will be based on system complexity and energy per blade. The complexity leads to higher risk, and in turn mass and cost. The energy leads to a larger and heavier deployment system (e.g. spring), as well as impacts Touch and Go (TAG) operations since the spacecraft would need to somehow account for larger excavation forces.

The sampling approach would be the same irrespective of the number of blades and can be described as follows. As the spacecraft descends to the comet's surface, the robotic arm is unstowed from its launched configuration. At this point, the TeCoS/PyCoS is in the stowed configuration as shown in Figure 1.43. After touchdown on the surface, contact sensors within the TeCoS/PyCoS trigger the pyrotechnic release of the blades. Once all of the blades have reached the end of their travel, they form an enclosed sample collection chamber, which collects over 500 cm^3 of sample within the required duration (less than 3 s). The inverted tetrahedron or a pyramid shape of the sampler helps to prevent over-penetration into the subsurface in the case of very weak surface strength. The TeCoS/PyCoS is then removed from the surface with the captured sample inside.

The concept in Figure 1.43 shows the PyCoS deployed using a robotic arm in the same manner as the sampler on the OSIRIS-REx system.

Such sampler shape has a number of advantages. The V shape acts as an arrow piercing into the comet surface at a steep angle. The opposing blades offset tangential forces, meaning that only vertical forces would need to be reacted during impact. These forces could be minimized by making the height low (and in turn the shape would be flatter).

After sample collection, the spacecraft maneuvers away from the comet's surface and retracts the robotic arm. The inverted shape of the sampler naturally makes it difficult for the sampler to become lodged in the surface, compared to harpoon-type samplers which need to overcome sidewall friction during retraction. This design is less prone to getting stuck since the sampler gets progressively smaller with depth, and as the sampler is retracted, it will gradually move into a larger volume (tube samplers, even double wall ones, have to retract through the same volume unless the tube samplers are tapered).

The final design for PyCoS (Figure 1.44) has an approximately 700-cm^3 sampling chamber once the 4 blades have been closed. The size of the sampling chamber can be adjusted based on mission requirements. The pyramidal (in the case of PyCoS) shape continues above the sampling chamber with a number of impact plates, which help to minimize over-penetration into the subsurface.

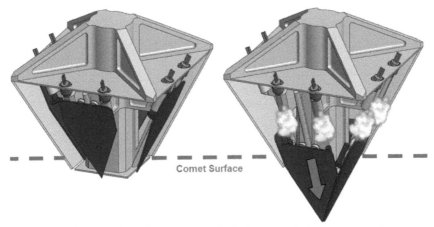

- Sampler impacts surface of comet
- Upper pyramid and flat bottom surface prevent over-penetration

- Redundant pyrotechnic charges on each blade are fired. Shear pin is broken
- Blades close off the collection chamber
- Passive locking mechanisms keep blades in place

FIGURE 1.44 Pyramid Comet Sampler operation. *Source:* Courtesy Honeybee Robotics.

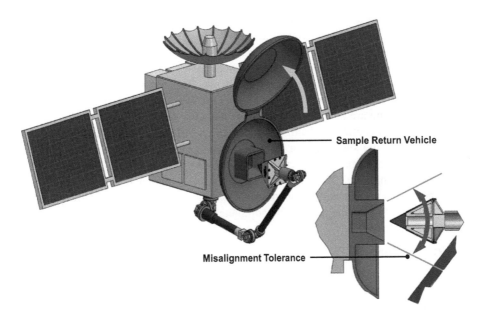

FIGURE 1.45 Pyramid Comet Sampler inserted into the Sample Return Vehicle after obtaining sample. The V-shape sampler and sealing container guide the insertion process without the need for a high accuracy arm. *Source:* Courtesy Honeybee Robotics.

Each cutting blade uses redundant pyrotechnic charges to actuate the sampling process. The charges would be designed to fire simultaneously; however for each cutting blade, one charge would be enough to actuate the blade.

Once the spacecraft is free of the comet, the robotic arm repositions the sampler and places the entire sampler into a Sample Return Vehicle for eventual return to Earth, as shown in Figure 1.45.

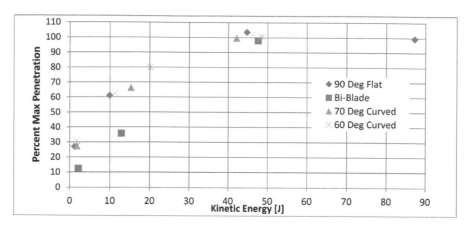

FIGURE 1.46   Penetration into FOAMGLAS as a function of blade assembly impact energy. *Source:* Courtesy Honeybee Robotics.

On the sampler delivery side, the pyramid (or tetrahedron) shape allows significant angular and axial misalignments. Tapered sides act as an alignment feature to more easily insert the pyramid into its docking station—hence the arm does not need to be extremely stiff.

We used three materials that represent a range of upper strengths of a comet: 600-kPa FOAMGLAS, 1-MPa Firebrick, and 4-MPa Aircrete (only data for FOAMGLAS is shown). A large number of different types of steel blades were designed and fabricated to provide knowledge with respect to blade geometry. The blades included BiBlade, a 90° flat blade for the Tetrahedron Comet Sampler and 3 blades with different apex angles for the Pyramid geometry. The BiBlade replica would give us a good reference point to compare with the JPL-developed BiBlade sampler, and the curved blades let us compare the effects of tip angle and curvature.

Figure 1.46 shows penetration depth (as a percentage of max penetration before the bottom of the carriage strikes the surface) as a function of the blade assembly's energy in the 500-kPa FOAMGLAS. The percent maximum penetration of greater than 100% is achievable when the carriage itself penetrates material. This occurred on most trials when energy exceeded 45 J. Given the geometry of the blades a penetration depth of only 70% was required for the blades to form a closed shape and in turn to consider the penetration complete and satisfactory.

1.5.11 ROSETTA PHILAE SD2

The goal of the European Space Agency's Rosetta mission was to study comet 67P/Churyumov–Gerasimenko. The Rosetta mission consisted of two spacecrafts: the Rosetta orbiter and the Philae lander. In November of 2014, the Philae landed on the comet and performed several science investigations of its surface.

The lander was equipped with a sampling drill called Sample Drilling and Distribution (SD2) as shown in Figure 1.47; however, possibly due to unexpected position of the lander with respect to the surface (one of the three legs was not in contact with the surface), the drill did not manage to penetrate the surface (Figure 1.48).

The SD2 weighed 5 kg and was designed to penetrate up to 230 mm depth (Finzi et al. 2007). After reaching the desired depth, a sampling tube was to design to extend from the drill bit to pick up the sample. The drill is then moved back to its home position, ready to deliver the sample to one of Philae's 26 ovens, which are mounted on the rotating carousel. Once the drill is at its home position with the sample, the carousel is rotated to put the assigned oven under the drill. The drill moves

FIGURE 1.47 Rosetta Drill, SD2. *Source:* Courtesy of ESA.

FIGURE 1.48 Philae has been identified in Rosetta's OSIRIS narrow-angle camera images taken on September 2, 2016 from a distance of 2.7 km. Philae's 1-m-wide body and two of its three legs can be seen extended from the body. *Source:* Courtesy of ESA.

downward to place the sampling tube on the oven opening and then pushes the sample into the oven. The carousel is then rotated to deliver the sample to the instruments. Two kinds of ovens were available: 10 medium-temperature ovens (maximum temperature: 180°C) and 16 high-temperature ovens (maximum temperature: 800°C). By heating the samples to different temperatures, a range of different gases would be released for analysis by COSAC and Ptolemy instruments.

The mechanical unit of SD2 consisted of a carbon fiber toolbox, a drill, and a rotating carousel. The drill was designed to penetrate material with strength ranging from fluffy snow to materials with a strength approaching a few MPa. As measured on Earth, the average drilling power between material extremes was in the range of 10 W. The drill could also withstand storage temperatures down to −160°C and could operate at temperatures as low as −140°C. As the comet neared the Sun, it was hoped that the Philae batteries would be able to recharge again and another drilling attempt could be undertaken.

1.5.12 Sample Acquisition System (SAS) for CAESAR

The Sample Acquisition System (SAS), developed by Honeybee Robotics, was designed to collect comet nucleus material during an approximately 5-s Touch-and-Go (TAG) event, as part of the New Frontiers CAESAR mission (Squyres 2018). The SAS hardware with a cross section indicating critical elements is shown in Figure 1.49. It consists of Ripper Tines that, during the contact with the comet surface, break up and loosen up material; gas nozzles that direct high purity N2 gas towards the surface and in the direction of sample container; and a CANCAM camera that takes pictures of samples being collected and in turn witnesses results of the sampling operation. In addition to the sample collection verification using CANCAM, SAS includes sample mass measurement using a load cell. The SAS has been designed for three TAG events.

After it has been verified that the SAS captured sufficient sample mass, the system was designed to detach the sampling hardware and expose the sample container. A robotic arm would then insert the sample container into the sample containment system (SCS). The SCS uses a knife-edge seal and copper gasket to hermetically seal the container with a sample inside it. After the sealing operation is verified, the sample would be warmed allowing volatiles to sublime. These volatiles would then be passively cryo-pumped into a gas containment system (GCS), which is kept below −60°C. After volatile transfer, the GCS-SCS valve would be closed and the SCS would be vented to space until just before Earth return.

SAS underwent significant test campaigns with pathologically worst-case simulants. The final performance tests at zero gravity and in vacuum at the NASA Glenn Zero Gravity Research Facility collected consistently over 300 g of comet simulant. SCS and GCS have also undergone significant testing, bringing the entire system to Technology Readiness Level (TRL) 6.

1.5.13 The Sample Acquisition and Transfer Mechanism (SATM) Drill

The Sample Acquisition and Transfer Mechanism (SATM) is a 4-axis, instrumented drilling system that features a sample preparation and handling system, as well as sample return containers

FIGURE 1.49 Sample Acquisition System (SAS) for CAESAR New Frontiers mission. *Source:* Courtesy Honeybee Robotics.

FIGURE 1.50 (Left) Artistic impression of SATM on Champollion spacecraft. Image: NASA. (Right) SATM Drill. *Source:* Courtesy Honeybee Robotics.

(Zacny et al. 2008a). A prototype was developed and tested by Honeybee Robotics to validate the performance requirements for the NASA ST/4 Champollion mission (Figure 1.50). The drill was designed to acquire subsurface samples from a comet at selectable depths to 1.2 m. The mass is 9 kg and volume is 60 × 60 × 138 cm. Laboratory tests in cryogenic regolith simulant have shown that a total energy of 25 Whr is required to sample at a rate of 0.88 cm/min with an auger speed of 194 rpm, a WOB of 55.6 N, and a drilling torque of 325 mNm.

1.5.14 SAMPLE RETURN PROBE

This concept includes several sampling probes that travel to a small body onboard a parent spacecraft (Chu et al. 2014a). Once the spacecraft arrives at the small body, one of the probes detaches from the parent spacecraft, spins to stabilize using an attitude control system, and propels itself towards the surface. Upon impact, the probe collects and transports a sample into its upper stage where it will later be hermetically sealed. The upper stage of the probe with the collected sample then detaches from the rest of the probe body and takes off from the surface. The probe with a sample inside it then docks with the parent spacecraft and hands off the hermetically sealed sample (Figure 1.51).

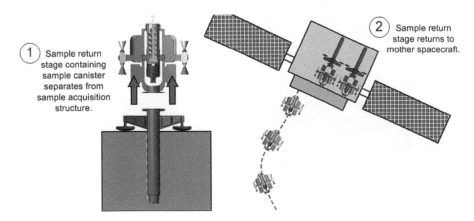

FIGURE 1.51 Honeybee Sample Return Probe. *Source:* Courtesy Honeybee Robotics.

Multiple probes could be used to ensure mission success or to sample multiple locations. The main advantage is that the sampling system is independent of the spacecraft; in turn the dangers associated with proximity operations are eliminated.

1.5.15 Touch and Go Surface Sampler

The "Touch and Go Surface Sampler" (TGSS) consists of a high-speed sampling head attached to the end of a flexible shaft—see Figure 1.52. The sampling head rotates its counter-rotating cutters at speeds of 5000–8000 rpm and consumes 20–30 W of power. The mass of the prototype is 450 g, with a volumetric envelope of 50 mm × 75 mm × 150 mm. The TGSS was demonstrated to sample regolith at a rate of 30 cc/s and consolidated chalk with strength of 10 MPa at a rate of 0.5 cc/s. A number of microgravity tests have shown that the TGSS can sample both consolidated and unconsolidated material at zero-g.

1.6 SURFACE SAMPLERS

In the majority of missions, a sample is acquired from either loose regolith or a more competent material from depths driven by science goals. This section provides several examples of past and present surface sampling tools that can achieve this task. Scoops, which also fall within a surface sampling category, have been covered earlier in Section 1.2.

1.6.1 Dual-Rasp

The Dual-Rasp is a surface sampler developed initially for a potential future Enceladus surface lander mission (Backes et al. 2020), as shown in Figure 1.53. The Enceladus environment provides unique challenges for sampling including low gravity, vacuum, cryogenic temperatures, and science preference for very shallow surface samples. The material of interest would have been deposited on

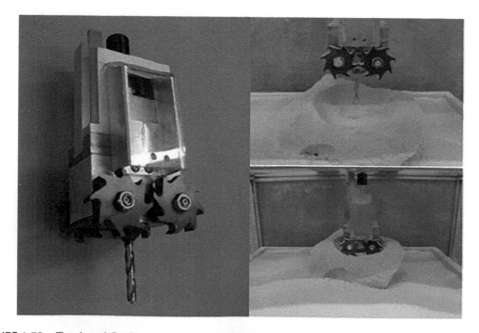

FIGURE 1.52 Touch and Go Surface Sampler (TGSS). *Source:* Courtesy Honeybee Robotics.

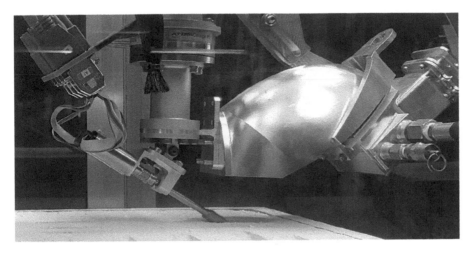

FIGURE 1.53   Dual-Rasp sampler prototype. *Source:* Courtesy of NASA JPL.

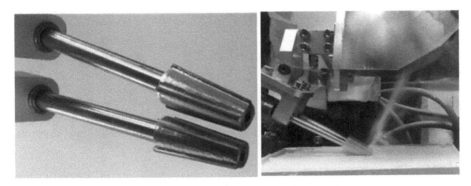

FIGURE 1.54   Counter-rotating rasps acquire and throw sample up between them. The sample is directed by a guide into a sample cup. *Source:* Courtesy of NASA JPL.

the surface from plume ejecta that originated in the subsurface ocean. The surface material has a wide range of potential material properties of approximately 400-kPa to 12-MPa cone penetration test strength (Choukroun et al. 2018, Molaro et al. 2019).

The Dual-Rasp sampler has counter-rotating rasps that break apart surface material and throw it up between them for capture within the collection system, as shown in Figure 1.54. Rasps have the benefit of being able to acquire samples over a wide range of loose to strong material while requiring only very low reacted forces. The rasps also enable careful acquisition of only the shallowest material. The Dual-Rasp is anticipated to be deployed from a lander by a 2-degree-of-freedom robotic arm which has azimuth and elevation actuators at its base. This allows for sampling across an arc in front of the lander while enabling pneumatic sample transfer through rigid pneumatic tubes.

The guide is shaped as an ellipse with the rasps at one focus and the second focus at the entrance of the sample cup, as shown in Figure 1.55. Cuttings that bounce off of the guide are directed into the sample cup.

A grid pattern at the entrance of the sample cup, not shown, prevents the sample particles from bouncing out of the sample cup. Samples would be transferred to science instruments on the lander using pneumatic gas flow through transport tubing. After the sample has accumulated in

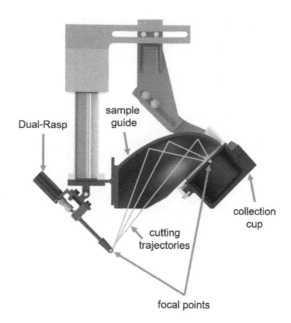

FIGURE 1.55 Sample guide elliptical shape directs the sample stream into the sample collection cup. *Source:* Courtesy of NASA JPL.

the sample cup, the cup would be closed and compressed gas from a tank on the deployment arm would lift and transfer the sample from the collection cup, through a rigid tube down the deployment arm, through passive pneumatic joints at the arm base, and to a sample handling system for science instruments on the lander.

1.6.2 ELECTROSTATIC AND MAGNETIC REGOLITH TRANSPORT AND CAPTURE

Regolith drilled on the Moon, Mars, and asteroids must be captured and transported to a container for scientific analyses or for In Situ Resource Utilization. Since electrostatic and magnetic handling methods have several advantages, such as simple configuration, high reliability, low power consumption, and no requirement for consumables, the following research and development have been conducted to realize reliable and efficient regolith handling technologies.

1.6.2.1 Electrostatic Capture

A high AC voltage is applied between the parallel screen electrodes set at the end of a tube (see Figure 1.56) (Adachi, Maezono, and Kawamoto 2016; Kawamoto 2014). The regolith particles beneath the lower screen electrode are agitated by the alternating electrostatic field in the vicinity of the electrodes and ejected by passing them through the openings in the upper screen electrode. It was demonstrated that not only regolith but also crashed ice particles were captured using this system (Kawamoto and Yoshida 2018). When the system is applied in a microgravity environment, the captured particles are transported to a collection capsule located in the upper part of the tube because of their own momentum. It was demonstrated that not only small particles (lunar regolith simulant) but also large stones with diameters on the order of millimeters were captured in the microgravity environment (0.01 g) reproduced by the parabolic flight of an aircraft (Adachi and Kawamoto 2017). However, when the system is applied on the Moon or Mars, the captured particles must be transported by separate electrostatic, magnetic, or vibrational means, as described in later sections.

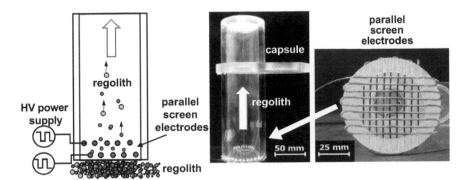

FIGURE 1.56 Electrostatic capture system.

1.6.2.2 Magnetic Capture

Because regolith on the Moon and Mars is magnetic (Goetz et al. 2007; Liu et al. 2007), a magnetic sampler that uses a coil gun mechanism was developed (see Figure 1.57) (Adachi et al. 2018). When an electric current is supplied to a solenoid coil, the resultant magnetic field acts on the particles close to the lower end of the coil. The particles are magnetically polarized and attracted into the coil. If the current is turned off immediately after the particles are attracted by the magnetic force, the particles are transported upward owing to their own inertial force in the low gravity environment. A configuration of multi-coils in series is suitable to transport regolith for a long distance on the Moon or Mars (Eimer and Taylor 2007). An LCR circuit is used to generate a large pulse current.

1.6.2.3 Electrostatic Transport

The captured particles can be transported by the electrostatic traveling wave. A conveyer consists of parallel electrodes printed on a plastic substrate, and a four-phase voltage is applied to the electrodes to transport particles on the conveyer (see Figure 1.58) (Kawamoto and Shirai 2012). Not only regolith particles but also crushed ice and ice mixed with regolith can be transported. In addition to the linear and flat transport, various paths were developed, such as a zig-zag path (Kawamoto et al. 2016a, b), a curved transport path in which the direction of the particle motion was changed by varying the direction of the traveling wave, a circular transport path, transport through a tube, and gathering path. These designs can be combined to generate many forms of transport.

1.6.2.4 Vibration Transport

Vibration transport is suitable for the transport of a large amount of regolith on an inclined path on the Moon and Mars (Mantovani and Townsend 2013), because a large amount of regolith cannot

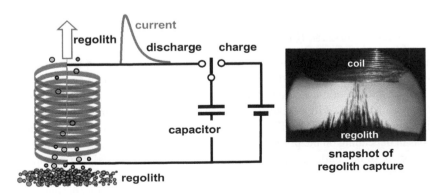

FIGURE 1.57 Magnetic capture system.

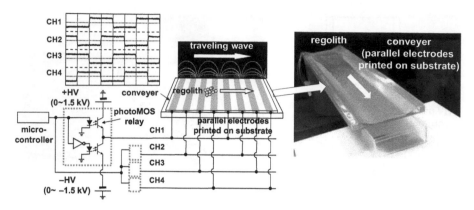

FIGURE 1.58 Electrostatic transport system.

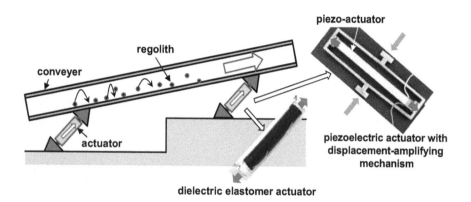

FIGURE 1.59 Vibration transport system.

be transported upward solely by the electrostatic force. The mechanism of vibration transport of particles is the same as that of a vibration feeder, which is widely used for transport of granular materials in industries. A plate or tube that contacts particles vibrates a slant using an actuator, and the vibrating parts repeatedly push the particles in one direction (see Figure 1.59). Instead of the electromagnetic actuator, a dielectric elastomer actuator (Adachi et al. 2017) and piezoelectric actuator with a displacement-amplifying mechanism were developed to realize high reliability and low power consumption.

1.6.3 Europa Drum Sampler (EDuS)

Europa's surface topography is unknown on the scale of a lander and even less on the scale of the sampling system itself. This requires the sampling system to be compliant to variable surface features. The Europa surface could be composed of cryogenic water ice of different densities (very dense to very porous), salt, or frozen sulfuric acid. As such, the sampling system needs to be able to work with any of these materials. The strength of cryogenic water ice can be equivalent to the strength of basalt depending on the porosity. As such, sampling time, forces, and life of the sampling system could prove to be challenging. Since the local gravity is 1.3 m/s², the maximum force the lander could provide to a drilling system is highly limited.

The Europa Drum Sampler (EDuS), shown in Figure 1.60, is based on a continuous miner/road-header design, used in road construction and mining (Zacny et al. 2016a). The advantage of the

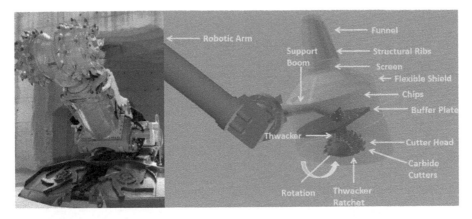

FIGURE 1.60 Terrestrial roadheaders (left) inspired design of the Europa Drum Sample (EDuS) (right). Components of EDuS are shown. *Source:* Courtesy Honeybee Robotics.

roadheader cutter drum design is that it's compliant to highly variable surface roughness. It's also extremely robust and has been shown to cut through weak and hard rocks (continuous miners have been field tested in iron ore mines, cutting >200-MPa iron ore formations).

Using a roadheader cutter drum as a starting point, the Europa sampler requires several features to fit the mission profile and requirements. The Europa Drum Sampler (EDuS) consists of a Cutter Head with carbide-tipped teeth enclosed in a Flexible Shield. The Cutter Head engages the surface while spinning in a counter-clockwise direction. This has a tendency to pull the robotic arm down towards the surface, reducing required excavation forces. The Cutter Head has a passive Thwacker (a ratchet) that works whenever the drum spins. The Thwacker-induced vibrations provide micro-hammering to the cutters and in turn lower excavation forces. Excavated chips ballistically fly out and up—some of them hit the Flexible Shield and fall behind the Buffer Plate. The Shield is flexible to allow for the EDuS to penetrate deeper into material and while the Cutter Head moves forward, the Shield deflects and wraps around the surface features. The Shield could be made of transparent Mylar to allow visual assessment of sample volume. The Buffer Plate is used to prevent the chips from falling out during the excavation process and robotic arm motions. The Flexible Shield ends with a Screen and a Funnel. The Screen size will depend on the maximum particle sizes that the instruments require. As such, for now this is a placeholder. The reason for having a Screen is that it will be very difficult to control particle sizes during excavation—these can range from microns to cm-size chunks, depending on material properties. To aid in the sieving process, the Cutter Head will rotate to engage the Thwacker. The shape and size of the Funnel can be determined based on the location of the instrument inlet port, and kinematic position of the robotic arm. Figure 1.61 describes a notional sequence of operation.

EDuS experiments were performed in Honeybee Robotics' walk-in freezer, which typically maintains a temperature of −20°C (253 K). While this is far from the 100-K conditions expected on Europa, the walk-in freezer is preferable for testing at this phase. Initial tests were aimed at evaluating the general viability of EduS. A sample collection shell was attached to the front to demonstrate the collection and transfer of a sample, and a handle was attached to allow the user to manipulate the instrument as a robot arm might (Figure 1.62). The 48-V motor was commanded to spin in velocity control mode at 25,000 rpm, which after the gearhead, corresponds to a cutter speed of 1316 rpm. Tested in a bin of water ice, it was demonstrated that the Cutter Heads could shave ice from the surface when pressed down.

These tests revealed that the initial sample collection shell did not possess an optimal geometry for catching the generated cuttings. A wider mouth was needed, as only a few grams of

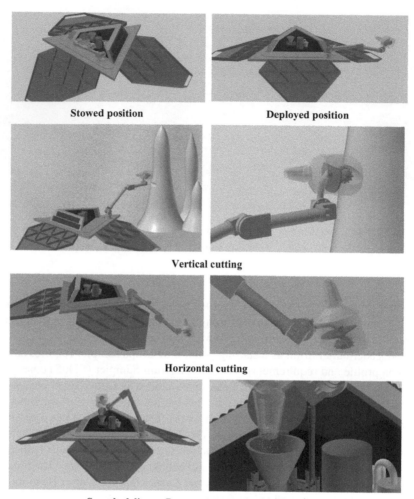

Sample delivery. Drum rotates to engage Thwacker.

FIGURE 1.61   Europa Drum Sampler (EDuS) sequence of operations. *Source:* Courtesy Honeybee Robotics.

sample were collected after minutes of scraping the surface. This did prove, however, that while the Thwacker subjectively didn't seem to have a significant impact on sampling, it did work to vibrate the device and help deliver the few grams of cutting down and through the outlet port on the shell. A Thwacker oriented vertically that could hammer the surface with greater energy might make a greater impact on sampling. One last demonstration proved that the rounded cutters allowed for lateral rasping across the surface, as planned. A well-controlled robot arm could drag EDuS across the Europan surface to collect more ice in one pass than could be collected in a single vertical penetration.

Following this evaluation, two types of tests were conducted: those investigating the excavation rate and power efficiency of EDuS, and those investigating the sample collection efficiency of EDuS. To perform these tests, a frame was fabricated and secured inside the walk-in freezer with a linear rail that could attach to the instrument. This creates a low-friction, passive Z-stage that can generate consistent and measurable results as EDuS penetrates vertically into the icy surface. Weight can be added to the instrument to vary the weight on bit during experiments, from 19.6 N (typical) to 29.4 N. During each test, EDuS was held above the ice surface and allowed to spin up to speed,

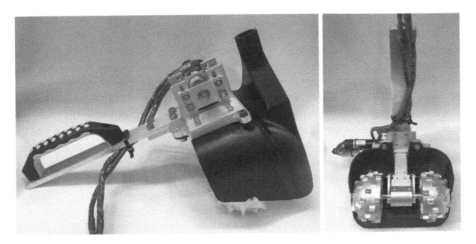

FIGURE 1.62 Hand-held configuration of EduS for general evaluation testing. *Source:* Courtesy Honeybee Robotics.

then was gently placed on the surface and allowed to proceed without human intervention. The experimental setup can be seen in Figure 1.63.

In order to investigate the excavation rate and efficiency of EDuS, we performed multiple tests on pure, solid water ice while varying the weight on bit (WOB). The sampler cut the ice for approximately 1 min in each trial and then the average current from that duration was taken. A few key results were learned from these tests. First, it was shown that increasing the WOB increased the electrical power consumption by a rate of approximately 1.8 W/N, within the range tested. While spinning in air with no load (functionally 0 WOB), the motor typically draws 56 W on average. The highest WOB tested, 30 N, drew an average of 116 W.

The solid ice excavation rate increased with weight on bit. For every Newton of weight added within the range, the excavation rate increased by about 3.3 g/min. At 30 N, EDuS was generating an average of 51 g of cuttings per minute. This means the sampler could excavate 1 kg of ice in 20 min of operation time.

FIGURE 1.63 EDuS: the experimental setup (left) before and (right) after an ice excavation test. The sampler rides up and down a passive z-stage and is weighed down by various masses. *Source:* Courtesy Honeybee Robotics.

By combining this data, we reveal how the excavation energy efficiency increases with weight on bit. While this trend may not be true of all weights on bit, within the range tested the mass of ice excavated per unit of energy consumed increased with WOB. The maximum efficiency tested was at 30 N, where an average of 7 g of ice cuttings were generated per kJ of energy consumed. This means EDuS would use 135 kJ to generate 1 kg of ice cuttings.

This excavation efficiency comes with a few caveats, though. First of all, it does not consider that cryogenically frozen Europa simulant is more difficult to cut through and will require a greater energy input to excavate the same amount. It also does not consider that a portion of that excavated mass will be salt and contaminants. And lastly, we must not confuse excavation rate with sample collection rate. Not all of the ice excavated is captured by the instrument. In fact, the tests in solid water ice had a collection efficiency of only 40%. That is, if 10 g were excavated, only 4 g were captured in the sampler for later use. The rest of the sample was kicked back or under the sample collection shell. The amount of sample collected in a typical test can be seen in Figure 1.64.

Some additional tests were conducted in loose and compact water ice snow. A range of media is important to test, as it is still uncertain what consistency the surface ice of Europa will be. The loose snow used had a density of about 0.5 g/mL, and the compacted snow about 0.650 g/mL. Neither media put up any resistance, as EDuS penetrated through each with ease. There was no significant increase in current draw as it proceeded through the snows, and the penetration rate was limited only by how fast the experimenter let the device proceed (approximately 1/3 inch per second). While the sample collection efficiencies were even worse at 4–16%, such a media would pose little challenge to the lander as it could run the sampler through the snow indefinitely until the desired amount of water was captured. It would likely be limited not by the power draw of the sampler, but by the power draw of the arm in such a circumstance.

These tests shed light on some key benefits of the EDuS sampler, as well as areas for improvement. It is excellent at grinding solid ice into a fluffy powder, which can be easily collected and transferred to other parts of the rover for future use. While 1 kg of water is a lot to demand of a 2-kg device, we proved that it can feasibly be done in under an hour of operation time and for as little as approximately 140 kJ.

However, improvements will be needed for sample collection efficiency if this excavated ice is to be used by the lander. Optimized bin geometries coupled with well-designed deflection plates and brushes to keep the sample in front of the cutters could drastically increase the sample collection

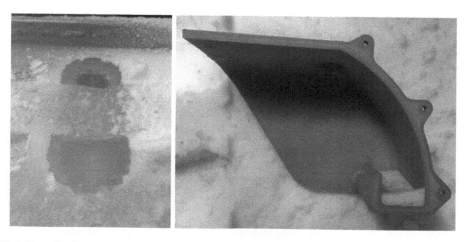

FIGURE 1.64   EDuS: (Left) The holes carved out by the sampler after a test and the resulting powder flung outward, versus (right) the amount of cuttings collected in a bin. *Source:* Courtesy Honeybee Robotics.

rate. It will also be important to demonstrate that such technology can still work as fast and efficiently in more Europan-like conditions (colder, vacuum, harder and saltier ice).

Additional tests were conducted in different icy media to observe the effect on power draw and excavation rate. In addition to the water ice at −20°C tested before, we tested in water ice at −86°C and a Europa simulant made of water saturated in Epsom salt at room temperature (257 g/L) then chilled down to −86°C. In conducting these tests, we hoped to learn how temperatures more like those on Europa would affect the cuttability of the ice, and how adding salt to the solution would change the result. The samples were chilled in a chest freezer overnight before quickly being relocated to the walk-in freezer for tests, all completed within 5 min of leaving the −86°C environment. A WOB of 30 was used, as it was the most efficient in −20°C conditions.

Power draw was relatively constant for the same WOB: 116 W in −20°C ice, 113 W in −86°C Europa simulant, and slightly lower at 95.6 W in −86°C ice. However, the excavation rate dropped significantly in both −86°C media: 16.8 g/min in −86°C Europa simulant and 16 g/min in −86°C ice, compared to 51 g/min in −20°C ice. Combining this information, we see that the excavation efficiency (Figure 1.65) in −86°C Europa simulant dropped to 2.5 g/kJ, while −86°C ice dropped to 2.8 g/kJ, compared to 7.4 g/kJ in −20°C ice. Despite the drastic reduction in efficiency, these tests still show that at 30 N WOB, EDuS could conceivably excavate 1 kg of Europa simulant using just 400 kJ.

1.6.4 NanoDrill for Drones

The NanoDrill uses a single actuator called the Auger-BO (Auger-Break/Off) which couples auger rotation, percussion and the ability to break off and capture a rock core (Zacny et al. 2013a). As the auger rotates in the drilling direction (i.e. coring bit in CW direction as viewed from top) the axis will simultaneously percuss at a rate of 5 blows per auger revolution.

In the reverse drilling direction (i.e. coring bit in CCW as viewed from top), percussion is disabled and the inner Breakoff Tube misaligns itself with respect to the auger which will cause a predrilled core to shear and become captured inside the Breakoff Tube. This principle is shown in Figure 1.66.

A second actuator which is smaller than the Auger-BO, is the Z-stage actuator used to deploy and retract the drill head. The Z-stage subassembly consists of a rack and pinion, which makes it possible for the actuator to sense WOB by measuring the change in its motor current. Another part of the drill assembly designed specifically for this effort is the Solenoid Lockout mechanism used to stow the drill and prevent the drill head from any movement along the Z-stage. The lockout mechanism

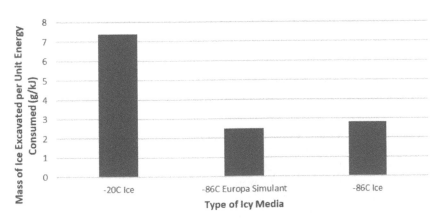

FIGURE 1.65 Excavation efficiency of EDuS in different icy media with WOB held constant at 30 N. *Source:* Courtesy Honeybee Robotics.

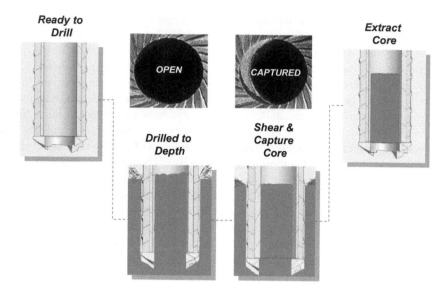

FIGURE 1.66 Honeybee Robotics Patented Core Breakoff Technology. *Source:* Courtesy Honeybee Robotics.

FIGURE 1.67 The NanoDrill is "seeking" the surface of a limestone block in preparation for a drilling operation to acquire a core. *Source:* Courtesy Honeybee Robotics.

consists of a solenoid that retracts a plunger when powered or naturally pushes it out with the help of a spring when no power is applied. The drill is stowed when the plunger is pushed out while the drill head is positioned so that the plunger is above the stop.

Indiana Limestone and Kaolinite rocks were used to perform three tests. Figure 1.67 shows the limestone block used for two tests and Figure 1.68 shows the third test in Kaolinite.

These two rocks are commonly used to test coring drills as they present different rock hardness and grain texture which affects the Rate of Penetration (ROP) differently. Limestone is harder than Kaolinite which will reduce the ROP. Kaolinite on the other hand consists of finer grain sizes which

FIGURE 1.68 Kaolinite rock sample with the NanoDrill retracted after performing a test. *Source:* Courtesy Honeybee Robotics.

TABLE 1.1

Drilling Performance Off Static Arm in Indiana Limestone and Kaolinite Under Atmospheric Pressure (760 Torr) at Room Temperature (~23°C)

Rock Type	Depth (mm)	RPM	Power (W)	WOB (N)	Energy (Whr)	ROP (mm/min)	Time (min)
Ind. Limestone	15	150	8.0	8.5	0.47	4.2	3.55
Ind. Limestone	20	150	7.8	9.8	0.62	4.2	4.75
Kaolinite	20	150	6.6	2.5	0.08	29.1	0.69

causes more friction and will slow down the ROP. The approximate compressive strengths measured with a Schmitt hammer were 40 MPa for the Indiana Limestone block and 15 MPa for the Kaolinite rock.

Table 1.1 summarizes the drilling performance data in Kaolinite and Indiana Limestone. The tests were conducted with a drill deployed from a static arm under atmospheric pressure (~760 Torr) and at room temperature (~23°C). All cores were found to be broken into four or five sections with the core diameter ranging from 6 to 7 mm per section.

The results presented in Table 1.1 are based on a sampling rate of 4–5 Hz while the drill was actively drilling in the target rock (Zacny et al. 2015a). The average WOB was under 10 N for all three tests; however considerable fluctuations occurred in limestone where a max value of 18 N was reached. Average WOB was much lower in Kaolinite than in limestone (2.5 N versus 8 N) which resulted in a much faster ROP (29 mm/min versus 4 mm/min) and low overall energy consumption (0.1 Whr versus 0.5 Whr). A high and varying ROP may be one factor that caused the cores to be broken up into many sections. This can be improved by optimizing the control loop used to set ROP based on the WOB and the Auger-BO motor current. Total power never exceeded 10 W and was relatively similar for all three tests. Figure 1.69 shows cores of limestone and Kaolinite inside coring bits.

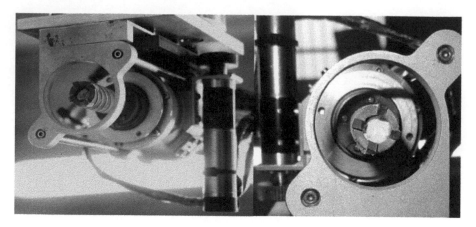

FIGURE 1.69   NanoDrill with cores for limestone (left) and Kaolinite (right). *Source:* Courtesy Honeybee Robotics.

After successful tests, the NanoDrill form factor has been changed to fit onto a Skyjib drone vehicle. The final integrated NanoDrill, and mounting structure are shown in Figures 1.70 and 1.71. As shown, a pair of lightweight carbon fiber panels position the NanoDrill center of mass in the geometric center of the Skyjib by attaching to the existing quadcopter leg frame. The frame used a system of linkages and preloaded bolts to create a stiff structural attachment to the Skyjib legs, while allowing the entire NanoDrill structure to be easily installed and removed from the quadcopter. A simple carbon fiber mounting plate was cantilevered off of this structure to allow for mounting of avionics instrumentation such as a laser range finder and a camera.

The NanoDrill housing also contained an integrated, custom electronics board which controlled all aspects of the drill operations. All onboard drill control was conducted via a single

FIGURE 1.70   Skyjib Quadcopter with NanoDrill. *Source:* Courtesy Honeybee Robotics.

FIGURE 1.71   Preliminary Testing of NanoDrill in Patio Bricks and Aircrete. *Source:* Courtesy Honeybee Robotics.

battery-powered printed circuit board assembly mounted to the back of the drill housing. The controller contained circuitry for two brushless motor drives with current and Hall-effect commutation feedback, a 900-MHz communications radio, and a single STM32F3 ARM microcontroller to provide all motor control functionality. A dedicated 22.1-V/900-mAh lithium-polymer battery pack was used to power all onboard drill electronics. High-level control commands were relayed to the drill from a PC base station using the 900-MHz LoRa long-range RF module coupled to an external omnidirectional antenna. The UAV side antenna was coupled via coaxial cable to facilitate flexible placement on a UAV body for improved link margins.

Microcontroller firmware to control the motors was developed in C utilizing ST vendor libraries for motor control. The firmware implemented high-speed capable field-oriented control of the motor to optimize both high-speed operation and motor efficiency. Control commands were relayed over a simple packet serial interface from the onboard radio.

Preliminary testing of the NanoDrill was performed in various types of paver stones, as these represented likely targets for an end-to-end demonstration, where the autonomously controlled Skyjib would fly to a "landing pad" made of these pavers, acquire a core sample, and return to the "home base" to jettison the core. Testing of the NanoDrill is shown in Figure 1.71.

1.6.5   NanoDrill and PowderDrill for Axel Rover

The motivation for the NanoDrill and PowderDrill was driven by a desire to accommodate mobility platforms which have very limited volume available for a drilling system. One such platform is the Axel Rover designed and built at JPL and shown in Figure 1.72 (Nesnas et al. 2012; Zacny et al. 2013b).

In any mobility platform, stowed volume is typically very limited. Demonstrating a reduced volume of a drill would give mission planners more options when choosing a mobility platform, which may prove increasingly interesting as the community attempts to accomplish more with less.

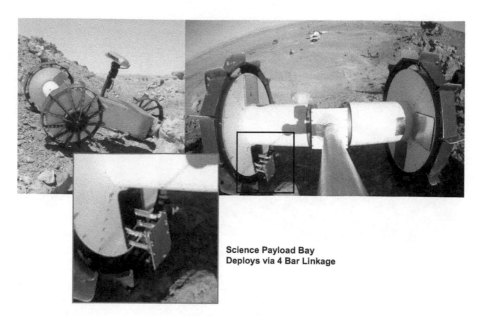

Science Payload Bay
Deploys via 4 Bar Linkage

FIGURE 1.72   Axel Rover and payload bay. *Source:* Courtesy of NASA JPL.

The Axel Rover has one of the smallest stow volumes among mobility platforms which could be considered for future missions. It is designed to explore areas which are inaccessible to conventional mobility platforms, such as cliffs, skylights, and crater walls, and to deploy payload instruments to these science-rich areas.

The NanoDrill and PowderDrill were designed to fit within the science payload bay of the Axel Rover. The four-bar linkage deployment mechanism greatly limits the volume and form factor of the drill. This design is particularly suited to a mission where a rock core is desired, but payload volume is very limited.

Both drills, shown in Figures 1.73 and 1.74, are dual actuator rotary percussive drilling systems which could be used on missions where a low number of samples are required, bit changeout is

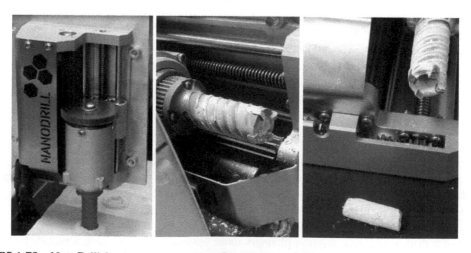

FIGURE 1.73   NanoDrill for Axel Rover. *Source:* Courtesy Honeybee Robotics.

FIGURE 1.74 PowderDrill for Axel Rover. *Source:* Courtesy Honeybee Robotics.

not needed, and a robotic arm or other positioning system may not be available to support the drill. The drills could be used on an exploratory mission such as an Axel-class mobility platform to scout an area for geologic features of interest in order to plan a future sample acquisition and caching campaign.

This mechanism implemented in both drills has the capability to drill and acquire a rock core or rock powder, using a linear actuator to control drilling depth and drill retraction from the borehole. In the case of an Axel Rover, the drill could be retracted into the payload bay, where the core and powdered sample could then be vibrated out of the drill bit and into an instrument for further analysis. In the current design, NanoDrill captures 25-mm-long and 7-mm-diameter core. To capture that core, the bit has an Outside Diameter (OD) of 11.5 mm and an Inside Diameter (ID) of 7.5 mm. However, this can be changed based on mission requirements. The PowderDrill captures samples in a hopper above the bit. The 4.8-mm-diameter bit is designed for 15 mm depth of penetration and can capture approximately 250 mm³ of sample—this can be changed depending on the mission requirements. The powder sample is discharged by reversing the bit direction or free drilling in vertical direction.

The NanoDrill weighs approximately 1 kg. It can fit into a 13 cm × 10 cm × 6 cm volume. The auger actuator was designed for the bit to spin at 225 rpm with a maximum torque output of 1.4 Nm. Since the auger motor runs the percussion as well, 0.6 Nm of the 1.4 Nm is attributed to the percussion and the remaining 0.8 Nm is set aside for bit rotation/cutting. Percussive energy was fine tuned to 0.1 J per blow. Percussive frequency is a function of bit rotation—at 225 rpm, percussion frequency is 675 BPM (Blows Per Minute).

PowderDrill weighs approximately 1 kg. The auger actuator was designed to spin the bit at 2020 rpm with a maximum torque output of 0.5 Nm. Just as with the NanoDrill, the same actuator drives rotation and percussion. To run percussion, a torque of 0.1 Nm is needed. This leaves 0.4 Nm for bit cutting torque. Percussive energy is 0.013 J per blow. The percussive frequency is also linked to rotation—at a maximum rotation of 2020 rpm, the frequency is 39,000 Blows Per Minute (BPM).

The key benefits of the two drills are the simplicity of the design, the light weight of the system and the reduced packaging volume. A dual actuator system means significantly lower development cost compared to multi-degree of freedom systems, and lower volume translates to greater flexibility in the mobility platform selection. This drill is ideal for exploratory missions where bit changeout and sample return are not necessary, and a low number of drilling operations are required.

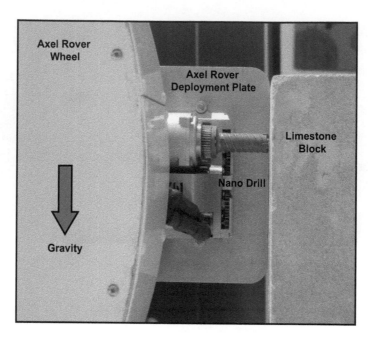

FIGURE 1.75   NanoDrill lab testing in limestone. *Source:* Courtesy Honeybee Robotics.

Figure 1.75 shows testing of the NanoDrill at Honeybee Robotics' test facility, drilling into 45-MPa Indiana Limestone. During the 15-min test run, NanoDrill captured a 26-mm-long core—this corresponds to a penetration rate of 1.7 mm/min. The average power was 15 W. As such, the energy required to capture the core was less than 4 Whr. During this test all drilling operations were performed successfully. The core was drilled, sheared and captured, and successfully ejected by rotating the bit clockwise again— i.e. as if it was drilling. During this process the drill rotated the Breakoff Tube within the auger back into the "drill" position allowing the core to slide out and at the same time engaged the hammering system to assist with core ejection.

To determine the performance of the PowderDrill, the drill was initially tested in Indiana Limestone and Briar Hill Sandstone, both having Unconfined Compressive Strength (UCS) of approximately 45 MPa. The difference between the two rocks is in the degree of abrasiveness (sandstone being made of quartz is abrasive, while limestone is nonabrasive) and the way drilled cuttings or powder behaved. Sandstone cuttings were non-cohesive while limestone cuttings were very cohesive.

The PowderDrill successfully reached 15 mm depth in 2 min and acquired powder (in both rock types). The average rate of penetration (ROP) was ~9 mm/min while the average power was less than 5 W. Hence, the energy required to capture the powder was approximately 0.2 Whr. The drill was used in vertical and horizontal (with respect to gravity vector) orientations to demonstrate powder acquisition at these two deployment orientations. To eject the powder, the drill was positioning down the gravity vector and the bit was rotated. Since percussion is linked to rotation, the hammering action helped with the flow of the powder out of the bit.

Both drills were integrated into the payload bay of the JPL Axel Rover and tested in 80-MPa Berea Sandstone rock at the JPL Mars Yard. The Axel Rover was driven up the vertical rock wall at 65°, and the drills were deployed, effectively drilling in the horizontal position. Successful placement and drilling operations were performed as shown in Figure 1.76. The drills reached the required depth within minutes. NanoDrill successfully sheared and captured the core, while PowderDrill captured rock powder. This demonstrated ability of both systems to capture samples during actual planetary mission.

FIGURE 1.76   NanoDrill Testing on Axel Rover in JPL Mars Yard. *Source:* Courtesy of NASA JPL.

1.6.6   THE PACKMOON

Operations in space are significantly different from those on Earth. As noted earlier, this is notably due to the nontrivial influence of gravity (which is typically lower than on Earth), limited mass and power, extreme temperatures, and vacuums. The lack of gravity is a particularly difficult problem during sampling or excavation. For example, Skonieczny (2013) provides evidence that low gravity has a nonlinear influence on continuous or discrete excavators. In recent years, the CBK together with other scientific institutes and universities has developed a new tool for extraterrestrial regolith sampling, called PACKMOON and shown in Figure 1.77 (Seweryn et al. 2014d).

The stability of the lander is a key concern during sampling in reduced gravity. Various solutions have been developed to prevent it from unexpectedly taking off, overturning, or changing its position. One is based on securing the lander to the surface, and includes devices such as anchors, harpoons, or thrusters. Another is to not land at all, and perform sampling "on the fly," minimizing (or eliminating) contact time between the lander and the surface. A third solution is to use a sampling tool that exerts minimal influence on the lander (either a force or momentum transfer). Drawing upon previous experience with Low Velocity Penetrometer (LVP) mechanisms (Seweryn et al. 2014c), where the idea is to perform movements using a hammering action, a novel, rotary hammering principle was invented. Unlike previous LVP devices that hammer themselves into the soil via linear movements, PACKMOON's spherical jaws are driven by a rotary movement. The hammer's energy and momentum are transferred to the jaws in consecutive strokes. It accelerates against supporting masses and, since the system is kinematically symmetrical, the direction of movement of the hammer and the jaw oppose each other. Similarly, reaction torques from the hammer's acceleration and the horizontal component of reaction forces are opposite and canceled. After a number of strokes, the jaws close and the sample is collected. As this technique is highly efficient, the device only needs a small amount of energy for sampling.

As Figure 1.77 shows, the system is duplicated; there are two hammers and two jaws. These components are mounted on a common shaft called the main axis. The hammers' drivers are connected through the support frame. As the jaws can be automatically opened after sampling, multiple samples can be collected.

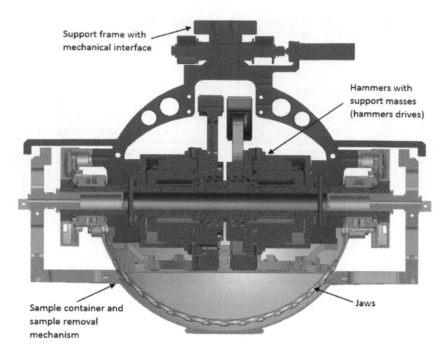

Support frame with mechanical interface

Hammers with support masses (hammers drives)

Sample container and sample removal mechanism

Jaws

FIGURE 1.77 Cross section of the PACKMOON subsystems. Two hammers and their support masses (shown in blue) are mounted on the main shaft. The two jaws and their bearings are shown in orange. The support frame and its mechanical interface are shown in purple, and the sampling container is shown in green.

Samples are preserved in a container that can be automatically dismounted and inserted into, for example, the Earth Reentry Capsule. The current PACKMOON specification is in line with the European Space Agency's Mars Moon Sample Return mission requirements. The device can acquire up to 150 cm^3 of samples of both loose regolith and consolidated material (up to 5-MPa compressive strength). Its key parameters are listed in Table 1.2.

The sampling procedure is shown in Figure 1.78. It starts with the deployment of the sampling device to its initial position. Once delivered to the correct point on the surface, the jaws are released, and hammering begins. After a certain number of strokes (dependent on soil properties), the jaws close and the sample is acquired. Then sample verification begins; this consists of deciding whether to repeat the process or to proceed to the next step. This stage is not done by the device itself. If the sample is not verified, the jaws open and the process of sample acquisition is repeated. If the

TABLE 1.2
Key Parameters of the PACKMOON Device

Parameter	Value
Mass	2.7 kg
Volume (sample container)	150 cm^3
Dimensions of sample container	Φ124 mm × 40 mm
Hammering energy	<2 J
Sampling time	<10 min

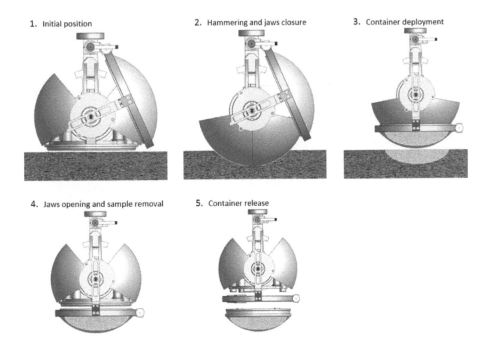

FIGURE 1.78 PACKMOON operational scenario, divided into five phases.

sample is verified, the sample container is released and rotated. Then the jaws open and the sample is removed by pushing it into the container. Then, a pushing plate locks with the container and serves as a lid, protecting the sample. The last step is to release the sample container, so that it can be then placed, for example, in the Earth Reentry Capsule.

An important advantage of this method is that the device has little contact with the specimen. Only the outer part of the sample is in direct contact with the jaws (Figure 1.79) and the sample remains intact. Furthermore, thermal coupling between the motors and the jaws is very low, and the pushing plate (placed between the motors and the jaws) serves as a radiation shield minimizing the thermal influence.

Breadboards of the PACKMOON sampling mechanism were developed to prove the concept of the rotary hammering method, to determine design parameters, and to validate the simulation's results.

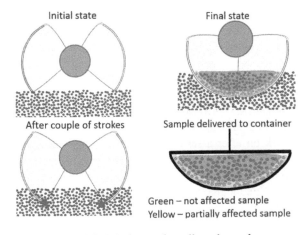

FIGURE 1.79 Impact of the PACKMOON device on the collected sample.

FIGURE 1.80 The PACKMOON prototype in its initial position (phase 1). The left view shows the jaw's teeth, the right view shows the sampling container.

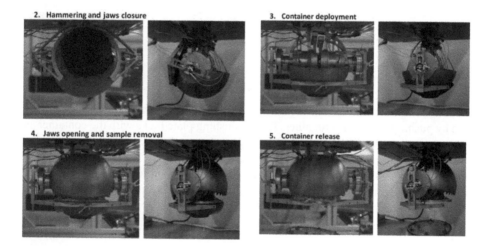

FIGURE 1.81 Other phases of PACKMOON operations during tests.

In the first, hammers were driven by torsional springs. This proved the concept of rotary hammering. In the next breadboard, the spring drive was replaced by a brushless DC motor drive. An extensive test campaign determined the performance of the device and defined parameters for the final model.

The final, functional model was developed in 2017 and is presented in Figure 1.80. The hammer drive is based on a brushless DC motor, connected directly to the hammers. This solution minimizes the complexity of the mechanism. Performance is optimized by using a supercapacitor battery to power the motor, and controlled pulse width modulation, which switches the current off after reaching a certain level and back on after a set amount of time. This type of current control needs a minimal amount of additional electronic components, and only one current sensor. This addition ensures that RMS remains constant, and thus maintains torque at the desired level. As the motors can generate torque of up to 1.4 Nm, the maximal energy of a single stroke for one hammer is about 2 J.

The phases of PACKMOON's operation shown in Figure 1.78 were verified in a test campaign. This confirmed the tool's operational timings shown in Table 1.2. Figure 1.81 shows the sequence of phases.

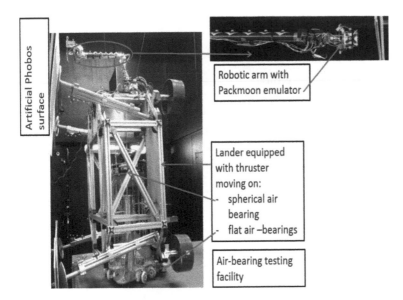

FIGURE 1.82   Test setup used in the ESA's SAMPLER project.

PACKMOON breadboards were tested on various materials, described in Seweryn et al. (2014b). Low gravity conditions were simulated using an off-loading spring system in Paśko et al. (2017). Finally, a dedicated test rig was developed to measure the reaction generated by PACKMOON's interfaces. The device was attached to a measuring unit, consisting of three force sensors, each able to record forces in three directions. These sensors were arranged in such a way that it was possible to calculate reaction torques. The results of these tests are presented in Paśko et al. (2017) in Figure 1.81 and these results were used to conduct further tests on an air bearing table (Rybus and Seweryn 2016) that included both a PACKMOON emulator and a lander mock-up (in the framework of the ESA's SAMPLER project led by Airbus UK). The main aim of the project was to study the interaction between the unanchored lander and sampling devices on zero gravity bodies. The test setup prepared by the CBK is shown in Figure 1.82.

The final prototype was tested on a test rig in analog planetary environments. In all cases, the number of impacts, reaction forces and torques at the mounting interface, sampling time and mass of the acquired samples were recorded. The most important data concerns vertical force components and torque. Results are presented in Table 1.3. This highlights the wide diversity in the number of impacts for Phobos soil simulant (prepared by Open University), caused by its heterogeneity. The volume of FOAMGLAS samples shows that, in most cases, the required $150\ cm^3$ was acquired. This volume depends on the distance between the initial position of the device and the surface. Sampling time is below 10 min (consistent with the parameters shown in Table 1.2).

The jaws are equipped with an encoder that tracks progress. Encoder recordings from the OU soil simulant test are shown in Figure 1.83. These results show the need for a control system that would symmetrize the operation of the jaws through the correct application of torque into the hammers.

1.7 PERCUSSIVE PENETRATORS

There are two categories of hammering devices: (1) with drive unit always located above the surface that is being penetrated (and therefore its penetration depth is limited to the length of the penetrating rod); (2) and tethered "mole-type" devices with a drive unit encapsulated in outer casing that can fully travel below the surface of the celestial body (and therefore its penetrating depth is limited by the length of a cable connecting the device and spacecraft).

TABLE 1.3
Final Test Results from the PACKMOON Model

Probe	Impact	Vertical Force Z Peak (N)	Vertical Force Z Mean (N)	Vertical Torque Peak (Nm)	Vertical Torque Mean (Nm)	Sampling Time (s)	Sample Mass (g)	Sample Volume (cm³)
M-A1	26	89.9	59.6	19.7	12.4	208	233.1	-
M-A2	20	104.6	56.8	18.2	13.2	160	294.9	-
M-A3	19	86	61.2	17.9	12.8	152	266.3	-
M-A4	50	131.9	65	18.4	13.3	400	268.9	-
R-A1	4	73.8	60.7	13.6	11.9	32	276.5	-
R-A2	3	91	67.2	16	10.6	24	218	-
R-A3	3	71.3	57.6	18.6	11.9	24	267.9	-
F1-A1	21	108.1	67.1	19	14.7	210	20.2	~165
F1-A2	22	87.8	49.5	19.5	13	176	19.3	~158
F1-A3	21	120	76.6	24.7	15.7	168	18.27	~150
F2-A1	55	103.2	64.6	15.5	9.9	440	32.2	~187
F2-A2	47	149.3	70.4	19.5	12.1	376	25.3	~147
F2-A3	50	105.4	76.7	19.3	10.4	420	27.2	~158

M results are for OU soil simulant, which is a Phobos soil simulant; R results are for AGK2010, which is a lunar regolith analog; F1 and F2 results are for lesser and denser FOAMGLAS. The properties of these analogs are given in Seweryn et al. (2014b).

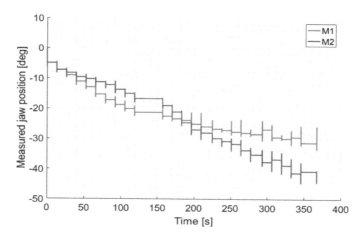

FIGURE 1.83   Progress of jaws (M1 and M2) during M1 and M2 sample acquisition.

The first category originates from the MUPUS device (Spohn et al. 2015). Typically, they have an electromagnetic drive unit (e.g. linear solenoid actuator) powered by a current accumulated in a capacitor, which possess a significant diameter that cannot be inserted into the regolith. They are used when the penetration depth is rather shallow (centimeters or a fraction of meter). They also possess an advantage of mechanical simplicity and possibility of operation in various power settings allowing for adjustment to various regolith types.

The second category originates from the concept of mole-type penetrator presented in (Gromov et al. 1997). In most of the cases, a spring is used to accumulate hammering energy, which allows for compact design. The best example of mole-type penetrator is DLR's HP³ for NASA InSight mission (Grygorczuk et al. 2011, 2016a; Spohn et al. 2018). Alternatively, in the EMOLE device (Grygorczuk et al. 2016c) we managed to implement the concept of electromagnetic drive into mole-type penetrator, by removing the energy storage (capacitor), outside the mole. As a result, it is the first mole-type penetrator with capability of interchangeable power settings.

The most essential and promising achievements described in this section include:

- MUPUS device for Rosetta mission (the main mission objective was to insert a thermal probe into the surface of comet 67P) (Spohn et al. 2015).
- CHOMIK device for Phobos-Grunt mission (the main objective of this mission was to collect a sample of the regolith of Phobos) (Rickman et al. 2014).
- EMOLE, a new type of mole penetrator with an electromagnetic drive unit (Grygorczuk et al. 2015).
- EMilia, the most energetic hammer drive built up to this moment.
- Two Lunar Drill percussions drives: the first based on MUPUS-like electromagnetic drive with digitally controlled discharging process to optimize its efficiency; the second with a crank mechanism.
- A concept of a dual hammer mole-type penetrator.
- A combination of hammering and drilling features in a single electromagnetic drive system (as in e.g. patent PL 229850) (Grygorczuk et al. 2018) that would allow to not only penetrate granular matter, but also drill into solid materials.

1.7.1 MUPUS

The MUPUS experiment on the lander Philae of the Rosetta mission was designed to measure temperature and thermal conductivity profiles of a cometary nucleus down to 40 cm below the comet surface. The investigations were supervised by Professor Tilman Spohn and the German Aerospace Center (DLR). The penetrator was a hollow tube made of space-qualified glass-fiber composite that hosted 16 ring-like thermal sensors attached to the inner wall of the tube. The main challenge was to insert the penetrator into the nucleus in an almost gravity-free environment. In this case, each stroke of the hammering device that is used to insert the penetrator is followed by a recoil, so that the center of mass of the whole system (hammer + penetrator) returns to its initial position. To achieve insertion progress, the environment should be put to work. The idea is to use friction of the cometary material during the recoil, and design the dynamics of the hammering action in such a way that a strong and short stroke is followed by a slow backward motion of the penetrator with as small as possible reaction force that, ideally, is smaller than the friction (Figure 1.84).

In order to deliver a strong hammering impulse of few Joules, it is necessary to provide power to the hammer that is an order of magnitude larger than the average power of 3 W supplied by the lander. An energy storage was, therefore, required. Out of three possible energy storage forms: a spring (mechanical), a reaction wheel (rotational), or a capacitor (electrical). It was decided to implement a capacitor, since it had no moving parts and could easily be incorporated into a penetrator head. The consequence of this choice was to employ an electromagnetically driven hammer that was accelerated during the capacitor discharge. The electromagnetic circuit comprised of a hollow cylindrical electromagnet, inside which an iron hammer could move. It is closed with a metal plate from above and a titanium foot of the penetrator rod from below. The electromagnet is placed inside the hollow cylindrical capacitor (Figure 1.85). The modelling of the e-m circuit showed that an efficiency of such system could reach 28% but practically the achieved efficiency was 9% smaller (around 19%).

The friction-supported insertion works well when more than half of the penetrator is already progressed in the ground. In the earlier phase of hammering, a pushing force from the lander of

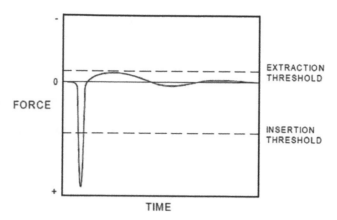

FIGURE 1.84 Time dependence of the force exerted by the penetrator on the environment. The peak corresponds to the hammering stroke.

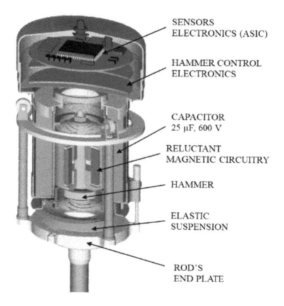

FIGURE 1.85 Main elements of the MUPUS insertion mechanism.

about 1 N is needed to support the insertion. In the case of MUPUS it was provided by the deployment device. In addition, a special design of a conical penetrator tip with elastic barbs resisted the backward motion of the penetrator.

In the mechanical design of the hammering device one can easily identify three cooperating, but independent masses: the penetrator, the hammer, and the penetrator head, i.e. a counter-mass (Figure 1.85). Intuitively, the counter-mass should be relatively heavy to move slowly backward during the recoil. The calculations led to (1,1,10) (almost) optimal ratio of (penetrator, hammer, counter-mass) masses. The real distribution of masses, (60 g, 30 g, 350 g), forced by design constraints, was not too far from the optimal. The masses were linked together by springs that kept the device together: (1) a membrane-type spring between the counter-mass and the penetrator, (2) a weak spring linking the hammer and the counter-mass and keeping the hammer in the upper position inside the e-m circuit before the stroke.

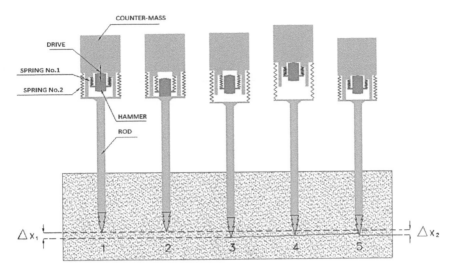

FIGURE 1.86   Five phases of the MUPUS insertion cycle (single stroke).

The hammering cycle is divided into several phases: (1) before the stroke, hammer in neutral position, (2) hammering action, transfer of momentum to the penetrator, (3) forward movement of the penetrator rod, (4) recoil, backward movement of the counter-mass, (5) return of the system to the neutral position (Figure 1.86).

Since the insertion device is deployed to a distance of about 1 m away from the lander, it was necessary to equip it with its own electronics that: (1) controlled the measurement process and converted the analog thermal measurements to digital signals that was sent to the lander, (2) controlled the hammering action, (3) kept the electronic compartment within the temperature limits. The measurement control and analog-digital conversion was realized by a dedicated ASIC chip. The hammering control allowed to program the insertion force by choosing the maximum voltage load of the capacitor. Four levels were equal to 200, 280, 400, and 620 V that corresponded to roughly two times energy increase between levels.

The MUPUS insertion device was a part of more complex systems comprising of a deployment device, two hold and release mechanisms, cable storage, supporting structure (Figure 1.87). The MUPUS operation scenario anticipated the following stages: (1) release of the deployment device (DD), (2) deployment of the insertion device (ID) to a given distance from the lander, (3) touchdown of the penetrator tip with the ground, (4) insertion to a depth of about 25–30 cm, (5) release of ID from the DD, (6) retracting of the DD to the lander, (7) full insertion of the ID, (8) thermal measurements. The insertion progress is measured by a potentiometer made of conducting paint on the outer surface of the penetrator tube, with two endings, one at the penetrator top and another, a moving one, on the clamp that links DD with ID, and after separation of both devices is in contact with the ground.

The functional tests of the insertion were performed on a horizontally suspended ID (to compensate for gravity). Since the maximum expected hardness of the icy cometary material was estimated to be 3 MPa, two solid foams and one foamed concrete with hardness of 0.6, 1.7, and 6 MPa, respectively, were used in tests. The results are presented in Figure 1.88, showing the (averaged) insertion progress after 4 strokes for different energy levels of hammering.

In November 2014 Philae, after two bounces, landed on the surface of comet 67P Churyumov–Gerasimenko. MUPUS was activated a few days after landing and realized the full anticipated operation scenario described earlier. Unfortunately, the penetrator apparently did not penetrate the comet's surface (Knapmeyer et al. 2018). While not unanimously accepted by the comet sampling

FIGURE 1.87   MUPUS PEN in stowed position.

researchers, the MUPUS scientific team attributed the lack of insertion progress to the larger-than-expected strength of cometary material (Spohn et al. 2015).

1.7.2 CHOMIK SAMPLING DEVICE

In 2010 CBK PAN built CHOMIK, a hammering penetrator with sampling capability for the Roskosmos Phobos-Grunt mission (Rickman et al. 2014). The device had three main functions: (1) firm or loose regolith sampling, (2) active and passive thermal measurements (Banaszkiewicz, Seweryn, and Wawrzaszek 2007), and (3) determination of geomechanical properties of regolith (Seweryn et al. 2014b). The retrieved sample from Phobos and in situ measurements could impact our understanding of the origin of the Martian moons and could bring extremely valuable insight to the study of origin and evolution of terrestrial planets. Unfortunately, the Phobos-Grunt mission had a critical failure in Earth's orbit and was not able to deliver the lander to the Phobos surface.

The penetrator CHOMIK is presented in Figure 1.89 and consists of an electromagnetic insertion device with a hammer, the composite rod, and the titanium sample container. Technical data is captured in Table 1.4. The rod transfers strokes from the hammer to the container and provides a

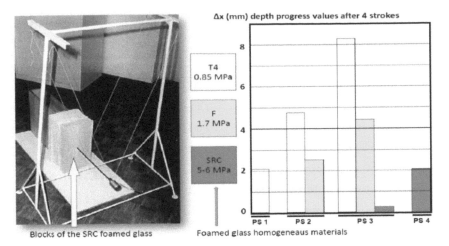

FIGURE 1.88 Insertion device in the test configuration (left panel) and the results of the insertion process for three different test materials (color codes) and four hammering energy levels, PS1 to PS4 (right panel).

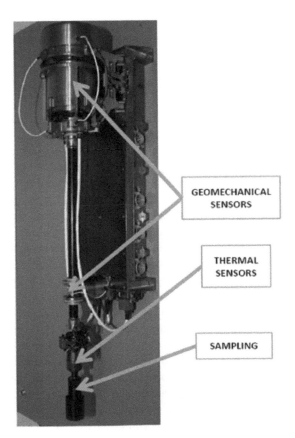

FIGURE 1.89 CHOMIK with indicated sampling container, thermal sensors, and geomechanical sensors.

TABLE 1.4
CHOMIK Technical Data

Parameter	Value
Size	340 × 110 × 100 mm
Mass	1.4 kg
Power consumption: average/peek	2.0/4.0 W
Operating temperature	−160° to +60°C
Operation in microgravity	yes
Force reacting on the manipulator arm	≤3 N
Energy of the hammer stroke	PS 1—1.0 J PS 2—0.1 J
Sensors	Heat flow, temperature, geomechanical

linear guide equipped with a depth sensor. Behind the sampling container there is a heat flow sensor as well as surface temperature sensor. Both the insertion device and the sample container are locked during launch and cruise phase. The penetrator is mounted on a manipulator arm developed by IKI RAN (Kozlov and Kozlova 2014).

CHOMIK belongs to a class of Low Velocity Penetrators (LVP) devices, which are low velocity, medium-to-high stroke energy, but low power, self-driven penetrators, designed as carriers of different sensors for in situ investigations of subsurface layers of planetary bodies. For given dimensions of the LVP, the maximum insertion depth is limited by the energy of a single stroke and by the soil resistance to dynamic penetration. A general description of the operation principle of the Low Velocity Penetrator was presented by Gromov et al. (1997), whereas a numerical model of the penetrator confirmed by experimental results was described by Seweryn et al. (2014d).

The LVP consists of three cooperating bodies: a cylindrical casing inserted into the soil, a hammer, and the rest of the penetrator body acting as support (or counter-) mass. The forward motion of the LVP is performed in cycles and each of them is composed of the phases analogous to those described in the previous section.

One of the most difficult challenges appearing at microgravity conditions and specifically concerning hammer-driven devices is isolation of the manipulator and lander from the reaction forces generated by the recoiled penetrator head (counter-mass). This problem was solved by designing a linear overload release coupling between penetrator rod and manipulator, and elastic suspension between rod and counter-mass. As a result, the reaction force is designed not to exceed 3 N, while the penetration force is at the order of 500–1000 N. Such a solution significantly reduces the tribological issues appearing in mechatronic devices working in vacuum environments. The detailed analysis of these aspects is described by Grygorczuk et al. (2011).

The most challenging sampling process was tested in terrestrial conditions in the IKI RAN laboratory (Seweryn et al. 2014c). The tests were done in Phobos soil analog. CHOMIK was mounted on manipulator arm and performed sampling operations. The sampling container was delivered to the dummy lander return capsule. These steps are presented in Figure 1.90.

Once the CHOMIK titanium container is separated, an extra titanium tool—a sharp tip—is exposed. Its task is crushing of the surface layers in order to make it easier for other devices on board the lander to collect samples. This function was extensively analyzed after CHOMIK delivery and was a major function of the HEEP penetrometer developed in CBK PAN (Grygorczuk et al. 2015). Most recently, a number of hammering devices like moles, drills actuators, scoops, or rotary hammering devices were developed in Poland. Some of them are described in following sections.

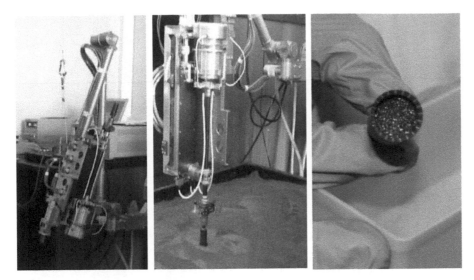

FIGURE 1.90 Sequence of CHOMIK tests. (From left) CHOMIK stowed on manipulator arm, CHOMIK during regolith sampling, sample container with soil delivered to return capsule.

1.7.3 EMOLE

EMOLE (Grygorczuk et al. 2016b, c) was the first mole-type penetrator which employs an electromagnetic linear drive system. Further, it is a prototype model and can be rebuilt to fulfil specific mission requirements. The EMOLE is lightweight and compact and can either be a complete device or, with its new electromagnetic direct drive, has the potential to become a flexible solution for space exploration missions providing a wide range of new possible applications (Figure 1.91).

The main structural novelty, i.e. the use of several electromagnets arranged in a stock as a direct hammer propulsion, gave twofold improvements. First of all, owing to the fact that the electromagnets do not need any drive transmissions in this case and they do not have any rotating parts as in a DC motor, the whole instrument was much simpler and more reliable. Secondly, the drive control has the ability to adjust the hammering energy during operation, which can contribute to energy savings and to the protection of scientific instruments from damage. Furthermore, to provide an additional mode in which the typical operation is superimposed with a high frequency and low energy mode, a new electronic control was developed. It is worth mentioning that for the magnetic circuit a new material—Permendur 49 (instead of soft iron ARMCO B)—was pre-tested and implemented into the final design. Thanks to all of the improvements, the EMOLE generates maximum stroke energy at the level of 2.2 J.

The penetrator has a diameter of 25.4 mm, length of 254 mm, and its overall mass is 0.704 kg (where hammer mass is 0.088 kg, counter-mass is 0.521 kg, and outer casing mass is 0.09 kg). The peak power consumption of the device is about 4 W.

The tests that were conducted proved that the penetrator is able to withstand simulated space conditions like vacuum and high and low temperature operation. Further, the higher stress levels generated during vibration testing sessions didn't damage the mole. After approximately 5 hr of operations and almost 8000 actuated strokes, the EMOLE remains fully functional, which suggests it is one of the more reliable penetrators.

1.7.4 EMILIA

EMilia is a prototype of an electromagnetic hammer head developed at Astronika (Warsaw, Poland). It is used for multiple purposes, including mimicking of various penetration devices and their

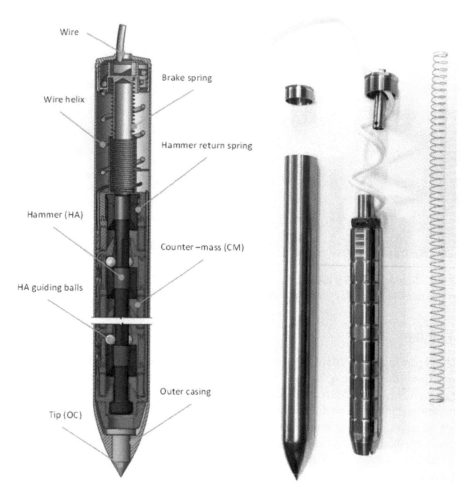

FIGURE 1.91   EMOLE cross section (left); EMOLE inside mechanism (right).

energies to test performance of penetration rate of a tip similar in shape to the outer casing of a HP³ mole penetrator. It was developed from lessons learned from multiple previous device designs and testing including CHOMIK, HEEP (Grygorczuk et al. 2013), and EMOLE. Thanks to a standardized interface and a wide range of possible hammer energies, the penetrating rod can be interchanged to other types and shapes allowing for a variety of penetrator rod experiments and studies with dynamical interaction with regolith. The cross section of the EMilia is presented in Figure 1.92 and the assembly is pictured in Figure 1.93. The characteristic parameters of the device are:

- Overall dimension of envelope: diameter 94 and 300 mm long.
- Overall mass of the hammer head: 0.326 kg (3.65 kg total with a 2.76 kg counterweight and a 0.57-kg rod).
- The hammer travel gap is 10 mm; however, thanks to the use of a 50-degrees cone, the magnetic gap is as low as 4.2 mm.
- The capacitor that stores the electric energy is placed outside of the hammer head.

The main advantages of this model that distinguish it from previous ones are:

- Multiple power settings with possibility to accumulate energy in a range 3–48 J.

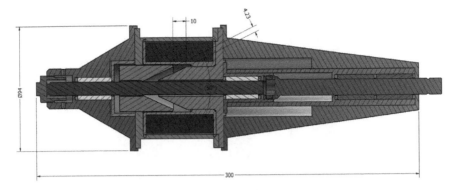

FIGURE 1.92 EMilia cross section. *Source:* Courtesy of Astronika.

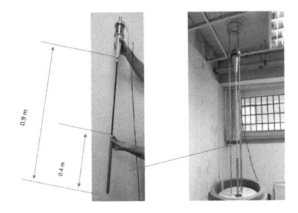

FIGURE 1.93 EMilia integrated with the penetrating tip. *Source:* Courtesy of Astronika.

- Stroke energy varies between 0.57 and 2.66 J.
- High energy to head unit mass ratio at c.a. 9.2–147 J/kg.
- Assumed efficiency of energy transfer on the level of 18%, similar to HEEP penetrator (Grygorczuk et al. 2013).
- Possibility to interchange the penetrating tips for various purposes depending on the need.

1.7.5 Dual Hammer Penetration Device

In the known solutions for a hammer percussion drive for insertion of the penetrator, a single drive is used. However, an exception occurs in nature as is demonstrated by the wood wasp, which to penetrate uses a two-valve ovipositor driven with a different method. During action, the wood wasp produces reciprocation movement and when one valve is protruding—the second one is retracted. After that the first one is retracted and the second is deployed. It is shown schematically in Figure 1.94.

A biomimetic design of this kind has been adapted in a device to confirm its applicability for drilling (Gouache et al. 2010) but the implementation of this idea into practical high-performance drive exceeding the regular single hammer was very challenging and difficult to realize. Therefore, it was proposed to intrinsically modify the nature-inspired idea of splitting the penetrator. In our solution, the penetrator would also possess a dual driving method but its main feature is that the

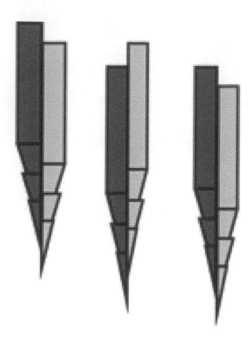

FIGURE 1.94 Schematic principle of operation of the wood wasp ovipositor.

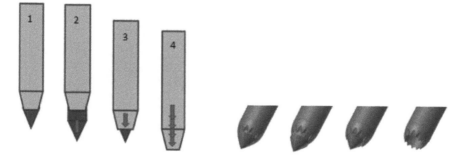

FIGURE 1.95 (Left) Schematic explanation of the principle of operation of the dual hammer penetration device: (1) Starting position, (2) The inner rod is driven, (3) The casing is driven, (4) The casing again is driven several times. (Right) Regarding to the left, the principle of action is shown on a CAD model. *Source:* Courtesy of Astronika.

moving parts are arranged coaxially and that the casing is driven by one hammer and the inner rod is driven separately by the second hammer. Additionally, both movements are not coupled and there is no necessity to keep one-to-one stroke order. This could provide some advantages, e.g. if only the casing is hammered many times then around the tip a hollow rod would be created, which could be useful for storage of a sample of soil (Figure 1.95). But the main advantage of a dual hammer penetration device is that the drive generates lower recoil and this is very beneficial in low-gravity applications.

1.7.6 THE HAMMERING DRILL (HILL)

HILL is Astronika's concept of a tool that provides drilling and hammering capabilities while minimizing the system's recoil in microgravity. The HILL concept (Hammering Drill) utilizes both

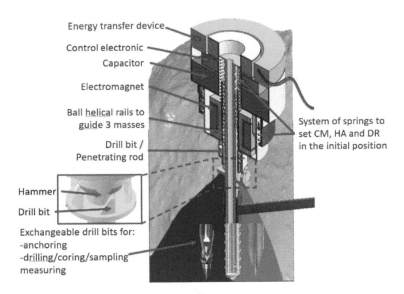

Energy transfer device
Control electronic
Capacitor
Electromagnet
Ball helical rails to guide 3 masses
Drill bit / Penetrating rod
System of springs to set CM, HA and DR in the initial position
Hammer
Drill bit
Exchangeable drill bits for:
-anchoring
-drilling/coring/sampling
measuring

FIGURE 1.96 HILL concept cross section. *Source:* Courtesy of Astronika.

axial and rotational impulses to drill in regolith under microgravity (typical for small Solar System bodies). In principle, a single hammering actuator generates both the rotational and the translational motion of the drill rod (Figure 1.96). Thereby, it combines both advantages of impulse hammering with drilling process—it shall consume less energy, require little lander/rover support, stay operational in microgravity, and be able to go through hard rocks without generating significant heat, which is especially crucial in volatiles prospecting.

The device uses a novel design of a hammer (HA) and a drill rod (DR), through which the hammer strikes. The HA's motion is turned into a composition of drill movements. The stroke generation is based on a modified reluctant electromagnet (EM) principle of operation. Thanks to this solution the device should be able to penetrate through hard rocks and compressed regolith while generating much less heat than, e.g., a conventional drilling system. This increases the performance and significantly mitigates the risk to applications in ISRU missions. A tool of this type is necessary during any operation on an asteroid or a planetary surface, as it allows to transfer sensors below the surface, couple them with the ground (especially important in a vacuum environment), provide anchoring for structures, measure mechanical properties (e.g. compressive and shear strength of the regolith), and perform rock sampling.

Furthermore, a set of different drilling rods with changeable bits will make the mechanism applicable to a greater range of mission types. A special drill bit is envisioned for collecting samples, a different one for assessing mineral and volatiles content of the subsurface, while yet another with heaters mounted on it to, e.g., investigate thermal conductivity of the regolith.

1.7.7 THE HEAT FLOW AND PHYSICAL PROPERTIES PACKAGE (HP³)

The Heat Flow and Physical Properties Package (HP³) was designed to measure the surface heat flow on Mars and includes an instrumented penetrator system to carry thermal sensors to depth. To measure the heat flow, the thermal conductivity of the Martian regolith as well as the subsurface thermal gradient need to be determined. The package uses a self-impelling, slow penetrator termed "the Mole" to emplace the sensors to a depth of up to 5 m (see Spohn et al. 2018, for a detailed description of the instrument package and Grott et al. 2019, for details on instrument calibration).

The first Moles have been developed by VINit Transmash in Russia and are described by Gromov et al. (1997). The Moles with electromechanic hammer mechanisms have been further developed at DLR (e.g., Richter et al. 2004) as sampling tools for the ill-fated Beagle II lander on MarsExpress and for potential heat flow probes on missions such as ExoMars and BepiColombo (Spohn et al. 2001). A second line of Mole development was carried through more or less in parallel at CBK, Warsaw (Grygorczuk et al. 2011; Seweryn et al. 2014d), who constructed a Mole based on an electromagnetic hammering system similar to the hammering mechanism of the MUPUS penetrator flown on the Rosetta mission (Grygorczuk et al. 2007; Spohn et al. 2007, 2015). The present HP³ Mole is a further development of the DLR Mole but with an electromechanic hammer mechanism from CBK and Astronika, Warsaw.

1.7.7.1 The HP³ Mole

The HP³ Mole is a small, low-velocity penetrator driven by a hammering mechanism in its interior (Krömer et al. 2019; Lichtenheldt et al. 2014a; Lichtenheldt, Schäfer, and Krömer 2014b; Wippermann et al. 2020). The hammer mass is accelerated by a spring and impacts periodically on an anvil connected to the Mole's hull, thereby driving the Mole forward. The drive spring is tightened by a cam and roller driven by a motor. The recoil generated by the impact on the anvil is absorbed by a counter-mass, which transmits the energy to the hull via a brake spring. This transmission of force is designed such that the resulting force is equilibrated by frictional forces on the hull. The distribution of masses and springs in the mechanism and the friction of the regolith on the Mole cause the hammering mechanism to function as a mechanical diode directing the Mole preferentially in the forward direction.

An image of the HP³ flight Mole as developed for the InSight Mars mission is shown in the top panel of Figure 1.97, while the interior of the HP³ Mole is shown in the bottom of the figure.

From left to right in Figure 1.97, there is a multi-purpose backcap (shown in red) which serves as the mechanical interface where the harness enters the Mole's interior and as the launch-lock interface. Further to the right is the payload cage (green) that contains tiltmeters and shock isolation springs specifically designed to protect the tilt sensors from hammer shocks. The payload cage connects to the brake spring and a helical cable (brown) that routes electrical lines further down. The suppressor (dark brown)—termed for its mass which is significantly larger than the hammer mass—contains the DC motor, gearbox, and drive train to operate the hammer (purple). The drive train connects to the hammer (made of tungsten) via a cylindrical cam and roller (blue) and the hammer is attached to the force spring (gray). The inner surface of the tip (dark green, made of titanium) provides the anvil against which the hammer is driven (see also Krömer et al. 2019). The shape of the tip, an ogive, has been optimized to reduce the resistance of the regolith against penetration.

The major engineering budgets of the HP³ Mole are summarized in Table 1.5, where mass, size, temperature range, hammering frequency, as well as typical depth progress per stroke are given.

Figure 1.98 illustrates the working principle of the Mole during one hammering cycle. The main masses are color-coded in the figure with the hammer in green, the suppressor in blue and the outer hull in gray. In addition, the payload cage and tiltmeters are shown in yellow and pink, respectively. The continuously operated DC motor drives the cylindrical cam that pulls the hammer away from the tip, thereby compressing the force spring. After the maximum compression is reached (Figure 1.98, leftmost frame), the force spring and the hammer are both released through a gap in the cylindrical cam and the hammer is accelerated towards the tip (Figure 1.98, frame 2). At the same time, the suppressor—suspended from the brake spring—is accelerated backwards (Figure 1.98, frames 2–5). Since the suppressor mass is 4.4 times the hammer mass, most of the force spring energy is transferred to the hammer's kinetic energy. The hammer separates from the spring after it is fully relaxed and continues to move freely before it impacts on the tip anvil for the first major stroke (Figure 1.98, frame 2–3).

After the brake spring has reached its maximum compression (Figure 1.98, frame 5), the suppressor mass is accelerated towards the tip (Figure 1.98, frame 6–7) driven by gravity and the relaxing

FIGURE 1.97 (Top) Image of the HP³ flight Mole in a transport fixture. (Bottom) CAD model showing the interior of the mole. (From left to right) The multi-purpose backcap including the feedthrough for the science tether, the payload cage which houses the static tiltmeters including the shock absorption springs, the brake spring assembly with the helical cable running to the motor, the suppressor and motor, gearbox and drive train, the drive shaft and cam and roller, the hammer, the force spring, and the Mole tip. The ogive shape of the tip minimizes the soil resistance against forward motion.

TABLE 1.5
Properties of the HP³ Mole

Property	Value	Units
Mole Mass	860	g
Mole Size	Ø 27 × 400	mm
Operational Temperature Range	228–323	K
Non-operational Temperature Range	148–323	K
Hammering Frequency	0.28	Hz
Typical Depth Progress per Stroke	0.1–1	mm
Average Power during Hammering	0.75	W

brake spring. Finally, the suppressor casing impacts the tip for a second major stroke to cause additional forward motion of the outer hull (Figure 1.98, frame 7). Due to internal rebounds, a few additional but much less energetic strokes can occur thereafter, but these are relatively weak and do not provide further significant forward motion. A full cycle takes 3.6–3.7 s depending on ambient conditions. Lichtenheldt et al. (2014a) give a detailed analysis of the cycle including an analysis of elastic waves propagating from the Mole and related effects.

FIGURE 1.98 Phases of a hammering cycle (after Spohn et al. (2018), left to right also cf. Lichtenheldt et al. (2014a, b)) showing the interaction between the hammer (green), the suppressor (blue), and the outer hull (gray). (1) Mechanism fully tensioned. (2) The hammer is released and impacts the tip, providing the first stroke. (3) Recoil movement of the suppressor, compressing the brake spring. (4) Recoil movement of hammer and suppressor away from the tip. (5) Suppressor point of reversal is reached. (6 and 7) The brake spring and gravity drive hammer and suppressor tipwards, providing a second stroke. (8) Mechanism at rest.

1.7.7.2 Instrument Testing

The Mole has been operated in various test stands at atmospheric pressure and temperature as well as under simulated Martian conditions to characterize its performance. In addition, the Mole has undergone the usual functional, vibrational, and thermal/vacuum testing during the flight qualification test program. Test stands used to characterize Mole hammering efficiency include setups where the Mole is fixed vertically and hammers against a spring, as well as devices simulating low gravity by hammering horizontally while measuring mole strength using a load cell. A deep penetration testbed was used at DLR Bremen that allowed for penetrating into various test sands to a depth of up to 5 m at ambient temperature and pressure conditions (Wippermann et al. 2020). This testbed is shown in Figure 1.99 and has an inner diameter of about 0.8 m. A similar testbed but capable of achieving Mars subsurface conditions of 213 K and Martian atmospheric composition and pressure was used at the Jet Propulsion Laboratory in Pasadena, California. The latter testbed contains regolith simulant up to 2 m height and the maximum depth of burial for the Mole was 1.7 m. That testbed was also used for a successful full functional test of the HP³ system. Other test sands included vacuum chambers at ZARM in Bremen with about a meter of sand with which the Mole performance at varying ambient pressure was tested. The main regolith simulants for these tests were a nearly cohesionless quartz sand (WF-34), Mojave Mars simulant (MMS, <2 mm size fraction), and a sand/dust mixture consisting of mechanically crushed commercially available (Syar, Inc.) basalt (see Delage et al. 2017; Vrettos et al. 2014, for details about the properties of the test sands).

The dynamic behavior of the Mole and the Mole-soil interaction are highly complex, and the rate of progress depends on soil mechanical properties such as compaction, friction angle, and density. Therefore, test results will to some extent depend on the preparation of the soil and the environmental conditions at the test stand. For the range of simulants tested during deep penetration testing, initial soil insertion was typically quite fast, with penetration rates exceeding 1 mm per stroke

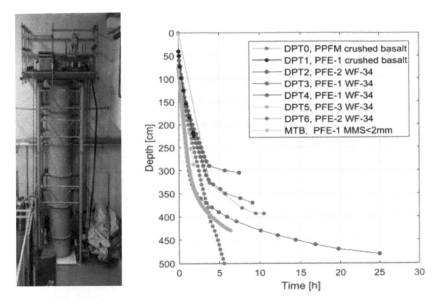

FIGURE 1.99 (Left) 5-m test cylinder at DLR Bremen used for deep penetration testing. (Right) Deep penetration test results in different regolith simulants (after Spohn et al. 2018, see text for details).

(compare Figure 1.99). Under Earth ambient conditions, penetration speed shown in Figure 1.99 typically started with several millimeters per stroke but decreased to values below 0.1 mm per stroke below depths between 3 and 4 m.

1.7.7.3 On Mars Operations

The InSight mission landed on Mars on November 26, 2018, and instruments were deployed onto the Martian surface after an initial phase of landing site characterization. The Mole and the tether with the temperature sensors as well as the tether connecting the instrument to the lander were housed in the support structure shown on the surface of Mars in Figure 1.100. HP³ started hammering on March 1, 2019, but the instruments' tether length monitoring system did not engage, indicating that penetration after the first 4000 hammering strokes was short of 45 cm. A contact sensor inside the HP³ support structure indicated a depth of penetration of at least 15 cm. A second hammering session of 5000 strokes a few days later did not result in further measurable depth progress.

Investigations into the cause of the Mole failing to penetrate deeper considered numerous possibilities including factors internal to the Mole (e.g., mechanism damage) and external factors (e.g., rocks). Over the course of recovery operations, most possibilities were ruled out as either not credible or not plausible. The three remaining root cause possibilities prior to support structure removal were:

1. The Mole had hit a rock sufficiently large to not be pushed aside (>10 cm).
2. The tether had somehow become snagged in the support structure.
3. The Mole did not have enough friction on its hull to balance the recoil from the hammering.

Areal density and size distribution of rocks on the surface along with models that projected the distribution to depth (Golombek et al. 2018) made the first possibility highly unlikely, but it could not be entirely discounted. Snagging of the tether was excluded after the controlled multi-day lift of the support structure to expose the Mole showed only smooth tether extraction. Only after the HP³ support structure was removed from the Mole on July 1, 2019 could the situation be assessed as shown in Figure 1.101. Following the lift (the support structure was lifted and re-placed on Sols

FIGURE 1.100 SEIS (The Seismic Experiment for Interior Structure) and HP³ (The Heat Flow and Physical Properties Package) on the surface of Mars. The image shows the support structure. The Mole is housed in the chimney-like part facing away from the camera. The tethers are housed in the containers behind the chimney.

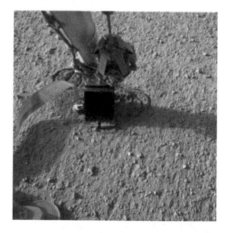

FIGURE 1.101 (Left) Instrument Deployment Camera (IDC) image showing the mole and surrounding pit before interaction with the robotic arm. The image was acquired on July 26, 2019 (Sol 235). (Right) Image of the mole after interacting with the robotic arm's scoop and hammering. The image was acquired on October 21, 2019 (Sol 320). *Source:* Courtesy of NASA/JPL-Caltech.

203–211), the most likely root cause remaining was #3: lack of friction on the hull to balance the recoil.

It was expected that the landing site regolith would consist mostly of cohesionless sand (Golombek et al. 2017), and this is what the Mole was designed to penetrate. Friction from cohesionless sand was estimated to be sufficient to balance the resulting recoil force of 3–5 N (the exact value depending on resistance of the regolith to penetration) if the Mole was buried to a depth of ~30 cm. Contrary to the test cases and other landing sites on Mars, a significant amount of the subsurface regolith at the InSight landing site turned out to be highly cohesive with considerable crushing strength estimated at 350 kPa (Spohn et al. 2019). From previous surface missions, a duricrust layer had been expected, but with a thickness of at most a few centimeters. Once exposed, the Mole appears to have carved a pit about twice as wide as the Mole and several diameters deep, suggesting that the duricrust was at least 7 cm thick. During initial penetration, when the Mole was still partly in the housing of the support structure, friction was provided by guiding springs in the support structure. Once the Mole left the lowermost guiding spring at a depth of ~35 cm, a significant portion of the length of the Mole was not in contact with the regolith, standing essentially in the air inside the pit. Without sufficient friction the Mole started to bounce (jostling the Mole and support structure and thereby widening the hole) and the Mole failed to penetrate further.

After assessing the situation and making some attempts to collapse the pit with the InSight robotic arm, the arm's scoop was used to push against the upper surface of the Mole (Figure 1.101, right). The applied horizontal and vertical force translated into increased friction with the regolith along the Mole's underside. This "pinning" technique was first used on Sol 308 resulting in forward (downward) progress of the Mole of about 0.5 cm. Subsequent hammerings with pinning allowed the Mole to penetrate nearly five further centimeters into the regolith until the backcap was nearly flush with the original regolith surface. This progress, aided only by added friction from the arm, simultaneously confirms root cause #3 (lack of friction) and rules out root cause #1 (obstructing rock). Pinning would only allow the Mole to go as deep as shown in Figure 1.101 (right), but even at this depth of nearly 40 cm the cohesive regolith and wide pit do not provide sufficient friction for the Mole to dig without help from the arm. Thus, some new technique is called for to help the Mole progress enough into the (presumably) unconsolidated regolith below the duricrust layer so it can penetrate unassisted. At the time of writing, efforts to assist the Mole using the scoop are continuing, and options to pin from another direction, to push directly onto the Mole's backcap, and to scrape soil into the pit are being evaluated.

1.7.8 PERCUSSIVE HEAT FLOW PROBE

Lunar heat flow measurements are important for studying the global heat flow budget for the Moon and the distribution of heat-producing elements in the crust and mantle (National Research Council 2011a). To determine heat flow properties, the probe needs to measure thermal conductivity and thermal gradient (Nagihara et al. 2018). In order to obtain the best measurements, the sensors must be extended to a depth of at least 2 m, i.e. beyond the depth of thermal cycles.

We investigated a percussive approach for deployment of a heat flow probe (Zacny et al. 2013e). A percussive approach utilizes a high frequency hammer to drive a cone penetrometer into the lunar regolith. Initially, we investigated Top-Down and Bottom-Up architectures of sensor deployments (Table 1.6). In the Top-Down approach sensors were deployed as the penetrometer was penetrating into the regolith, and in the Bottom-Up approach, sensors were deployed as the rods were being pulled out of the hole. The Top-Down approach was studied previously, and while sensor deployment was proven to be successful, complexity of the system significantly reduced its efficiency. The Bottom-Up approach seemed to have mitigated some of the problems encountered by the Top-Down approach, such as collapse of borehole, and elimination of additional actuator for sensor deployment. However, during the preliminary cone probe deployment tests, the retraction was found to be rather implausible by an automated system and alternative options were investigated for a solution.

TABLE 1.6
Summary Trade Study

Top Down	Bottom Up	Modified HP3 Mole (ModMole)
		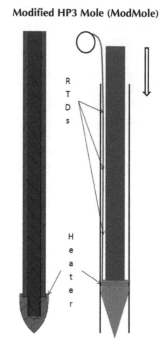
Sensors deployed on the way down	Sensors deployed as probe retracts.	Cone with T sensor and heater penetrates down. Additional sensors are tethered to the cone.

Since neither of these options were robust, the third option was considered—Modified HP³ Mole approach (ModMole), whereby the Mole is pulling a thermal tether behind it.

In the ModMole approach the hammer is placed above the ground and hence it can deliver greater blow energies at higher frequency. In addition, the cone diameter is reduced and as such the probe area which drives the penetration resistance is reduced. These changes mean that the probe can be deployed in less time and the probability of hitting a rock that might stop the probe from further penetration is significantly reduced. The drawback to the ModMole approach is that it requires a Z-stage system and a carousel for feeding the drive rods. However, since the rods have very low thermal conductivity, they can remain inside a hole—this greatly simplifies rod joints and reduces mass and volume along with the possible failure modes.

Figure 1.102 shows the probe mounted on a lander. There are only two actuators besides the percussive hammer within the whole system. One controls the linear motion of the hammer and the other feeds the rods within the carousel. The carousel utilizes tapered friction hold to keep the rods in place. The carousel also serves as a rod guide for the borehole and a guide rail for the Z-stage's rotational motion. The rod guide feature allows for more efficient transfer of percussive energy to the penetrometer. It also allows for misalignment. A Current Best Estimate (CBE) mass of the system is 15 kg. It fits within a box, 25 cm × 25 cm × 127 cm.

To perform penetrometer tests, we used the LITA hammer drill system in hammer-only mode. The hammer provided 2 J/blow at up to 30 Hz. The hammer and its Z-stage were bolted on to the structure fixed into the wall. Modifications were made to the rod guide and the hammering interface

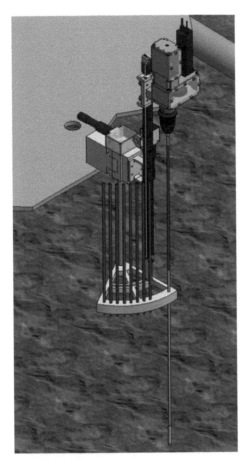

FIGURE 1.102 Concept of the percussive heat flow probe. *Source:* Courtesy Honeybee Robotics.

to accommodate the new ModMole design. We performed a number of tests to determine whether the borehole would remain stable during penetration (Figures 1.103, 1.104, and 1.105). We used a 1.27-cm cone diameter and steel rods. The NU-LHT-2M soil was vibratory compacted for 10 min to achieve 2 g/cc. Our tests showed that the borehole in fact collapses. This is a different result from our prior tests with JSC-1a soil simulant. It seems JSC-1a is more cohesive than NU-LHT-2M and in turn produces a stable borehole.

Although Apollo astronauts did find that boreholes remained open after drilling, this might not be the case if the hole is drilled near the crater rim which has more fluffy soil. As such, we should expect the hole to collapse or remain open depending on soil properties (density). It should be noted that if soil collapses, this would increase thermal conductivity to the tether and in turn result in potentially more accurate thermal data.

Figure 1.106 shows penetration tests with several cone-needle designs. Steel tube 1.9 cm OD with 0.21-cm wall thickness was used for penetration and retraction testing. Different cone geometries were used to determine needle integrity and penetration rates. With a 1.9-cm rod and cone, a depth of 839 mm was reached in 765 s. Increasing cone diameter (while keeping the rod at the same 1.9 cm diameter) resulted in a lower rate of penetration. It was also determined that in some cases, thermal needles below the cone used to measure thermal conductivity of soil, would tend to bend and even break (see second cone from the left in Figure 1.106). Hence this design is susceptible to possible stalling in penetration rate and catastrophic failure of thermal needles (which make the probe unusable).

FIGURE 1.103   Bin with NU-LHT-2M simulant. *Source:* Courtesy Honeybee Robotics.

Since a dedicated thermal needle is prone to breakage, it was decided to combine a cone and a needle into one. To make the cone look more like a needle (and in turn improve its characteristics as a thermal probe), the cone diameter was made much smaller (10 mm) and the cone length was made much longer (80 mm). This aspect ratio of 8:1 makes it almost an ideal thermal probe.

We tested two cone geometries as shown in Figure 1.107. Figure 1.108 shows a fully inserted penetrometer. Both geometries showed similar penetration rates until ~600 mm depth. Beyond that depth, geometry A penetrated deeper than B in the same amount of time. The rate of penetration was 2.7 and 2.3 mm/s for geometry A and B, respectively. For this reason, cone A was chosen for the final probe design.

We also performed penetration tests with several extension rods to determine the length effect of the rods on the penetration rate. Deployment rods were made of G-10/FR4 Garolite with low (negligible) thermal conductivity of 0.288 W/m-K. The tests showed little to no change from the penetration tests with just one rod.

1.8   SUCTION-BASED SAMPLERS

Honeybee Robotics developed several stand-alone soil sampling systems for use with small-scale drones that could one day explore Venus and Titan—planetary bodies with significant atmosphere— and Mars.

FIGURE 1.104 Hammer drill with one of the penetrometers. *Source:* Courtesy Honeybee Robotics.

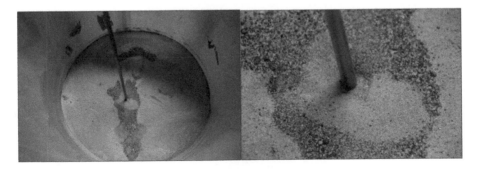

FIGURE 1.105 Percussive heat flow probe borehole collapse experiments. *Source:* Courtesy Honeybee Robotics.

Initially, research was done to determine the best methods for separating particles from air for this application. The most reasonable methods were the settling chamber, a settling chamber with baffles, and the cyclone separator. Simple representations of these technologies are shown in Figure 1.109. The settling chamber slows the flow by increasing the cross-sectional area of the duct allowing particles to fall from the stream under the acceleration of gravity. The baffle chamber works in the same fashion, but the flow is diverted using baffles. These barriers make the flow rapidly change direction and separation occurs because particles resist change in direction due to their inertia.

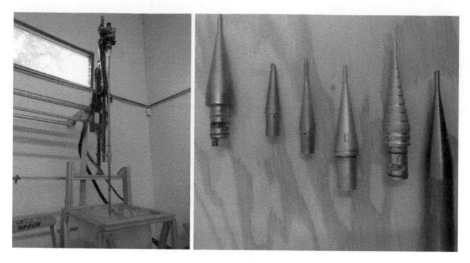

FIGURE 1.106 Percussive heat flow probe penetration tests with different cone-needle geometries. *Source:* Courtesy Honeybee Robotics.

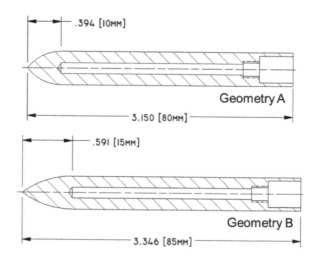

FIGURE 1.107 Different cone geometries tested for rate of penetration. *Source:* Courtesy Honeybee Robotics.

The cyclone separator works by inducing a vortex and flow direction reversal to achieve inertial separation.

Each method has advantages and disadvantages. The settling chamber is generally good for collecting larger particles and creates a lower pressure drop in the system. A settling chamber could consume a large amount of space, however. Adding baffles reduces the special requirement, but increases the pressure drop. The cyclone separator is generally better at collecting a larger range of particle sizes and it can be a more compact system. A cyclone separator has a higher pressure drop though because the flow must change directions many times.

Prototypes for these three separation methods were developed. The concepts were tested using medium-grain sand. Initial calculations were performed to determine the correct dimensions for each design. Pickup and saltation velocities were researched and calculated to get an approximate

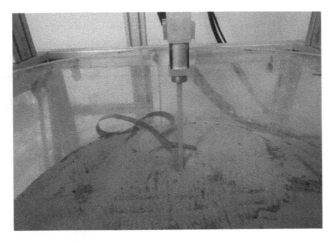

FIGURE 1.108 Non-adhesive Kapton tape was used to simulate the flex circuit being tethered down to the borehole. *Source:* Courtesy Honeybee Robotics.

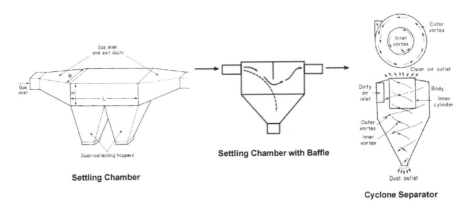

FIGURE 1.109 Methods to remove soil particles from air. *Source:* Courtesy of Flagan and Seinfeld (1988).

estimate of the flow velocities needed to pick up particles and transport them. Also, two fans, 50 and 80 mm, were selected for prototyping.

The first settling chamber prototype was designed for the 50-mm fan. The collection efficiency for a given particle size can be calculated given the settling chamber dimensions and fluid flow properties (Flagan and Seinfeld 1988). The dimensions of the chamber were specified to collect particles 50 μm and greater. The efficiencies are higher for larger particles. The system was designed and 3D printed for testing. Additionally, baffles were made to insert in the settling chamber for testing the baffle chamber concept. Figure 1.110 shows the sampler prototypes. Note that the collection containers were not printed and are not pictured. Also, an attachment was left for the sampling nozzles, which were not developed at this time.

The cyclone separator has been used often at Honeybee Robotics. A high-throughput design was chosen for the application. The cyclone separator was also 3D printed and assembled.

Tests demonstrated that all the concepts worked. The settling chamber collected the largest particles with the largest being around 1 mm. Fine particles were sent through the fan, however. Adding baffles reduced the amount of fine particles lost through the fan. The cyclone separator was better at

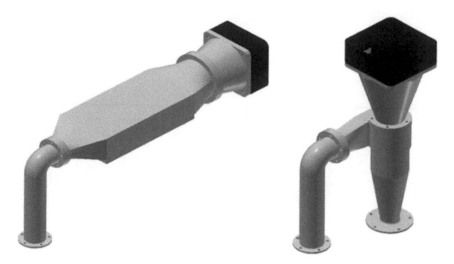

FIGURE 1.110   Suction-based sampler concept prototypes. *Source:* Courtesy Honeybee Robotics.

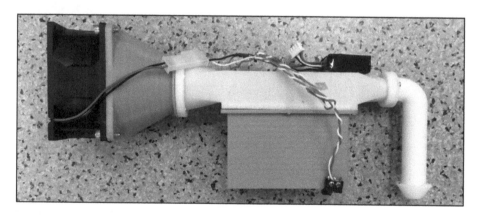

FIGURE 1.111   Pneumatic suction sampler prototype. Suction inlet is on the right, suction fan is on the left (black rectangle), and container for captured sample is a blue rectangle in the middle. *Source:* Courtesy Honeybee Robotics.

collecting the fine particles, but struggled to pick up particles larger than 710 μm. The larger 80-mm fan was also used to test the prototypes. For the baffle chamber, more particles were sent through the fan because the chamber dimensions were undersized for the increased flow velocity. For the cyclone separator, performance improved and larger particles were collected.

Honeybee Robotics decided to move on with the baffle chamber concept because it is more efficient in collecting large particles with less power. The inlet tube was the same for both prototypes and reducing its diameter should increase the flow velocity at the inlet. This should help in collecting the larger particles. Next work will focus on designing the inlet and nozzle. Various tube sizes will be analyzed and tested to gain understanding in the trade-off between inlet flow velocity and available surface area for collection.

Figure 1.111 shows a more complete prototype using the baffle chamber separation concept. The prototype includes a limit switch which will trigger the fan once the drone lands. The fan is powered by a small onboard 12-V battery. A nozzle attachment was added to the end of the inlet tube.

FIGURE 1.112   3DR Solo with suction sampler. *Source:* Courtesy Honeybee Robotics.

Honeybee chose a baffle chamber-type sampler because the pressure losses in the system were less than the cyclone separator, and it collected a larger portion of the material than the settling chamber. The baffle chamber design worked as expected.

After the first prototype was built, effort was dedicated to building a functional prototype of the pneumatic soil sampler for a commercial drone. Honeybee has the 3DR Solo and DJI Phantom 4 drones on hand. The 3DR Solo was selected as the craft for the initial prototype because it has more room onboard for mounting accessories. In fact, there is a 3DR Solo Development Guide online that provides information for integrating equipment with the drone. After studying the development guide and disassembling the 3DR Solo, Honeybee decided to try to run the sampler from the drone's battery and control the sampler using the drone's remote controller.

First, the sampler was resized to fit under the Solo, and the mechanical mounting solution was designed. Figure 1.112 shows the sampler on the 3DR Solo.

A Zortrax M200 3D printer was used to make the sampler components. The mount was designed to clip onto the underside of the Solo and provide a mounting surface for the baffle chamber. The mount was printed from Zortrax's yellow ULRAT plastic. The baffle chamber and its inlet and outlet nozzles were printed in black ULRAT. Most parts were printed from ULRAT with the exception of the sample container.

One of the landing gear legs of the Solo was replaced with a specially designed sampling leg. Figure 1.113 shows the custom sampling leg.

The foot of the sampling leg has an opening for attaching a sample collection nozzle. The sample collection nozzle is the yellow ULRAT part attached to the foot of the leg. The sample collection nozzle has inlet holes located on the top surface of the cone. These inlet holes allow air to make its way through the system even if the bottom of the cone is buried in soil. Figure 1.114 shows how the baffle chamber is attached to the mount with four flathead screws.

The sample container can be seen clearly in Figure 1.115. The sample container was printed from Zortrax GLASS plastic which is semi-transparent allowing the user to see if material is captured in the container. It has clips on either side to allow for easy removal and reattachment.

FIGURE 1.113   3DR Solo with sampler, close-up of sampling leg. *Source:* Courtesy Honeybee Robotics.

FIGURE 1.114   3DR Solo with sampler, bottom view, without sample container. *Source:* Courtesy Honeybee Robotics.

A 10-mm ID tube is routed from the sample collection nozzle to the baffle chamber inlet. The baffle chamber has slots for inserting up to 3 baffles. The fan used is a Mechatronics 80-mm 12-V fan. The baffles can be added or removed to change the range of particle sizes collected.

The fan is powered from the 14.8-V 5200-mAh onboard battery. Also, an I/O pin on the drone's PixHawk 2 flight controller was used to turn the fan on and off. The Mission Planner flight controller application was used to program the drone remote controller to turn the fan on and off by pressing the "pause" button. The I/O pin from the flight controller was connected to a MOSFET gate which operates a relay driving the fan.

FIGURE 1.115   3DR Solo sample container. *Source:* Courtesy Honeybee Robotics.

The breadboard was installed at the back of the drone under where the main board typically sits. After the prototype was built, the sampler was tested. A container was filled with slightly compacted playground sand. The drone was dropped from a height of 4 in into the sand, and the fan was turned on. The fan took approximately 5 s to reach a constant speed. The fan was run for 15 s before it was turned off. The sampler collected 30 g of sample.

The next steps for this portion of the project would be to further test the sampler in different soils with varying conditions, iterate on the current prototype, and develop a commercial product. Work could also be given to developing a smaller sampler that could integrate with the DJI Phantom 4.

1.8.1   Drone Deployable Pneumatic Sampler for Mars

Honeybee developed four designs of a pneumatic soil sampling system that could be used on the surface of Mars. It is assumed that an aerial vehicle would have enough lift capacity for carrying of a sampling system.

The first pneumatic sampling concept is shown in Figure 1.116. This concept would be suspended from the bottom of the vehicle using three cables in a tripod configuration. Suspending the sampler from the vehicle will allow it to conform to the local surface normal, creating a better seal around the lower circumference of the sampler. This will help to maximize the amount of sample acquired.

The components of Pneumatic Sampler Concept #1 are shown in Figure 1.117. A Commercial Off the Shelf (COTS) pressurized CO_2 canister is inserted by the user into a housing. The canister is held in place using a threaded end cap. These one-time-use CO_2 canisters have a thin metal foil which seals the end of the canister, preventing the gas from escaping. This foil needs to be punctured before gas can be released. Threading the end cap into the housing compresses the CO_2 canister against a puncture mechanism (not shown) and holds it in place for flight.

Once the CO_2 canister is punctured, the gas then passes into a Miniature Solenoid Valve. A rechargeable battery (not shown) will provide the power for the solenoid valve, rather than having to rely on power from the main vehicle. When a sample is desired, the solenoid valve is opened, allowing gas to flow into the central collection tube. The pneumatic system uses the Venturi effect

FIGURE 1.116   Pneumatic Soil Sampler Concept #1. *Source:* Courtesy Honeybee Robotics.

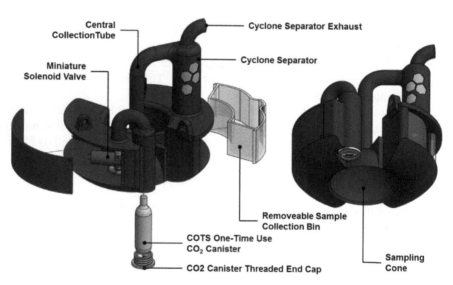

FIGURE 1.117   Pneumatic Soil Sampler Concept #1 components. *Source:* Courtesy Honeybee Robotics.

to create a flow of gas from the sampling cone into the cyclone separator. Fine particles under the sampling cone will be caught up in the flow, and sent to the cyclone separator.

Within the cyclone separator, as fine soil particles flow in, they will hit the interior wall of the cyclone, losing momentum and eventually falling downward into a Sample Collection Bin, which can be easily removed by the user after the vehicle returns. The remaining gas will flow out of the cyclone separator exhaust, which is directed to the side, away from the main body of the aerial vehicle.

FIGURE 1.118 Pneumatic Soil Sampler Concept #2. *Source:* Courtesy Honeybee Robotics.

This system is designed primarily for the collection of fine-grained material such as sand. Larger particles such as gravel will most likely not be collected using this method. This concept will be traded against the other four concepts to determine whether to move forward with a prototype.

The second Pneumatic Soil Sampler Concept is shown attached to aerial vehicle in Figure 1.118. Similar to the previous concept, it is suspended using three cables from the underside of the main vehicle body in such a way that landing the vehicle will release the tension in the cables and allow the sampler to conform to the surface to be sampled.

The components of Pneumatic Sampler Concept #2 are shown in Figure 1.119. Compared to other concepts, this system has a significantly larger footprint area, with the goal of acquiring larger

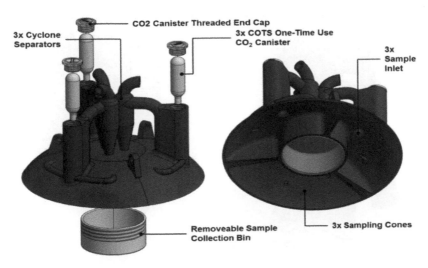

FIGURE 1.119 Pneumatic Soil Sampler Concept #2 components. *Source:* Courtesy Honeybee Robotics.

FIGURE 1.120   Pneumatic Soil Sampler Concept #3. *Source:* Courtesy Honeybee Robotics.

amounts of sample compared to the other sampler concepts, with the tradeoff being a larger, potentially heavier sampler.

This concept utilizes three separate COTS CO_2 canisters, connected to three independent cyclone separators. As in the previous concept, each set of canisters and cyclones use the Venturi effect to create a suction force to pull fine soil samples from the sampling cones into the cyclone separators.

A third pneumatic sampling concept is shown in Figure 1.120, hanging from the main body of an aerial vehicle. The three cyclone separators (Figure 1.121) deposit material into a single sample

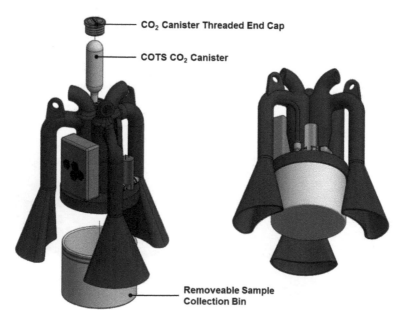

FIGURE 1.121   Pneumatic Soil Sampler Concept #3 components. *Source:* Courtesy Honeybee Robotics.

FIGURE 1.122 Pneumatic Soil Sampler Concept #4 with Skyjib Vehicle. *Source:* Courtesy Honeybee Robotics.

container. After recovery of the aerial vehicle, the user can simply unscrew the container from the Pneumatic Sampler. Not shown in this conceptual design is the solenoid valve which controls the gas flow, as well as the battery required to power the valve. This system would require a wired connection to the aerial vehicle in order to provide a triggering signal.

This system is designed primarily for the collection of fine-grained material such as sand. Larger particles such as gravel will most likely not be collected using this method. This concept will be traded against the other three concepts to determine whether to move forward with a prototype.

This concept is similar to Concept #2, in that it utilizes three separate sampling cones and three different cyclone separators which deposit samples into a single centralized Sample Collection Bin. This design is significantly reduced in footprint compared to Concept #2 and uses a single COTS CO_2 canister to provide pressurized gas to all three cyclone separators. This system will also use the Venturi effect to create a suction force on the soil.

This system is designed primarily for the collection of fine-grained material such as sand. Larger particles such as gravel will most likely not be collected using this method.

A fourth pneumatic sampling concept is shown suspended from an aerial vehicle in Figure 1.122. Similar to the previous conceptual designs, when contacting the surface to be sampled, the cables or ropes will allow the sampler to conform to the local surface prior to sampling. This helps to increase the efficiency of pneumatic sampling by preventing wasted gas from "leaking" out between the base of the sampler and the surface.

Figure 1.123 shows a cross-sectional diagram of Concept #4. This system uses a centralized COTS CO_2 canister, which is inserted from the top of the sampler, and held in place with a threaded end-cap.

The CO_2 canister would connect to a small solenoid valve (not shown), which would allow gas to flow into multiple spray nozzles in the center of the sampler. These nozzles would point downward at an angle in order to help "stir up" soil particles and move them up over the "Sample Retaining Wall" and into the Sample Collection Area. Filter material on the outer wall of the sample collection area would allow gas to flow out of the collection area, while trapping particles. It is expected that this pneumatic sampling concept would have the ability to sample both fine particles such as sand, as well as slightly larger particles such as fine gravel. After return of the aerial vehicle, the Sample

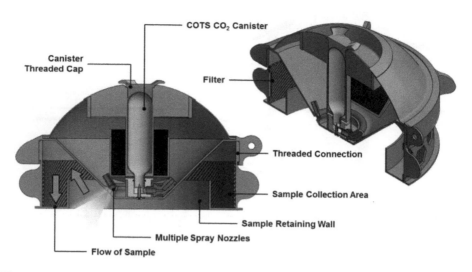

FIGURE 1.123 Pneumatic Soil Sampler Concept #4 components. *Source:* Courtesy Honeybee Robotics.

FIGURE 1.124 Pneumatic Sampler mock-up on UAV. *Source:* Courtesy Honeybee Robotics.

Collection Bin would be unthreaded from the main Pneumatic Sampler body in order to retrieve the sample.

The sampler with piston and universal joint is shown in Figures 1.124 and 1.125. This illustrates the sampler in contact with an uneven surface (relative to the base of the landing legs), showing that the universal joint helps to conform to the uneven terrain.

1.8.2 DRILL FOR ACQUISITION OF COMPLEX ORGANICS (DrACO) FOR DRAGONFLY

Drill for Acquisition of Complex Organics (DrACO) is a sample acquisition and delivery system designed for the New Frontiers Dragonfly mission to explore Titan (Turtle et al. 2017a, b; Lorenz et al., 2018; Zacny et al. 2019). The purpose of DrACO is to provide Titan surface and

FIGURE 1.125   Pneumatic Sampler conforming to uneven surface. *Source:* Courtesy Honeybee Robotics.

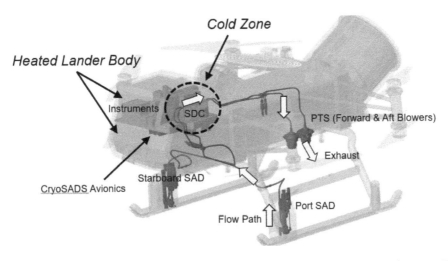

FIGURE 1.126   Example end-to-end sampling system configuration of DrACO. *Source:* Courtesy Honeybee Robotics.

subsurface samples to two sample interfaces for the Dragonfly Mass Spectrometer (DraMS, Trainer et al. 2017): Laser Desorption Mass Spectrometry (LDMS) and a Gas Chromatograph Mass Spectrometry (GCMS). The end-to-end configuration of the sampling system within the Dragonfly spacecraft is shown in Figure 1.126 and the details of the pneumatic acquisition and delivery are shown in Figure 1.127.

DrACO consists of four major subsystems: the Sample Acquisition Drills (SAD), the Pneumatic Transport System (PTS), the Sample Delivery Carousel (SDC), and associated Avionics. Drill cuttings are suctioned directly from the cuttings pile generated by the drill bit, and pneumatically conveyed in a fast-moving stream of ambient Titan air to minimize any temperature rise, reducing risk of sample alteration or fouling of the transport system. In contrast to traditional systems, the DrACO pneumatic architecture is gravity agnostic. The end-to-end system features two redundant sampling

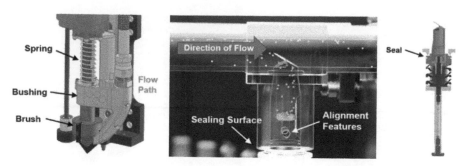

FIGURE 1.127 Details of sample acquisition and delivery (from left to right): SAD/PTS—drill bit with pick-up foot, PTS/SDC—sample cup inserted in sample collection chamber, SDC/Instrument—sample cup sealed against instrument analysis chamber. *Source:* Courtesy Honeybee Robotics.

drills and two redundant suction blowers in a "cross-strapped" configuration: either drill can deliver samples to either LDMS or GCMS using either blower.

Because the subsystems are connected by simple bent tubing and standard fittings, they can be located anywhere aboard the lander, giving mission designers the flexibility to trade off overall tubing length against other high-level vehicle configuration objectives, such as thermal control zones and vehicle center of mass. In the example provided, the entire sample transport path is exposed to the Titan ambient environment to minimize temperature rise. Due to the impact on pressure drop, increased tubing length comes at the cost of increased blower power consumption.

A Sample Acquisition Drill (SAD) consists of a drill head, feed stage, and a drill bit. The drill uses a rotary-percussive action to efficiently penetrate through the strongest Titan material expected, water-ice-cemented regolith ~80 MPa in strength. The rotation and percussion are driven by separate actuators, allowing for drilling flexibility using rotation with or without percussion. Percussion, particularly in hard materials, allows efficient drilling which reduces heating and alteration of sample. The heart of the drill head is an extremely robust and efficient cam-spring-based hammering system with Apollo Lunar Surface heritage. Honeybee Robotics has adapted this approach to over a dozen drills, ranging from near surface to 100s of meters deep, and for planetary conditions ranging from Venus to the lunar surface.

Two fully redundant SADs are mounted on the insides of opposing landing leg trusses, placing the drills close to the surface, and reducing length of the feed stage. Mounting the drills 10 cm above the bottom of the lander feet prevents damage during landing. And with a range of 20 cm the feed stage can place the bit up to 10 cm below the footpads. This enables drilling to at least 3 cm depth even in a situation where there is a 7-cm pit directly underneath the drill.

The 4.6-cm diameter drill bit has been designed to produce a sufficient volume of material to meet sampling volume for the mission. The bit is conical in shape to avoid getting stuck and has hybrid-grade carbide cutting teeth (an ideal combination for both rotary and hammer drilling). A pilot drill at the bit's center contacts the surface first, allowing for preliminary ground interrogation for hardness and cohesion via imagery and measurement of motor current. These images and other observations of the landing site inform whether or not to drill and/or ingest sample at a particular landing site.

The Pneumatic Transfer System (PTS) operates on the same principle used in commercial vacuum cleaners; suction blowers rapidly convey samples from the surface to the collection ports in the SDC (Sparta et al. 2019). Titan's 94-K air temperature, 4.4× higher air density, and 1/7th gravity result in pneumatic transport requiring less power than on Earth. Lifting cohesive particles on Titan thus requires 3× less power and transporting a particle on Titan requires 36× less power.

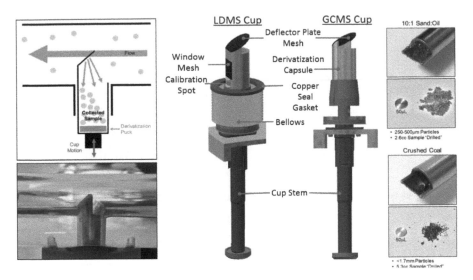

FIGURE 1.128 DrACO's one cup-one sample approach eliminates the need to clean cups. Pneumatic sample delivery is gravity-agnostic and keeps the sample in motion until coming to the rest in a cup, minimizing risk of fouling the pneumatic system. Excess sample is exhausted. Tests with a range of Titan analog materials confirm robust, repeatable sample capture. *Source:* Courtesy Honeybee Robotics.

Pneumatic sample collection has five advantages: (1) the delivery point can be at a distance from the drill, because the connecting tube can be routed around the vehicle. (2) Sample temperature remains cold during transfer because the medium is 94-K Titan air. (3) Sample transfer does not rely on gravity, a strong advantage if moving potentially cohesive material. (4) Cross contamination is significantly reduced by running air through the pneumatic lines prior to and after sample transfer. (5) Risk of clogging is minimized because the sample remains in motion until captured by the LDMS and GCMS sample cups (Figure 1.128) or exhausted outside the lander.

The PTS is physically decoupled from the drill, such that loose, unconsolidated samples can be acquired without using the drill, providing a contingency operating mode in the highly unlikely scenario that both drills cease functioning. This also enables interrogation of cohesiveness of the surface material.

The input to the PTS is a suction nozzle with 180° coverage around the drill bit. The opening is designed to exclude >1.5-mm particles that could foul the pneumatic system. The inlet is continuously cleaned by the rotating bit, which carries wire mesh brushes to scrub away debris. The PTS uses a traditional centrifugal blower design, with impellers designed for Titan air. A Diverter valve at the blower exhaust allows the system to both suck and blow; blowing is available potentially to unclog holes from the suction foot and to clean transfer lines.

The Sample Delivery Carousel (SDC) inherits significantly from the highly successful Sample Manipulation System (SMS) developed by Honeybee Robotics for the MSL Sample Analysis at Mars (SAM). As on SAM, the carousel's two rings rotate together: the inner ring carries sample cups to feed the GCMS inlet and the outer ring cups feed the LDMS inlet, manipulated by elevators at each of four stations around the carousel.

The LDMS and GCMS cups collect solid particles from the moving airstream by redirecting them into the interior of the cup with a deflector plate that extends into the flow (Costa et al. 2019). The cup is self-metering (cannot be overfilled) and nominally single-use, thereby avoiding complications of cleaning. Each LDMS cup features a fine "mesh" viewing window that contains sample

TABLE 1.7
DrACO Crilling Test Results

Simulant	Drilling Mode	Rate of Penetration (mm/s)	Drilling Power (W)	Specific Energy (Whr/cc)
Water Ice	RP	1.89	59.6	0.0052
Ammonia Solution 28–30%	RP	1.99	56.3	0.0047
Paraffin Wax	RP	1.12	65.1	0.0096
Water Ice	R	0.06	19.3	0.0522
Ammonia Solution 28–30%	R	0.13	34.1	0.0428
Paraffin Wax	R	0.27	22.4	0.0138

RP = Rotary Percussive, R = Rotary Only.

while preserving 87% of the viewable area, exposing the sample to the DraMS UV laser for compositional analysis.

Extensive testing has been performed to characterize the behavior of a range of Titan simulants (both room temperature and cryogenic analog materials) during drilling, transport, and collection (Rehnmark et al. 2018a; Sparta et al. 2018). Major accomplishments include tests of drilling into cryogenic material, pneumatic pickup and transport of room temperature simulants, sampling at Titan-like conditions, and end-to-end demonstration of the entire sampling chain.

Three materials (water ice, 28–30% ammonia/water solution, and paraffin wax) were identified as suitable Titan surface simulants for use in cryogenic drilling tests. Samples of these materials were prepared and cooled to Titan surface temperature prior to drilling by first placing them in a freezer at −80°C and then transferring them to a boiling liquid nitrogen (LN_2) bath at ~77 K for further cooling to reach a final temperature of <100 K. The samples remained in the LN_2 bath for the duration of each drilling trial. The drilling test results are summarized in Table 1.7.

The data clearly demonstrates the advantage of drilling with both rotation (auger) and percussion (hammer) vs. drilling with rotation only. For both ice and ammonia, ~90% less energy per unit volume is expended in penetrating the simulant and reducing it to fine particles when percussion is used. The rate of penetration is over 30x faster in ice and 15x faster in ammonia, for the same controlled weight on bit of 100 N. The difference is less pronounced in wax, with 30% less specific energy and 4x faster penetration when using percussion.

1.9 PNEUMATIC SAMPLERS

Pneumatic Samplers use compressed gas to stir up and convey material into a sample container for either in situ analysis or sample return missions. Gas is essentially used as a transport medium and since gas pressures and flow rates are high, the flow is not strongly gravity dependent. The pneumatic mining systems include actual pneumatic components (valves, tanks, pressure, and temperature sensors) as well as a sampling head, transfer tube, and delivery canister. Since pneumatic components can be used from propulsion systems, they would have flight heritage and in turn won't need any further technology development, unless specifically required to meet science mission objectives (e.g. cleanliness level).

Figure 1.129 shows three steps in pneumatic mining: sample acquisition, transfer, and capture. There are several options for sample acquisition, but in general, compressed gas is used to stir up and then convey material. The exact location and geometry of the sampling head would depend on

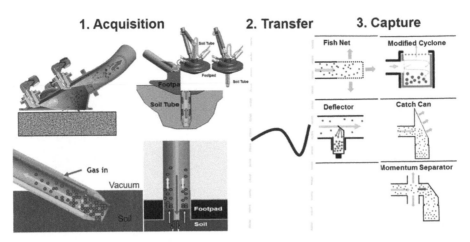

FIGURE 1.129 Various approaches to pneumatic sampling. *Source:* Courtesy Honeybee Robotics.

mission requirements (e.g. sample size). Sample transfer is essentially a tube that conveys sample to its destination. Concerning sample capture, there are numerous ways of separating particles from the gas stream. Additional features can be added to also allow sieving/sorting—capturing small or large particles.

While this architecture is not traditional, it solves many challenges. Pneumatic systems are significantly more powerful than gravity; everyone has moved dense particles including pennies up a vacuum cleaner hose against Earth's gravity. Pneumatic systems can also work well with dry and sticky materials as observed by powerful shop vacs easily cleaning up water and mud. Additionally, the flexible vacuum hose can be routed around any obstacle enabling a connection between the sampling head on the surface to the instrument on the other side of the lander. To reduce cross contamination, the system can be cleaned between the samples (using gas) and each sampling cup or canister could be used once (Zacny et al. 2004, 2008b).

Sullivan et al. (1994) evaluated pneumatic transfer for the movement of lunar regolith at lunar gravity conditions (and atmospheric pressure) on NASA's KC-135 reduced gravity aircraft. It was determined that choking velocity (velocity to keep particles afloat in vertical flow) is 3× lower at lunar gravity. Additional tests in vacuum conditions and lunar gravity (in reduce gravity flights) have shown that 1 g of gas at 60 kPa can successfully lift ~6000 g of soil particles at high velocity (Zacny et al. 2010b). This high efficiency ratio is attributed to the high-pressure ratio between the tank and outside vacuum.

Some of the pneumatic approaches have been covered in early sections—these include OSIRIS-REx Sampler and drone deployable samplers, for example. Here, we cover Pneumatic Samplers used on landed missions.

1.9.1 PLANETVAC

Honeybee Robotics built and tested a lander-based passive pneumatic sampling system known as PlanetVac (Figure 1.130). Testing was performed in a vacuum chamber and with two planetary simulants: Mars Mojave Simulant (MMS) and JSC-1A lunar regolith simulant. Demonstrations included a drop from a height of ~50 cm, deployment of sampling tubes, acquisition of regolith into a regolith box and the rocket, and the launch of the rocket (using compressed gas). In all tests, samples were successfully acquired to the regolith box and the rocket. The gas efficiency was measured to be ~1000:1; that is 1 g of gas lofted 1000 g of regolith.

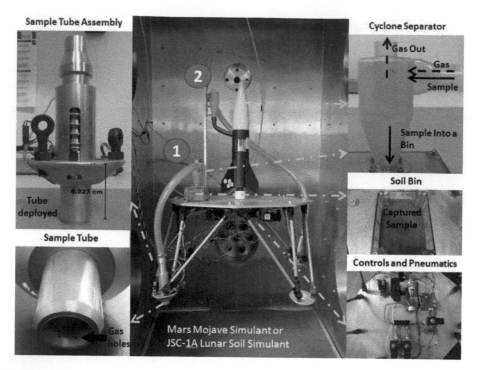

FIGURE 1.130   PlanetVac suspended in a chamber. *Source:* Courtesy Honeybee Robotics.

This was a first end-to-end demonstration of the pneumatic sample acquisition and delivery system in a vacuum. The next step was to demonstrate the system can actually work on a lander. For this purpose, PlanetVac was designed to double as a footpad and mounted on Masten Xodiac lander (Spring et al. 2019). In this particular iteration, nozzles in the lower section of the sampling head (blue in Figure 1.131) were used to spray gas across the surface of the material, stirring up regolith, and creating an area of high pressure. This area of high pressure caused a fluid flow towards the sample container (in gray) at lower pressure.

Depending on the composition of the planetary body's regolith, the sample makes its way to the sample container in one of two transport methods. If particles are small and light (generally subcentimeter size), they will tend to be swept up in the flow and follow the gas flow. If particles are larger (over 1 cm), their movement is dominated by momentum imparted from the gas hitting their surface and are forced towards the sample container. The dominant transport method is a strong driving factor in the design of the entire sampling system, and therefore the systems needs to be customized per specific mission.

From the sampling cone, regolith makes its way to the sample container. Depending on the configuration of PlanetVac, this could be a very short trip (as depicted in Figure 1.131) or a longer one into the spacecraft through a tube. In the case of a long sample transfer, additional nozzles might be added to assist in material transport. Once in the sample container, the sample needs to lose its energy and come to rest while the gas exits, usually through a filter. This process could take many forms but in the case of PlanetVac tested on Xodiac, the sample was captured inside a system that resembled a cyclone separator.

The PlanetVac Xodiac sampler can be separated into two distinct parts: the sampler cone, and the sample container (Figure 1.132). The sampler cone (1) is where the sampling process occurs. Pneumatic nozzles (2) push high-pressure gas into the regolith in order to stir up the surface and move sample towards the back of the cone, into the sample container. These nozzles have been

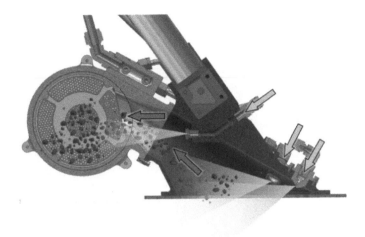

FIGURE 1.131 Cross section of PlanetVac Xodiac. Yellow cones and arrow depict gas flow. *Source:* Courtesy Honeybee Robotics.

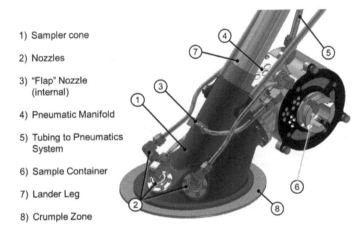

1) Sampler cone

2) Nozzles

3) "Flap" Nozzle (internal)

4) Pneumatic Manifold

5) Tubing to Pneumatics System

6) Sample Container

7) Lander Leg

8) Crumple Zone

FIGURE 1.132 Key components of the PlanetVac Xodiac sampler. *Source:* Courtesy Honeybee Robotics.

specifically designed to maximize sample movement towards the container. An additional nozzle (3) pointed towards the sample container further helped direct the sample into the container and assisted in opening the Kapton flap which is inside.

The PlanetVac was tested on Masten Xodiac lander during three ground-to-ground tests, with the tethered lander flying to the simulant bin and back (Figures 1.133 and 1.134). One major concern was the potential for the lander to tilt while sampling, due to pneumatic erosion of regolith under the footpad, making it unsafe to lift back off from the simulant bin. Upon the initial landing and after sampling, measurements were taken of the lander's tilt angle to ensure it was within acceptable bounds, and there were no problems reported, allowing lander to fly back and complete the test. In all three instances, the lander lifted off, flew to the simulant bin, landed, PlanetVac took a sample, the lander lifted off, then landed back in the initial position. There was no damage to the PlanetVac equipment during any of the three tests, and each sampling event obtained three times more sample than the 100-g requirement (Figure 1.134). As designed, the sample container was easy to remove

FIGURE 1.133 Masten Xodiac landing in the sample bin with PlanetVac on all 4 ft. *Source:* Courtesy of Masten.

FIGURE 1.134 Sample container with 332 g of Mars Mojave Simulant. *Source:* Courtesy Honeybee Robotics.

by hand after the launch. The sampling system showed no signs of leaks, and the electronics worked flawlessly, remotely triggering the sampling events.

These tests validated the PlanetVac in the rocket lander environment: surviving the vibration and thermal environments, surviving landing loads, and collecting and retaining sample. Other aspects of the system that were tested included surviving in-flight vibration and impact, as well as a full

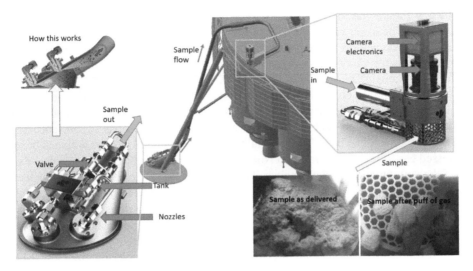

FIGURE 1.135 PlanetVac is scheduled to launch to the Moon in 2023. *Source:* Courtesy Honeybee Robotics.

remote electronics checkout. Landing in the simulant bin had previously been tested as well, but not with the PlanetVac installed, nor had the effects of the weight of the lander on PlanetVac while sampling been tested. These were all validated through successful ground-to-ground tests.

This testing therefore demonstrated a number of conditions directly relevant to flight applications, including a realistic thermal environment, in-flight dynamics, and landing impact loads. Minor departures from flight conditions include the gas pressure required for sampling, and a uniform fine-grained simulant. In a microgravity and reduced atmosphere environment, significantly lower gas pressure and flow rate is required to capture the same amount of material on Earth.

PlanetVac has been selected by NASA to fly to the Moon as part of the NASA's Commercial Lunar Payload Services (CLPS) initiative. The 19D mission is scheduled to launch in 2022 to Mare Crisium. The lander provider will be selected in September of 2020 from the 14 CLPS teams. The requirement for the PlanetVac is to demonstrate the sampling technology and in turn increase its flight readiness for future missions that require samples (e.g. missions with instruments or sample return missions). PlanetVac, shown in Figure 1.135, would deliver lunar regolith to a sample canister and take a photograph of the delivered sample.

1.9.2 Pneumatic Sampler (P-SMP) for Martian Moons eXploration (MMX)

The goal of the Martian Moons eXploration (MMX) mission, led by the Japanese Aerospace Exploration Agency (JAXA), is to make remote sensing and in situ observations of Phobos and Deimos as well as return with samples from Phobos (Kawakatsu et al. 2019). To increase science return objectives, MMX employs a dual sampling approach. The Core sampler, known as the C-Sampler, captures material from the subsurface, while the Pneumatic Sampler (P-Sampler), captures samples from the surface and near subsurface using burst of compressed gas (Zacny et al. 2020).

The P-Sampler uses a high-pressure N_2 gas to excavate and loft surface material into a sample return canister. As such, the gas acts as a sweeping element (e.g. brush) so that there are no mechanisms directly in contact with the sample. This makes the sampling architecture extremely simple and robust. Since the gas is ultra-pure, there is little danger of contamination, which otherwise might be present in mechanical sampling systems.

As shown in Figures 1.136 and 1.137, the P-Sampler is integrated onto one of the fixed struts of the landing gear. The lower end of the P-Sampler has the sampling head, while the upper end has the Control Box that contains all the plumbing (tanks, valves, etc.) and the sample canister. The two elements are connected via two tubes—one for the gas (Gas Delivery Tube) and one for the sample and carrier gas—Sample Transfer Tube.

The sampling head (see Figure 1.138) has three sets of nozzle pairs. The Excavation Nozzles are pointed down—the purpose of these is to kick up surface material—some of this surface material will no doubt be scattered around the local space and impact some parts of the spacecraft, while some will enter the sampling head. The material that enters the sampling head will then be motivated by compressed gas from the two Transport Nozzles, which transfers the material through the Sample Transfer Tube and into the sample container. There are also two Retro-Nozzles—these were added to negate the impulse created by the sampling nozzles which would otherwise affect the attitude of the spacecraft. The Retro-Nozzles are pointed directly opposite the sampling nozzles and are sized to equally distribute the gas flow. As the sample moves up the Sample Transfer Tube it passes through two sets of Beam Breakers (comprised of an IR LED directly opposing a phototransistor) that are used for sample verification. Directly downstream from the Beam Breakers the tube has a set of vents to separate the gas from the sample. The two vents allow gas to escape in opposing directions (to provide force neutral operation). Once the sample reaches the Sample Canister, it is

FIGURE 1.136 MMX P-Sampler on a landing gear. *Source:* Courtesy Honeybee Robotics.

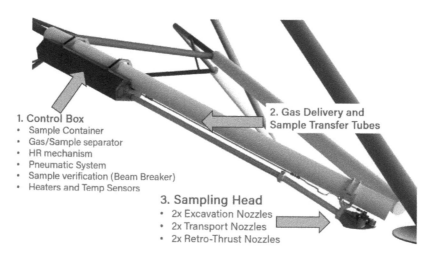

FIGURE 1.137 MMX P-Sampler consists of three elements: Control Box, Tubes, and Sampling Head. *Source:* Courtesy Honeybee Robotics.

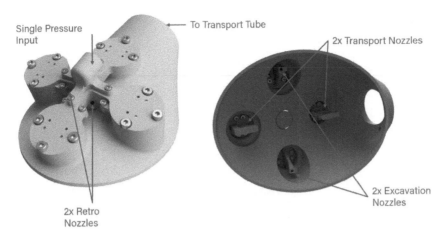

FIGURE 1.138 Details of the MMX P-Sampler Sampling Head. *Source:* Courtesy Honeybee Robotics.

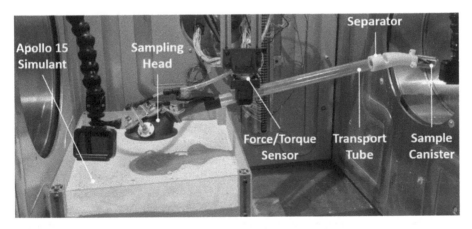

FIGURE 1.139 MMX P-Sampler test setup inside a vacuum chamber. *Source:* Courtesy Honeybee Robotics.

trapped inside it by a set of labyrinths. After sampling and once the spacecraft reaches orbit, the Sample Canister is transferred into the SRC via the robotic arm.

During the preliminary development phase, over 50 tests were conducted inside a vacuum chamber to determine the parameters affecting successful sample collection. The test parameters included simulant, gas flow rate, pulse (gas flow) duration, and sampler height above the surface. All tests were conducted with air (78% N2/21% O_2) and in a vacuum. The test setup shown in Figure 1.139 included the P-Sampler prototype mounted onto a Force/Torque sensor to determine forces during sampling events. The sampling head was placed above the soil bin; the height was adjusted for different tests. To date over 200 tests have been conducted in a variety of simulants and in off-nominal scenarios. So far in each scenario the P-Sampler was able to capture material.

1.9.3 FROST: Frozen Regolith Observation and Sampling Tool

Frozen Regolith Observation and Sampling Tool (FROST) captures several promising architectures into a single, fully integrated system (Figure 1.140). It includes a drill to drill below the surface and

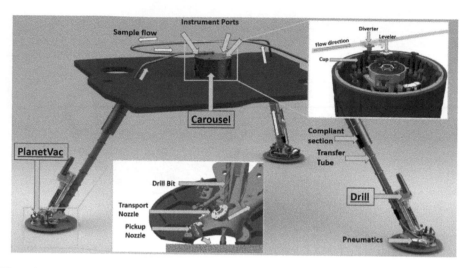

FIGURE 1.140   FROST: Frozen Regolith Observation and Sampling Tool consists of a drill, PlanetVac, and Carousel. *Source:* Courtesy Honeybee Robotics.

bring material to the surface, PlanetVac to capture and pneumatically move material, a Diverter and a cup for metering, and a Carousel with leveler to deliver samples to instruments. Since the drill and PlanetVac are mounted on the footpad/leg of a lander, the deployment is eliminated, reducing system mass and risk. Since PlanetVac can be used as excavation system for granular material, it offers system-level redundancy. It should be noted that a drill has an ideal aspect ratio (long and small diameter) for capturing samples from greater depths. But if there is a requirement to capture material from shallower depths, another excavation system can be used.

Figure 1.141 shows the upstream section of FROST: drill coupled to the PlanetVac. PlanetVac by itself is an excavation system and capture system wherein a sample of fines and clasts can be delivered from a surface to a cup with a single puff of gas. Utilization of PlanetVac makes the FROST architecture very robust as samples can be acquired without equipment deployment (or any other actuation) since PlanetVac is mounted on the footpad of a lander, placing it already on the target planetary surface.

PlanetVac alone, however, cannot perform excavation if material is competent (hard) and if the requirement calls for getting deep sample. For that reason, FROST couples a drill to the PlanetVac system to provide the additional capability of sampling hard/component materials and capability to bring material from greater depths.

In general, there are three types of materials likely to be encountered on planetary bodies: granular (fines), granular (coarse), and competent (e.g. rocks, ice, icy soil). In granular material with some fraction of fines, PlanetVac Pickup Nozzles are sufficient to loft surface material into a sample delivery tube (Figure 1.141, left). If surface material is coarse gravel, Vertical Nozzles are fired to bring up fines from below the ground (Figure 1.141, center). If material is competent (icy soil, ice, or rock) the drill is used first to break up that material into fines, and then the PlanetVac's Pickup Nozzle lofts the material to the transfer tube (Figure 1.141, right). Granular materials can also have different degrees of cohesion (stickiness)—this can be dealt with by increasing (or decreasing) gas pressures.

Figure 1.142 shows an example of a capture system that can meet sample presentation requirements for most instruments. As material is flying inside the tube, some of the material hits a Diverter plate and is directed into the cup, while the remainder of the material travels pass the Diverter and out. During this step, we will use Beam Breaker to determine that some sample has passed through the transfer tube and correlations to determine how much sample was captured.

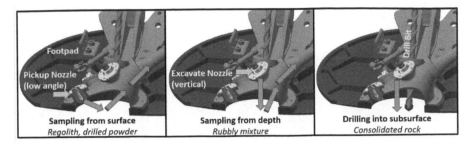

FIGURE 1.141   FROST: Three modes of sampling on. *Source:* Courtesy Honeybee Robotics.

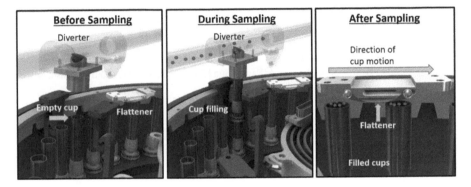

FIGURE 1.142   FROST: Sample capture and delivery. *Source:* Courtesy Honeybee Robotics.

After enough sample has been captured, cups are passed underneath a Flattener to smooth out the sample surface.

Figure 1.143 shows the experimental setup which includes all parts of FROST: drill for cutting into competent material, PlanetVac for excavation of loose soil and pneumatic transfer of samples, and Carousel for capturing of sample into cups and presenting them to various instruments. The drill and PlanetVac were attached to a mock-up lander leg and footpad and connected via a 1-cm transfer tube to a carousel above. The entire system was placed inside a vacuum chamber. The drill was a rotary-percussive with >20 cm depth capabilities. The rotation and percussion were driven by separate actuators to enable testing three modes: rotation, percussion, and rotation-percussion.

The PlanetVac was 3D printed and integrated into a footpad. The 2-m-long sample delivery tube was routed approximately 1.5 m up into a Diverter above a carousel. A mock-up carousel was fabricated with several cups—these cups were manually inserted into a Diverter.

Table 1.8 shows data from vacuum chamber tests in six different materials. JSC-1a is a lunar mare simulant with significant fine fraction, Aircrete is aerated concrete—very porous material with a bulk density of 0.6 g/cc—good gravity analog. Aircrete A15 is aerated concrete crushed and blended to fit Apollo 15 particle size distribution. Aircrete fines on top of an irregular surface were used to demonstrate that even in these extreme conditions (powder on an irregular block of rock) some material can be delivered. Aircrete coarse/fines were used to demonstrate that some material can be captured if bulk material is coarse gravel. 45-MPa limestone was used as room temperature strength proxy for cryogenic icy soil (Atkinson and Zacny 2018). In each test the sampling time was 1.5 s. Thus, to increase sample volume (assuming there is enough sample underneath footpad) the sampling time could be increased.

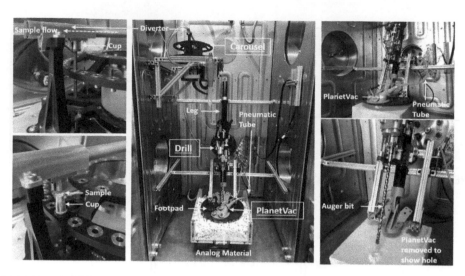

FIGURE 1.143 FROST testing. (Left) Aircrete sample delivered to a cup. (Center) Experimental setup inside vacuum chamber. (Right) Test on a block of 45-MPa Indiana Limestone. *Source:* Courtesy Honeybee Robotics.

TABLE 1.8
FROST Vacuum Chamber Tests with Various Materials

JSC-1A	Quartz Sand	Aircrete, A15	Aircrete Fines on Top Rough Surface	Aircrete Coarse/ Fines (10:1) Mixture	45-MPa Limestone
0.26 g	0.48 g	0.26 g	0.06 g	0.21 g	0.07 g
53 mm³	95 mm³	151 mm³	35 mm³	122 mm³	29 mm³

The cups are 1 cm in diameter.

These tests have shown that the system is highly robust to a range of material types and surface conditions.

1.9.4 NIBBLER FOR EUROPA AND ENCELADUS

Nibbler is a modified needle scaler combined with a PlanetVac system (Figure 1.144) designed for capturing ices on planetary bodies such as Europa and Enceladus. The needle scaler is typically used to remove rust and scale from steel structures in preparation for painting or welding. The scaler uses an array of loosely constrained steel rods, or "needles" which are hammered on using a pneumatic piston. The combination of the percussive action, flexibility of the needles, and loose constraint of the needles allow the tips of the needles to "wander" slightly on the surface, helping to scrape off old paint and rust. This same needle-to-surface interaction can be used on hard planetary surfaces to generate fine material without significantly altering the temperature of the sample.

To test the needle scaler in an end-to-end sample acquisition and transfer operation, a series of tests were conducted on cryogenic ice inside of a vacuum chamber. The setup is shown in Figure 1.145. The testing involved cooling the ice in liquid nitrogen (LN_2) inside the chamber and keeping the pneumatic gas cold in the walk-in freezer so that nothing would melt the generated ice chips. The walk-in freezer is kept at $-25°C$. A highly dynamic Presto refrigeration unit was used to pump refrigerant at $-90°C$ around cooling shrouds in the vacuum chamber to aid in keeping the needle scaler and surrounding equipment cold, below the melting point of briny ice. Thermocouples were installed in the ice sample and on the scaler to monitor the temperature changes.

Testing was performed in a thermal vacuum chamber (Figure 1.145) to demonstrate an end-to-end sample generation and sample transfer operation from an icy surface consisting of cryogenic briny Europa ice, to an elevated sample container within a thermally controlled vacuum chamber.

The following general procedure was followed to conduct an end-to-end test in the vacuum chamber.

1. Place Europa simulant in liquid nitrogen (LN_2) until thermocouple reads below $-170°C$.
2. Begin to chill the vacuum chamber cooling shrouds at $-90°C$.

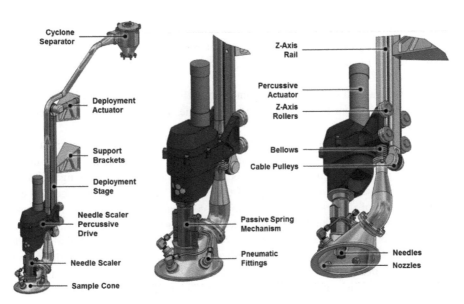

FIGURE 1.144 Needle Scaler Pneumatic Sampler components. *Source:* Courtesy Honeybee Robotics.

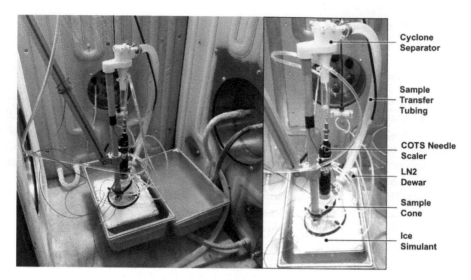

Cyclone
Separator

Sample
Transfer
Tubing

COTS Needle
Scaler

LN2
Dewar

Sample
Cone

Ice
Simulant

FIGURE 1.145 Nibbler experimental setup inside vacuum chamber. *Source:* Courtesy Honeybee Robotics.

3. Place sample in smaller tub of LN_2 inside the vacuum chamber.
4. Install needle scaler and pneumatic transfer cone on the sample and lock in place.
5. Record test setup.
6. Pump vacuum chamber down to an absolute pressure of 200 Torr.
7. Let experiment sit for about half an hour until the thermocouple on the needle scaler reads below $-20°C$ and the ambient temperature reads below $-15°C$.
8. Begin video.
9. After 3-s countdown, open needle scaler run valve, followed by the pneumatic transfer valve (exact order/durations may vary from test to test).
10. Stop vacuum pump, open vent to re-pressurize chamber.
11. Measure mass of gathered sample, back-calculate sample volume.
12. Measure volume of hole created by needle scaler by filling from a graduated cylinder.

Steps 6 and 7 were added to allow the equipment to chill down in an inert nitrogen environment, to prevent condensation from forming and freezing on the needle scaler, as this was found to be detrimental to performance.

Figure 1.146 shows a summary of some of the tests. Early tests (001–005) were used primarily to debug the system. In Test 006 a total of 8.1 cm³ of material was collected. No frost developed on any of the sampling equipment, indicating that our cooling method was working as intended, and we would not have "false positive" results from frost buildup. Test 007 was a successful end-to-end test. A total of 9.1 cm³ of material was collected. Test 008 resulted in a <1 cm³ amount of material being collected. This could be due to multiple reasons such as the sampler rim was set high on surrounding ice ridges or the needler was set in a particularly low spot reducing the available stroke of the needler. Test 009 resulted in no material collected. This was due to the needler freezing from condensation built up between tests. This problem was solved in future tests by simply running the needler between tests to blow out any settled moisture. Test 010 collected a significant 11.4 cm³ of material. This is due to a larger-than-estimated hole developed by the needler during testing. It appears that the ice fractured as the needler was penetrating the surface. Some of the liberated pieces of ice in the fractured section were too large to be pneumatically transported. Many pieces likely

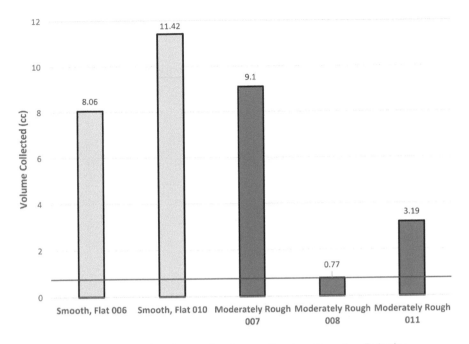

FIGURE 1.146 Nibbler end-to-end testing results. *Source:* Courtesy Honeybee Robotics.

fell into the needles during operation resulting in a larger than average amount of material being produced. Test 011 was a successful end-to-end test. A total of 3.2 cm³ was collected.

As demonstrated, even in rough surface conditions, the needle scaler and pneumatic sample transfer system are able to deliver some sample to the container. There has been no test that resulted in no sample. This demonstrates that this method of generating and transferring sample is robust to variations in surface geometry.

1.10 10-CM CLASS SAMPLERS

This section presents sampling systems in the 10-cm class. This range is ideal for small bodies sampling as well as for sampling at the end of robotic arms. As the sampling depth becomes greater, the predominant sampling approach becomes a drill which has ideal aspect ratio: small diameter and long. As such drills require less power, weight on bit, and energy to penetrate deeper. Not surprisingly this section consists of various examples of drilling systems developed for Solar System exploration.

1.10.1 VENERA DRILL

Venus is considered to be Earth's sister planet and hence we can learn a lot about Earth by investigating Venus tectonics, volcanism, and atmosphere. As opposed to Mars which lost most of its atmosphere but retained a lot of water, Venus has an extremely dense and hot carbon dioxide atmosphere (95% CO_2, >90 bar pressure, T ~ 480°C) and lost most of its water. One day our Earth could end up looking just like Venus or Mars. Mars has been mapped by a multitude of spacecrafts and we learn more about that planet each year. However, our understanding of Venus is relatively incomplete.

Table 1.9 shows a list of past Venus landers. The longest surface survival time was just over 2 hr, and the most recent mission to touch the surface of Venus was launched in 1984. The most

TABLE 1.9

Venera and Vega Venus landers

Surface Landed Missions	Launch	Surface Survival Time (min)	Pressure Vessel	Thermal Control	Surface Sample Acquisition	Notes
Venera 7	1970	23	No	No	No	Parachute failed, probe survived.
Venera 8	1972	50	No	Yes	No	
Venera 9	1975	53	No	Yes	No	
Venera 10	1975	65	No	Yes	No	
Venera 11	1978	95	No	PCM	No	
Venera 12	1978	110	No	PCM	No	
Pioneer	1978	60	Yes	No	No	
Venera 13	1981	127	Yes	PCM	Yes	Drill, XRF
Venera 14	1981	57	Yes	PCM	Yes	
Vega 1 Lander	1984	56	Yes	PCM	Drill activated during descent due to turbulences.	
Vega 2 Lander	1984	56	Yes	PCM	Yes	

PCM = Phase Change Materials.

successful Venus missions were in the Soviet Venera program. For example, Venera 13 launched in 1981 was a soft lander that survived at least 127 min on the Venusian surface. The exact time is not known since the 127 min is actually the time the relay spacecraft was in view of the lander. The lander was equipped with a 26-kg drill for acquisition of samples for XRF analysis.

A critical payload element of the Venera 13 and 14 and Vega 1 and 2 spacecraft was a drill referred to as the GZU soil sampling drill as shown in Figure 1.147 (Barmin and Shevchenko 1983). The 26-kg drill system was capable of drilling up to 3 cm into igneous rocks. This was the only mechanism with moving parts and hence required development of a high-temperature electric motor, lubricants, and new alloys. A very clever engineering approach helped to solve a major hurdle associated with the different coefficients of thermal expansions of two materials in close proximity (e.g. gears and shaft). These machine parts were in fact designed to function properly only after thermal expansion at 500°C.

The drill used a series of pre-coded steps—there was no control feedback. The drilling operation started as the telescoping drill head was lowered to the surface. The same 90-W electric motor lowered and advanced the drill into the formation, and rotated the drill bit at 50 rpm. The drill operated for 2 min, and then a number of pyrotechnic charges were used to transfer the sample into x-ray fluorescence (XRF) chamber. The first pyro charge broke a series of seals that allowed the high-pressure atmosphere of Venus to flow into an assembly of tubes. Soil was carried in stages, into a soil transfer tube and onto a sample container. The second pyro closed the soil feeding system to the outside atmosphere. The third pyro broke the seal between the vacuum tank and the soil feeding system—the pressure dropped to 0.06 atm which was required for XRF measurement. The fourth pyro moved a sample tray through an airlock and into the XRF chamber. During the forward motion, the tray hit a hard stop and the sample was sprayed into XRF cups. The drill telemetry revealed that in both cases (Venera 13 and 14), the drill penetrated to the target depth of 30 mm. In addition,

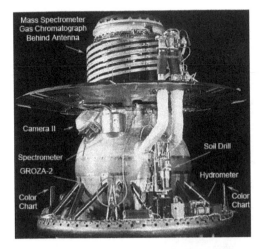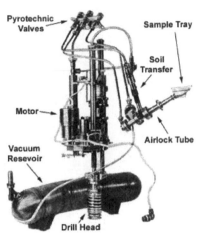

FIGURE 1.147 Venera 13 and 14 and Vega 2 drill penetrated 3 cm into the subsurface and transferred sample into the XRF chamber. (Left) Venera 13 lander (see location of the drill and other outside mounted instruments). (Right) Venera drill. *Source:* Courtesy of USSR Academy of Sciences.

the analysis of the drill telemetry indicated that the physical strength of the drilled formations corresponded to that of weathered porous basalt or compacted ashy volcanic tuff-type material. The surface sample at the Vega 2 landing site was found to be an anorthosite-troctolite rock, rarely found on Earth, but present in the lunar highlands, leading to the conclusion that the area was probably the oldest explored by any Venera vehicle.

1.10.2 VENUS IN SITU EXPLORER (VISE) DRILL

The VISE drill is based on the IceBreaker and LITA rotary-percussive planetary drills (Paulsen et al. 2010; Zacny et al. 2015b). The drill consists of two actuated Z-stages (one to place the drill on the ground—deployment stage, and the other to penetrate below the surface—feed stage), a rotary-percussive drill head, and a sampling bit. The drill head uses two actuators: for rotating the bit (Spindle) and for actuating the hammer system (Percussion)—see Figure 1.148 (the deployment and feed stages are not shown).

The design includes a hollow drill stem extension through the bit and hammer drive assemblies up to the top of the drill body, where it interfaces with the pneumatic sample transfer system. The sample flows directly from the bit, through the hollow transfer tube into the rest of the sample transfer plumbing (not shown).

To capture the required sample volume to 5 cm depth and allow for pneumatic transfer, the bit is hollow and uses a full faced cutter (Figure 1.149). The VISE bit does not need flutes (since it uses pneumatics to transport the cuttings). The hollow bit was designed with an external diameter of 1.75 cm and an internal diameter of 1.26 cm. This bit can generate approximately 2 cc of sample for each 1 cm depth, assuming a fluff (volume expansion) factor of 2, and up to 50% sample loss.

VISE drill uses Brushless DC motor (BLDC) with Pulsed Injection Position Sensor (PIPS) for commutation and a two-stage planetary gearbox to convert high speed, low torque output from the motor into either higher speed, lower torque or lower speed, higher torque. Thus, the same motor can be used for Spindle, Percussive, Feed, and Deployment stages just by employing different gearboxes. To characterize the torque-speed curve of the motor and measure its temperature rise due to self-heating, the actuator was placed in an oven at 462°C and connected to a dynamometer via an extension rod (Figure 1.150).

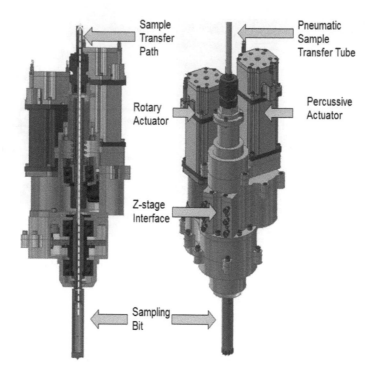

FIGURE 1.148 Main subsystems of the VISE drill. Deployment and feed stages are omitted here. *Source:* Courtesy of Zacny et al. (2017a).

FIGURE 1.149 Custom Venus drill bit. *Source:* Courtesy Honeybee Robotics.

Performance characterization curves were captured at room temperature (RT), 100°C, 200°C, 300°C, 400°C, and the target operating temperature of 482°C, representing the surface temperature of Venus plus a 20°C test margin. The tests have shown that the maximum power output (current-limited) dropped from 200 W at RT to 130 W at 482°C (Rehnmark et al. 2017). The drop is attributed mainly to an increase in wire resistance (3× higher resistance at high temperature). The output power of 130 W is more than sufficient for each of the drill's 4 degrees of freedom (Rehnmark et al. 2018b).

Drill functional tests were initially conducted at room temperature in 120-MPa Saddleback Basalt. The drill successfully penetrated to 4.5 cm depth in 9.5 min and captured 25 g of sample (~25 cc). In addition, several Venus tests were conducted in NASA JPL's Venus chamber—465°C,

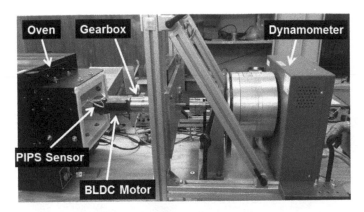

FIGURE 1.150 Venus actuator dynamometer tests. *Source:* Courtesy Honeybee Robotics.

FIGURE 1.151 Drilling tests inside Venus chamber at NASA JPL. *Source:* Courtesy of NASA JPL.

92 bar pressure, and CO_2 atmosphere (Figure 1.151). The drill successfully penetrated to a target depth of 5 cm in approximately 10 min and pneumatically transferred sample. This demonstrates that drilling and sample delivery under Venus conditions is possible and as such it paves the way for future Venus missions.

1.10.3 THE MARS SCIENCE LABORATORY DRILL

Launched in the fall of 2011, the Mars Science Laboratory is part of a long-term effort of robotic exploration of Mars to assess whether Mars ever was, or is still today, an environment able to support microbial life.

FIGURE 1.152   (Left) Curiosity Rover self-portrait at Big Sky; (right) SA/SpaH RA and turret from Sol 14.

The MSL rover, Curiosity, features the most advanced robotic Sample Acquisition, Sample Processing and Handling (SA/SPaH) subsystem ever sent to another planet (see Figure 1.152). The major elements of the SA/SpaH subsystem are a robotic arm (RA) with a tool and instrument laden turret. The tools are a Sample Acquisition Drill, scooping, sieving and portioning device called CHIMRA described here: Sunshine (2010) and the Dust Removal Tool (DRT). The turret-mounted instruments are the APXS and MAHLI. For more detail description of the SA/SPAH subsystem and development, please see Jandura (2010).

The SA/SpaH subsystem is responsible for both acquiring powdered samples from rock interiors via rotary percussive drilling and delivering these samples to the two analytical instruments on board the rover, SAM (Sample Analysis at Mars) (Mahaffy et al. 2012) and CheMin (Chemistry & Mineralogy) (Blake et al. 2012). These two instruments are the primary reason for the existence of the drill as they both require powdered sample for analysis.

To meet the objectives of the MSL mission and the needs of the science instruments the MSL Drill has the following sampling capabilities:

1. Sampling depth from 20 to 50 mm.
2. Acquisition of 8–14 cc of sample.
3. Sufficient quantity of rock cuttings that are <150 um.
4. Less than 40-ppb organic contamination in delivered sample.

The MSL Drill is comprised of 7 sub-elements depicted in Figure 1.153(a–g). The interface to the rock and sample is a bit assembly that cuts the rock and collects the sample (a). A chuck mechanism (b) engages and releases the new and worn bits. A spindle mechanism (c) rotates the bit through a drive coupling. Within the MSL Drill there is a percussion mechanism (d) which generates hammer blows to break the rock and create the dynamic environment used to flow the powdered sample. The aforementioned components are mounted to a translation or feed mechanism © which provides linear motion and senses weight on bit with a force sensor. A passive contact sensor/stabilizer mechanism (f) detects contact with the rock and, when preloaded, stabilizes the drill's position on the rock surface. Lastly, a flex harness management hardware (g) provides the power and signals to

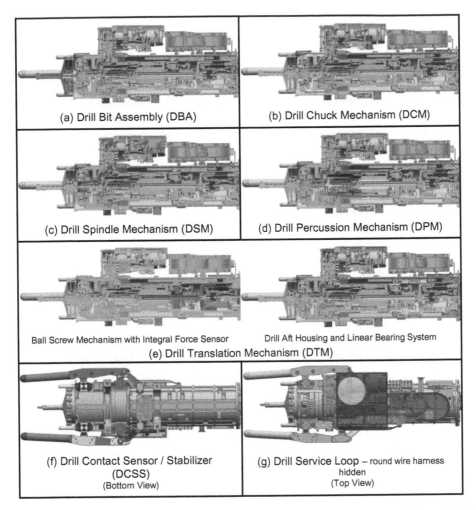

FIGURE 1.153 Curiosity Rover drill. Depictions of the drill subassemblies. (a) Drill Bit Assembly (DBA); (b) Drill Chuck Mechanism (DCM); (c) Drill Spindle Mechanism (DSM); (d) Drill Percussion Mechanism (DPM); (e) Drill Translation Mechanism (DTM); (f) Drill Contact Sensor/Stabilizer (DCSS); and (g) Drill Service Loop.

the translating components. A more comprehensive description of the MSL Drill hardware and its development can be found in Okon (2010).

1.10.3.1 Salient Hardware Details

The MSL Drill sample acquisition and handling is performed exclusively by the drill bit assembly. The drill consists of a 5/8" commercial rotary-percussive bit that has been modified with custom flutes. A sleeve that surrounds the fluted section enables the collection of the cuttings. The bit has two sample chambers that store the cuttings before they are transferred to CHIMRA via an exit tube to complete the processing and delivery processing. The chambers create a labyrinth that ensures the sample is not lost while drilling in a variety of orientations with respect to gravity. The MSL rover carries two spare drill bit assemblies to replace a worn-out or fouled bit.

The MSL Drill powders rocks using a percussive hammering mechanism that features a 400-g hammer operating at 1800 blow per minute. The percussion output has 6 levels ranging in kinetic energy at the impact from 0.05 to 0.8 J per blow. The percussive blow is transferred to the bit via an anvil rod that is in contact with the back of bit shank. The percussion mechanism is decoupled from the spindle mechanism which allows the hammer energy to be independent of the bit rotational speed.

The MSL stabilizers feature 2-point articulating, sharp-toothed spherical tips which provide good purchase onto rock targets. The articulation allows the drill to be stabilized over a range of rock surface topographies while locking out 4 degrees of freedom (3 translational DOFs and a rotational DOF about the bit axis) between the drill and the rock target.

The drill's mechanism, primary structure, and bit assembly are designed to survive large external loads. This makes the MSL Drill robust to shifts in the rover or rock target during operations.

1.10.3.2   Algorithm and Software Overview

The MSL Drill software low-level algorithms contribute functionality that is used in many parts of the basic and advanced behaviors. These include: force sensor signal filtering and processing to accurately measure weight on bit (WOB), force control algorithms to apply a static WOB and to regulate the WOB using force control during drilling.

The MSL Drill software basic behaviors are used in nominal command sequences to perform simple tasks between the advanced behaviors. They can also be used in off-nominal conditions to provide flexibility for diagnostics or unforeseen required functionality. These behaviors are responsible for dispatching motion requests for individual and simultaneous actuations to common rover motion control software.

The advanced behaviors of the MSL Drill software are the building blocks of the operational drilling sequence. These include: Seek Surface, Start Hole, Hardness Test, Drilling, and Retraction. A more comprehensive description of the MSL Drill software and its development can be found in Helmick et al. (2013).

1.10.3.3   Drill Concept of Operations

The robotic arm preloads the drill on a rock target and the robotic arm actuators are powered off and held in position with a power-off brake. The drill seeks the surface of the rock to locate the zero-depth position.

To increase the likelihood of a successful drill operation in the face of rough and angled rock surfaces, a hole-start operation is performed to create a pilot. The bit makes a series of cuts in the rock at discrete rotational positions using percussion without bit rotation. After each cut the bit is retracted, rotated slightly, and then advanced to contact with the rock at a prescribed force. Once a full bit revolution is achieved, a brief rotary-percussive operation is performed to clear out the divot. This series is repeat until a prescribed depth is achieved.

Following Start Hole creation, but prior to actual drilling, a Hardness Test also is performed on the drill target. It consists of several techniques used in succession to create an indentation at the bottom of the Start Hole. First, the drill bit is pressed into the rock using 130 N (~13 kg) of force. Second, a tap test is performed by creating a single percussive impact with the voice coil. Finally, a series of percussive bursts are applied at each percussion level from 1 to 6.

Once the Start Hole is complete, the MSL Drill is commanded to drill to a specified depth. The weight on bit value and spindle rotation rates are command-able, but typically fixed. The drilling algorithm is parametrizable allowing the operator to choose behavior that can minimize the drilling duration or operate at a reduced percussion level. The percussion level starts at a prescribed value and then is increased, up to a specified maximum level, if the rate of penetration (ROP) falls below a specified threshold value. Conversely, if the ROP is higher than a defined threshold the percussion level is decreased. The drill operation can be duty-cycled to reduce the actuator and sample temperature rise.

When the desired depth has been achieved, the drill feed is stowed. Then, the robotic arm repositions the drill and CHIMRA with respect to gravity to begin the sample processing and sample handling activities. The sample is transfer from the drill bit to the CHIMRA by a series of pose changes by the RA wrist and turret and dynamic environment generated by the drill's percussion mechanism and the CHIMRA vibration mechanism.

Often in MSL operations a shallow test hole is drilled into a rock prior to committing to a full-depth sampling. The depth of this hole does not engage the bit sleeve into the rock and thus does not acquire sample. This operation is intended to produce powder the operations team can assess to ensure it behaves as a dry powder or assess if there is any clumping or sticking present that could clog the sieve and pathways within the sampling system.

A more in-depth discussion of what is necessary to acquire samples, including the steps taken to ensure sampling hardware is safe when drilling into a target (i.e., evaluation of rock type, rover stability, prior testbed experience, etc.) and the drilling parameters used to acquire these samples can be found in Abbey et al. (2019).

1.10.3.4 Drilling Mission Performance and Results

The Curiosity Rover completed its prime mission in the first Martian year, 669 Sols (687 Earth days) on June 24, 2014. In that time the rover successfully drilled three full-depth drill holes into the Martian surface and analyzed the recovered material using onboard instruments, giving us new insights into the potential habitability of ancient Mars. These drill targets are known as "John Klein" (Sol 182) and "Cumberland" (Sol 279), which lie in the mudstones of the Yellowknife Bay formation, and "Windjana" (Sol 621), which lies in the sandstones of the Kimberley formation. The graphs in Figure 1.154 show the sample acquisition data for the full and mini-drill operation on the targets. Please see Abbey et al. (2019) for additional details about the drill performance. Figure 1.155 shows the holes drilled through August of 2019.

Starting the drilling operation at the highest allowable percussion level created some problems under certain conditions. Rocks that were poorly embedded or too weak moved or fractured under the initial higher percussion levels. Following an abandoned drill opportunity at the Bonanza King target on Sol 724, when the rock moved under percussion during a presampling operation (mini-drill), the engineering team worked to develop a second, more adaptive "reduced percussion" configuration. Weeks of testing in terrestrial rocks resulted in the development of the reduced percussion algorithm, which monitors ROP and WOB thresholds, guiding the VCL (Voice Coil Level) control algorithm toward the lowest VCL possible while still maintaining adequate ROP into the rock. Reduced percussion initiates the drill operation sequence at the lowest percussion level of VCL-1 and only increases percussive energy when the ROP falls below 0.05 mm/s. The system will continue to increase VCL up to a maximum of VCL-4 and will eventually stop the drilling process (i.e., fault out) if the ROP falls below 0.025 mm/s while at VCL-4.

After a mini-drill attempt on Sol 867 at the Mojave target resulted in a fractured slab, the decision was made to put the reduced percussion algorithm into service. During the next drill attempt, on Sol 882, reduced percussion was successfully used to sample the Mojave 2 target.

The reduced percussion algorithm had been used to drill the next ten rocks. The use of this algorithm has enabled the development of a methodology that uses the drill to indicate the uniaxial compressive strengths of rocks through comparison with performance of an identically assembled drill system in terrestrial samples of comparable sedimentary class. For additional details about the method and the results see Peters et al. (2018).

The mission and the drill, after almost a decade on Mars, are still operational.

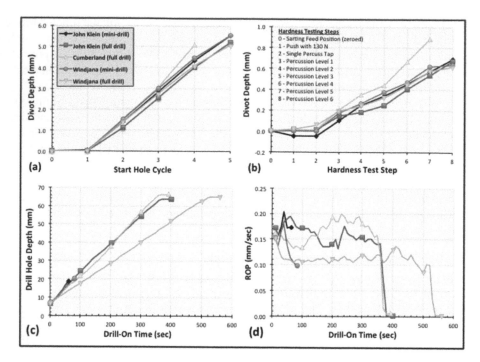

FIGURE 1.154 Curiosity Rover. Graphs of drilling performance data. (a) Divot depth by Start Hole cycle, (b) divot depth by hardness test step, (c) drill hole depth by drill-on time, (d) rate of penetration by drill-on time.

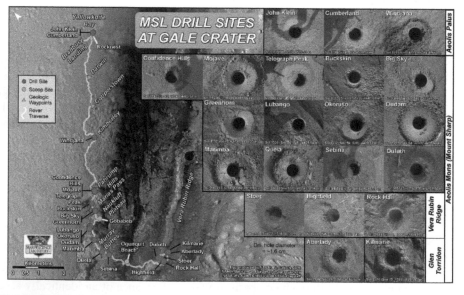

FIGURE 1.155 The locations of the sites where NASA's Curiosity Mars rover collected its rock and soil samples for analysis by laboratory instruments inside the vehicle and images of the drilled holes where 21 rock-powder samples were acquired as of August 2019.

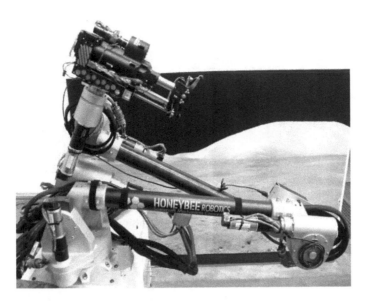

FIGURE 1.156 RoPeC Drill on robotic arm. *Source:* Courtesy Honeybee Robotics.

1.10.4 ROTARY PERCUSSIVE CORER (RoPeC)

The Rotary Percussive Corer (RoPeC) Drill was developed in order to advance a technology area that is necessary to sample acquisition for many mission concepts (Figure 1.156).

The RoPeC Drill was designed to be deployed by a 5-degree-of-freedom robotic arm. The drill obtains various types of drill bits from a Bit Station—this is akin to terrestrial systems where one drill is used with various types of drill bits, depending on a specific application (Chu et al. 2014; Zacny et al. 2014a, b). For RoPeC system, the following drill bits have been developed: coring bit to capture core samples, PreView bit to allow the core to be inspected in situ, Powder Bit to allow for capture of regolith and rock cuttings, and Rock Abrasion and Brushing Bit (RABBit) to brush and grinds rocks (Figure 1.157). The RoPeC Drill was assembled, fully integrating all motors with connectors, as well as the custom flex cable. Additional electronic items such as proximity switches, limit switches, and magnetic encoders were also integrated into cabling harnesses. The following images show the completely integrated RoPeC Drill. Figure 1.157 illustrates the various attachments used by the RoPeC Drill.

The Core PreView bit is used to acquire and examine a short 2–3-cm core from the side before deciding to use the Core Acquisition Bit to acquire a full-sized core (~5 cm in length) for further analysis or caching. The Core PreView bit has a window which closes when the core is sheared off from the base rock. It is kept closed while the drill retracts, is moved into position, and is ready to be examined by a camera or other instrumentation. Just before the window is opened, the drill is tilted with the help of the arm while the drill percussion actuator is enabled. This causes the core to slide down in front of the window. The window is then opened by rotating the inner sleeve relative to the outer auger sleeve.

The PreView bit was tested in Indiana Limestone and Travertine under atmospheric conditions. Tests were performed with and without the stabilization tines (a.k.a. Butterflies) which are typically used to preload and stabilize the drill while drilling.

ROPEC with Rock Abrasion Brushing Bit

ROPEC with Powder and Regolith Acquisition Bit

ROPEC Drill with Core PreView Bit

ROPEC Drill with Coring Bit

FIGURE 1.157   RoPeC Drill attachments. *Source:* Courtesy Honeybee Robotics.

The coring bit can be used to acquire a full-size core. After the core is broken off, the drill retracts with the captured core. The core can be deposited to any desired location or the entire bit with the core can be detached and cached.

1.10.5   Percussive Rapid Access Isotope Drill (P-RAID)

The acquisition of subglacial bedrock samples from the polar regions has been a long-standing aim for researchers in the field of glacial geology and paleoclimatology (Clow and Koci 2002; Flowerdew et al. 2012; Spector et al. 2018). By obtaining samples from multiple sites, data on the last exposure of the bedrock can be used to deduce the evolution of the ice sheet, with emphasis on the thickness and extent of the ice sheet over time.

While the drilling of polar ices is well established, the polar science community does not possess the same level of capability with bedrock drilling due to a variety of uncompromising logistical and technical limitations. It is for this reason that novel approaches to this task are particularly valued. The fact that it may not be possible to predetermine the terrain to be drilled, and that drilling operations typically take place in cold, dry environments that are relatively inaccessible suggests that the planetary drilling approach may offer a suitable solution. The development of a sampling mechanism which is influenced by the design of planetary drilling systems, whereby parameters such as mass, volume, power consumption, and weight on bit are minimized, may offer a practical solution to this challenge.

1.10.5.1   Logistical Limitations

Obtaining subglacial bedrock samples is difficult due to the numerous technical and resource challenges associated with deep drilling activities, exacerbated by the extreme conditions present at the drilling locations. Furthermore, the need to constrain deep field operations to the polar summer months ensures that light aircraft are typically utilized for the rapid transportation of equipment and personnel. As light aircraft operating in these regions typically operate at the edge of their performance envelopes, it is essential that the development of any new deep drilling system must be approached with a holistic view of the problem, placing the supporting logistics effort as one of the major constraints on the solution that is to be developed. It is for this reason that drilling systems which have a low resource requirement are likely to be favored over industrial-scale drilling rigs.

In order to reduce the resources required at deep field sites, it is essential that any new drill development must be power efficient as only small-scale electrical generators, capable of a few kilowatts, may be made available. As these generators typically utilize petroleum-based fuel, an efficient system will reduce the volume of fuel, which is required, extending the operational time available for a given available fuel volume. Furthermore, while it is common for many drill systems to operate "wet," making use of cutting fluids and antifreeze, these fluids also have to be transported to the field site and the use of large quantities of heavy, potentially polluting fluids also has implications for the preservation of drilling sites. Organizations such as BAS tend to operate "leave no trace" policies, in keeping with the Protocol on Environmental Protection to the Antarctic Treaty 1991. Thus, any system which can operate "dry," without the need for these fluids, is highly favorable.

1.10.5.2   Technical Challenges

The development of subglacial bedrock sampling technologies faces technical challenges similar to those encountered when designing planetary drilling hardware. Predominantly, uncertainty surrounding the target terrain to be sampled means that the system must be capable of extracting material from a host of potential terrain types. The subglacial environment may consist of sequential layers of debris-rich basal ice, an unfrozen layer, till, debris beneath the unfrozen layer and bedrock, or some combination of these layers. It is for this reason that a high degree of robustness is required in order to ensure that the developed system is capable of breaching multiple layers of material before extraction of a bedrock sample can be achieved.

It has been suggested that the topography and geology of subglacial Antarctica remains less understood than the surface of Mars (NRC 2011b). The subglacial bedrock may include sedimentary rocks (e.g. siltstone and sandstone), metamorphic species (e.g. gneisses and metamorphosed granites), or igneous formations (e.g. granites and basalts) (Talalay 2013; Talalay et al. 2014). As each of these rock types will not respond equally to the drilling process, any new development would seek to mitigate against this risk by designing for the worst-case scenario. While harder rocks, typically of the igneous or metamorphic variety, may be the most difficult to penetrate, softer rocks which are more easily drilled may increase the failure rate in drilling operations through the onset of drill choking or other associated error states.

The ability to sample various types of bedrock at a range of depths is essential in order to ensure that any system can be utilized throughout the region of interest. Wireline operations, whereby the drill is attached to an umbilical spool, may allow for depth extremes to be reached provided that the drill forces and torques are low enough to be reacted downhole.

Secondly, the nature of subglacial drilling limits real-time visibility of the drilling process, so systems with a high degree of autonomy (and fault tolerance) are preferred, particularly where this autonomous control can be affected downhole. Autonomous systems also reduce the need for human-in-the-loop control which is favorable when drilling in adverse weather conditions.

Clearly, the development of a lightweight subglacial bedrock sampling solution capable of rapid and low resource drilling is highly desirable. While systems capable of bedrock sampling do exist, deployment of these systems in the field are logistically intensive and the use of such systems may be excessive for shallower targeted depths (Goodge and Severinghaus 2016; IDDO 2015). A deployable, wireline rock drilling system is therefore an attractive proposal.

1.10.5.3 BAS Rapid Access Isotope Drill (RAID)

The Rapid Access Isotope Drill (RAID) is a drilling system developed and operated by the British Antarctic Survey. Designed to sample ice cuttings from medium depths of approximately 600 m (the depth limit for dry drilling), the RAID offers a new approach to the task of penetrating the ice sheet as it attempts to reduce operational complexity while increasing drilling speed at the expense of high-resolution core sampling. Inspired by many drills but mainly developed from the BAS medium-depth ice corer (Mulvaney et al. 2002), RAID is able to acquire multi-meter "pecks" of loose, icy cuttings from the icepack utilizing an anti-torque device which reacts torque from the rotating cuttings barrel. The cuttings rotate around and climb the ice auger until the peck depth is reached, allowing the drill to be raised to the surface and emptied.

Having succeeded in operating at multiple Antarctic field sites over the course of a number of seasons in service with BAS, a degree of confidence in RAID has been reached which provides strong foundations upon which the drill can be modified, reducing the need for a costly new development (Rix et al. 2018). Using this approach, RAID can both be used in its standard configuration to reach the subglacial bedrock and then adapted for the task of bedrock sampling.

While the RAID system provides an excellent baseline for the development of a bedrock sampler, its adoption imposes a number of constraints upon any new developments.

Firstly, in order to minimize the cross-sectional area of the borehole, the diameter of the RAID system is fixed at 80 mm. While this acts to reduce the total volume of ice that must be removed during drilling operations, it also imposes a limit on the development of the bedrock penetrating system because any development must be capable of utilizing the ice borehole created by RAID.

Secondly, as the existing RAID device receives electrical power from a surface-level petrol generator of the kilowatt class, the electrical umbilical cable is rated only to this level. This prevents the use of a more powerful generator, thus limiting the power available to a bedrock sampler.

Finally, the existing RAID system imposes mechanical limitations on the bedrock sampler development due to the inherently lightweight design of the system and its low torque demand and reaction capability. In a related manner, the axial force available to penetrate the bedrock is limited by the self-weight of the drill, and so a clear limit is imposed on any development due to the necessarily lightweight nature of the system.

These limitations of weight on bit, torque, and power are such that the option of utilizing pure rotary drilling as a means of penetrating the bedrock is diminished. However, percussive drilling, whereby the process of penetrating the bedrock target is assisted by the application of a repeated hammering load to the rock face, may prove to be a viable alternative to conventional rotary drilling in a low resource setting (Wang et al. 2015). In fact, it has been demonstrated that the use of rotary-percussive drilling may be in excess of seven times faster than pure rotation (Melamed et al. 2000).

To date, there have been no attempts made to incorporate a percussive actuator into a bedrock sampling drill of any form. While the reasons for this are not explored in the literature, it is assumed that industrial-scale bedrock sampling rigs do not require percussive action due to the availability of high forces and torques at the cutting head. In the case of lightweight electromechanical cable-suspended drilling, it is foreseeable that the constraints imposed by the need to house the percussive actuator within the narrow housing bore of the drill has restricted the application of a hammering system.

Nonetheless, in response to the drivers facing our specific design, this is the architecture we have adopted for our subglacial bedrock sampling tool.

1.10.5.4 The P-RAID (Percussive RAID)

The P-RAID (percussive RAID) prototype was designed in a modular manner such that multiple, individual elements could be brought together to form a functional assembly, but each section could be removed and maintained with ease. This was deemed essential in order to allow the performance of the rig to be assessed while ensuring minimal assembly and disassembly complexity. The footprint of the prototype was also designed to be consistent with the RAID borehole diameter constraint so as to ensure that the kinematics of the percussive actuator were of the correct order of magnitude as any future field-ready models.

Upon performing a trade-off study, it was decided that a spring-cam actuated hammer mechanism would be utilized due to its simple working principle and robustness. Such systems have heritage in a number of planetary drilling developments (Chu et al. 2014; Zacny et al. 2013c) and offer a reliable means of generating percussion using a single motor actuator. Spring-cam systems also benefit from being easily tunable in both the hammer energy generated and the percussive power which can be delivered by altering either the cam geometry or the rotary speed of the motor used. This ensures that a system can be designed which is capable of penetrating even the hardest rock types foreseen. Figure 1.158 depicts the architecture of the P-RAID system.

The P-RAID design has two central features: the cam-hammer-driven percussive actuator and the multi-motor gearbox assembly. The latter of these features consists of a gearbox assembly which is driven by three Maxon brushless DC motors. The decision to use three smaller diameter motors driving a single shared sun gear, as opposed to one larger motor, was a result of the need to maximize the torque deliverable to the drill bit while ensuring that the design was compatible with the narrow-gauge borehole.

The P-RAID system has been tested in a variety of settings in order to characterize its performance. Perhaps the most arduous of tests occurred in the British Antarctic Survey Cold Facility, whereby the complete assembly was tested at −25°C for one week. These tests provided both baseline performance results when drilling in multiple terrain types under laboratory conditions and also acted as a means of proving the robustness of the system at reduced temperatures.

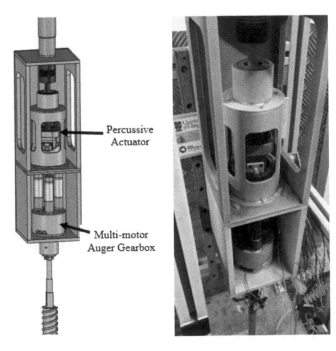

Percussive Actuator

Multi-motor Auger Gearbox

FIGURE 1.158 P-RAID architecture.

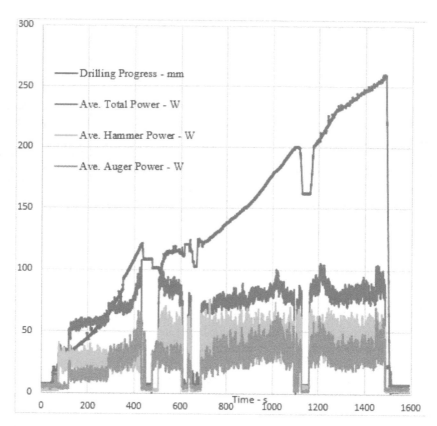

FIGURE 1.159   P-RAID. Performance data from a successful drilling run in a limestone sample.

In order to establish performance, a metric known as specific energy may be used. Specific energy provides the designer with a means of establishing the energy required to penetrate a given volume of terrain, and thus normalizes against the rate of penetration used during the drilling operation and the physical geometry of the drill bit, allowing dissimilar to be compared. Figure 1.159 details the performance of the system during a typical run in limestone terrain. The average total power (rotation and percussion summed) is approximately 75 W.

Using a custom-designed core drilling bit with an outer diameter of 40 mm with an internal bore diameter of 25 mm, specific energies in the region of 300, 500, and 1800 MJ/m^3 are achievable for limestone (25 MPa), medium-strength sandstone (~50 MPa), and micro gabbro (in excess of 120 MPa). Though results for micro gabbro are unavailable, these results broadly resemble those found in the literature. These drill runs successfully created cores of varying quality, depicted in Figure 1.160.

Assured by the performance of the P-RAID proof-of-concept prototype tested in the laboratory, it was decided that the system would be redeveloped into a field-deployable architecture for deployment in West Antarctica with the British Antarctic Survey. Figure 1.161 depicts a cutaway of the system.

While the key components of the system are identical to the proof of concept version, each element of the design has been reworked allowing the system to be stretched axially in order to reduce the burden on the tight diameter constraint present. The new system measures approximately 5 m long, but with a diameter of only 80 mm as before.

Taking advantage of pre-existing field campaign with an objective of drilling through ice to the bedrock (albeit, not with the RAID drill but with a larger diameter ice coring system), the P-RAID

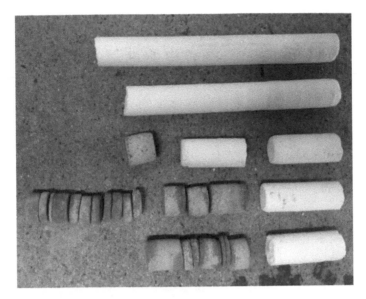

FIGURE 1.160 P-RAID. Rock cores captured during laboratory testing. Hartham Park Limestone (white cores), Locharbriggs Sandstone (reddish cores) and micro gabbro (gray core) are shown.

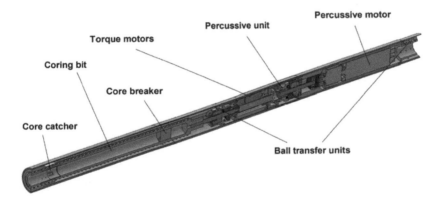

FIGURE 1.161 Field deployable P-RAID system.

team opportunistically deployed the system to Antarctica for initial shakedown testing in January 2019. Deployed to Skytrain ice rise (79° S, 78° W), the system was integrated with the top-end RAID and winch system to test mechanical and electrical integration in the field. Figure 1.162 shows this activity.

At the time of writing, a drill campaign is in planning for the 2019/20 summer season. The first targeted drill site is intended to be at Sherman Island.

1.10.5.5 Conclusions

The development of the P-RAID system presents the European planetary science and glaciological communities with a promising technology for ongoing exploration activities. The drill system has compared favorably with existing technologies and initial shakedown tests in Antarctica have bolstered confidence in the capability of the system going forward.

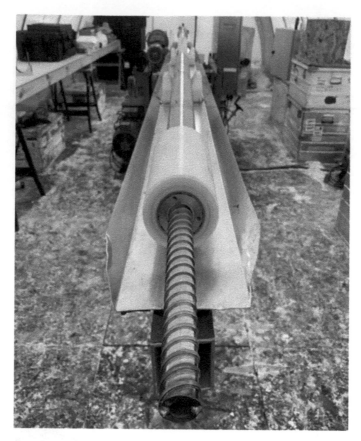

FIGURE 1.162   Field deployable P-RAID undergoing integration tests in Antarctica.

1.11   1-M CLASS SAMPLERS

This section focuses on 1-m class sampling systems. As mentioned earlier, as the excavation depth becomes larger, the most energy efficient sampling system has a small diameter and it is long. Drilling systems fall in this category. The section presents a number of drills developed by various space agencies and organizations around the world.

1.11.1   JAXA/SHIMIZU DRILL

Subsurface access by robotic lunar exploration may require a drill system which is low mass/low power consumption and capable of quick excavation. Despite the fact that the drill excavation in harsh environments of the moon is different from the ground, the data necessary for the design is not sufficiently obtained. This section gives an example of drill excavation experiments in the simulated environment, utilizing the lunar soil simulant. Table 1.10 shows the specifications of prototyped test drill, and Table 1.11 shows the simulated environmental conditions, compared with the actual lunar environmental data.

Figure 1.163 shows the drill driving unit. The driving unit is composed of an elevating mechanism for raising and lowering the auger and a rotating mechanism for rotating the auger. The 6-axis force/torque sensor, the rotation mechanism part, and the auger are all assembled to the elevating mechanism attached to the guide rail. Elevation is controlled by driving it with a trapezoidal drive screw. The 6-axis force/torque sensor measures the excavating force and the rotational torque. All mechanical parts are accommodated in the vacuum chamber as shown in Figure 1.164.

TABLE 1.10
Specification of Experiment Equipment

Item	Specification
Drilling depth	400 mm (max)
Excavation speed	10 mm/s (max)
Shooting force	50 N (max)
Rotational speed	2 r/s (max)
Rotation torque	1.9 Nm (max)
Deep soil depth	450 mm
Tab diameter	200 mm

TABLE 1.11
Experimental Environmental Condition

Item	Experimental Environment	Lunar Environment
Temperature	At normal temperature	-170 to $110°C^{*1}$
Soil vacuum degree[*2]	Less than 25 Pa	10^{-8} Pa
Gravitational acceleration	1 G	1/6 G
Excavation target	FJS-1[*3]	Lunar regolith

[*1] Lowest at night near the equator and highest at daytime.
[*2] Degree of vacuum of the lunar soil simulant and the lunar regolith.
[*3] Name of the lunar soil simulant manufactured by Shimizu.

Figure 1.165 shows the auger and drill bit. The shape of the auger was a helical with an outer diameter of 25 mm and a pitch of 10 mm. In addition, the diameter of the drill bit at the outer circumference was 30 mm, which is slightly larger than that of the auger, to reduce the frictional influence of the borehole later.

Inside the vacuum chamber, the drilling mechanism part and the soil bath filled with lunar soil simulant are arranged, and excavation and measurement can be performed by a command from an external PC. Also, to confirm the drilling condition, LED lights and cameras are installed inside so that the excavation state can be visually confirmed through the display.

The internal structure of the chamber is divided into upper and lower parts as shown in Figure 1.166. The outside of the lower part of the chamber is connected to the vacuum pump, and the soil tank is installed inside the chamber. The soil tank has a double structure of punching metal (4 in the figure) and a stainless-steel mesh (3 in the figure) to prevent outflow of the lunar soil simulant, and it exhausts the air inside the lunar soil simulant through the mesh part (arrows in the figure represent air flow). Since the air is discharged in the upper part of the chamber (1 in the figure) where the drilling mechanism part is installed as a filter with the lunar soil simulant as a filter, it is possible to suppress the boiling phenomenon of the lunar soil simulant caused by the vacuum drawing from the surface. The degree of vacuum at the top of the chamber (1 in the figure) was taken as the degree of vacuum of the lunar soil simulant.

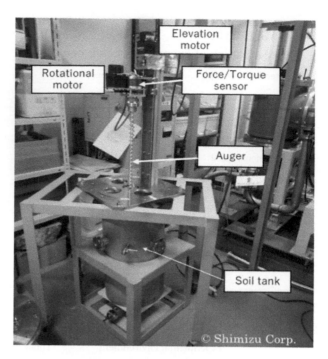

FIGURE 1.163 JAXA/Shimizu Experimental drill drive unit.

FIGURE 1.164 JAXA/Shimizu. Drill drive unit accommodated in a vacuum chamber.

FJS-1 is one of the lunar soil simulants, and was manufactured in Japan by Shimizu Corporation. The particle size distribution lies below 2 [mm]; the median particle size is between 70 and 75 [micro-m]; the shear strength, i.e. cohesion c, is 0 to 10 kN/m^2; and the internal friction angle, phi, is 30°–50°. The chemical composition is also close to Apollo samples with slight difference that FJS-1 contains more ferric oxide and alkaline components than Apollo samples.

The comparison of the excavation torques of 20 cases are shown in Table 1.12. The experimental conditions are as shown in the table. The bulk densities were 1.75 g/cm^3 (equivalent to the depth of

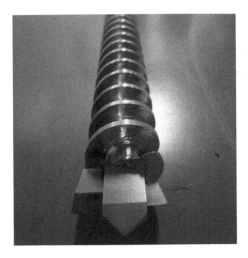

FIGURE 1.165   Drill bit and auger.

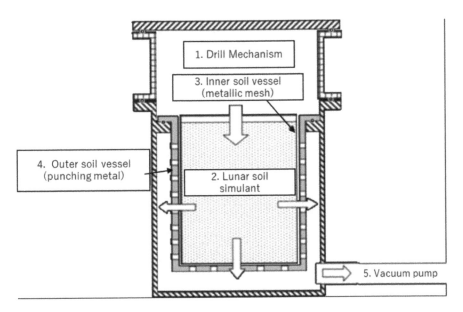

FIGURE 1.166   Internal structure of vacuum chamber part.

50 cm) and 1.90 g/cm³ (equivalent to the depth of 250 cm). The 2 degrees of vacuum were chosen as normal pressure and the pressure less than 25 Pa. The excavation speed was fixed to 2.0 mm/s. From this excavation speed, the auger rotation speed when excavating along the auger pitch of 10 mm (excavating like a wood screw) is 0.2 r/s. In the experiment, five kinds of rotation speeds were set with 0.2 r/s as a reference.

The experimental conditions are shown in Table 1.12, and the results of torque vs. depth are shown in Figures 1.167, 1.168, 1.169, and 1.170. All data is filtered through a low pass filter with the cut frequency of 0.1 Hz.

Based on the experimental results, it is possible to excavate with a small excavation torque regardless of each bulk density and vacuum environmental conditions if a sufficient rotational speed with respect to the pitch (for example, 0.75 r/s, 1 r/s) is maintained.

TABLE 1.12

Experimental Conditions for JAXA/Shimizu Drill

Exp. #	Bulk Density (g/cm³)	Rotational Speed (r/s)	Degree of Vacuum (Pa)
#1	1.75	0.2	Normal pressure
#2		0.3	
#3		0.5	
#4		0.75	
#5		1.0	
#6	1.75	0.2	11.7
#7		0.3	14.7
#8		0.5	10.8
#9		0.75	5.33
#10		1.0	6.67
#11	1.90	0.2	Normal pressure
#12		0.3	
#13		0.5	
#14		0.75	
#15		1.0	
#16	1.90	0.2	14.7
#17		0.3	17.3
#18		0.5	18.7
#19		0.75	22.7
#20		1.0	17.3

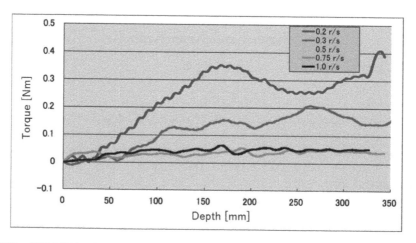

FIGURE 1.167   JAXA/Shimizu drill. Drilling torque vs. depth, vacuum (<25 Pa), 1.75 g/cm³ (#6–10).

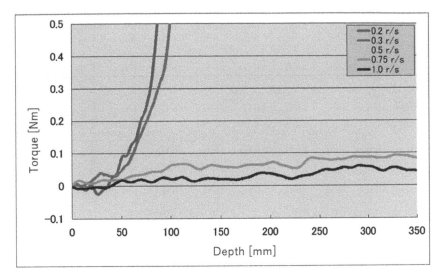

FIGURE 1.168 JAXA/Shimizu drill. Drilling torque vs. depth, normal pressure, 1.90 g/cm³ (#11–15).

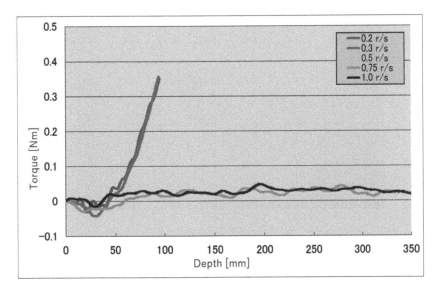

FIGURE 1.169 JAXA/Shimizu drill. Drilling torque vs. depth, vacuum (<25 Pa), 1.90 g/cm³ (#16–20).

For an auger of 25 mm in diameter and 10 mm in pitch, it can be thought of as 3–4 times the rotation speed corresponding to the pitch of 0.2 r/s is a sufficient speed. This is because, as shown in Figure 1.170, when there is a sufficient rotational speed, it is possible to discharge drilling slips (shavings) from the borehole. In the case of an auger, the larger the friction between the particles, the more advantageously it works. Therefore, under the condition that the shear can be discharged, it is considered that the larger the friction between the particles, the smaller the excavation torque can be.

On the contrary, if a sufficient rotation speed to discharge the shear is not obtained, the auger re-compresses the lunar soil simulant in the borehole. Thus, the resulting excavation force and torque increase. For example, as shown in the experiment #16 in Figure 1.170, it is impossible to discharge the shear for low rotational speed, and thus, a large torque is required. In addition, in the experiment of #16, there was a misalignment of more than the volume of the borehole. Since it could not be

FIGURE 1.170   JAXA/Shimizu drill. Comparisons of removed soil (#16–20).

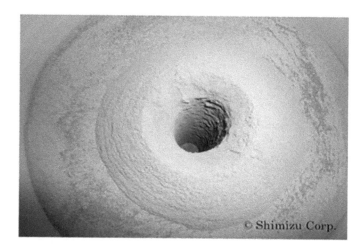

FIGURE 1.171   JAXA/Shimizu drill. Borehole after excavation (about 40 cm depth).

discharged, a stress was generated inside such that a crack would form in the vicinity of the borehole (due to excavating force and excessive torque, the experiment was interrupted).

Figure 1.171 shows the state of the borehole after excavation. After excavation, one can observe upright boreholes which cannot be achieved with ground soil. This phenomenon is presumed to be able to form similar boreholes on the lunar surface because there are records observed also in Apollo drill excavation, etc.

In terms of soil mechanics, lack of borehole collapse is considered to be the effect of apparent adhesive force, and interlocking of particles due to the shape of the particles and dilatancy etc. where the volume of soil expands at the time of shear failure.

1.11.2   EXOMARS DRILL

The ExoMars mission will deploy on Mars a Rover equipped with a sophisticated sample acquisition and analysis system, inclusive of a drilling unit for subsurface exploration (Figure 1.172). The

FIGURE 1.172 ExoMars Rover flight model undergoing tests. The drill is housed inside the black vertical box in front.

drill is designed to operate on Mars, acquire soil samples from its subsurface to a depth down to 2 m, and present them to a suite of mechanisms and instruments for in situ processing and analysis.

The drill flight model has been accepted for integration on the Rover flight model in February 2019, at completion of a long process of design, development, and qualification that lasted more than a decade.

ExoMars (EXM) is an ESA project based on international cooperation with Roscosmos, NASA, and other National Space Agencies, and involving the major European industries, with Thales Alenia Space Italy as overall Prime, and Airbus Defense & Space-UK as the main contractor for the Rover. It consists of two missions. The first one, launched in 2016, has carried to Mars a demonstration lander and the Trace Gas Orbiter (TGO) that is currently analyzing the planet atmosphere and will serve as data relay for the following Rover and Surface Platform (RSP) mission, scheduled for launch in 2022. The RSP lander (entry mass 2 tons) will bring to the Mars surface in 2023 an instrumented platform and a Rover, which will deploy and egress from the platform to perform at least 8 months of scientific operations.

The EXM Rover is the first if its kind for Europe. The scientific and technology objectives include on one side the environment characterization and the search for past and present life, and on the other the demonstration of key technologies, including locomotion, drilling, sample processing and automated operations. These are necessary steps to prepare the future planetary missions, on Mars and on the Moon. The Rover, working as a mobile scientific laboratory, will have to map the terrain, identify targets of interest, acquire soil samples from the surface and the subsurface and measure their content and characteristics, while moving with autonomous navigation capabilities and maintaining daily communication contacts with relay orbiters and Earth ground control.

The drill—developed by *Leonardo, Italy*—is therefore an essential element of the Rover mission, and its operations and performance are key to the mission success. Extensive resources have been used for its development, starting from dedicated breadboards, to testing with a variety of Mars simulant soils and qualifying in representative Mars environment conditions.

1.11.2.1 EXM Drill System Configuration

The EXM Rover Drill system includes a drill unit (or drill box), a positioner, a jettison device and a set of electronics and software, controlling also other Rover mechanisms and functions, located both inside the Rover (in a Central Electronic Unit—CEU—or warm electronics) and outside it, in a separate package mounted externally on the drill box structure (drill cold electronics). In summary, the drill unit consists of the following main elements:

- **Drill Tool**: A 700-mm-long augered steel tube, equipped with a drill-tip and a sample acquisition device, inclusive of a shutter to retain the collected sample, a movable piston to open the sample volume and push out the collected sample, position and temperature sensors; and inclusive of the Ma_Miss tip components (optical fiber, lamp, window, reflector) as shown in Figures 1.173 and 1.174. Its main function is to penetrate the terrain, acquire a core sample (reference is a cylinder core 1 cm in diameter × 3 cm in length), extract/retain the sample, bring it upwards, and deliver it into the inlet port of the Rover Analytical Laboratory, once the drill unit has been properly positioned.
- **Back-up Drill Tool**: Stored in the rod magazine, it is there to be engaged and used in off-nominal situations (drill main tool stuck in the terrain or malfunctioning). The back-up tool is not equipped with Ma_Miss optical head.

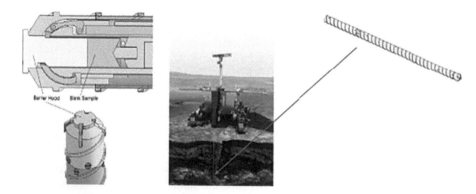

FIGURE 1.173 ExoMars Drill tool with sample acquisition capability.

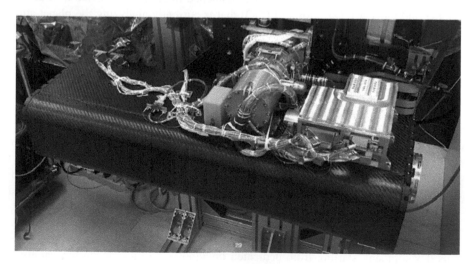

FIGURE 1.174 ExoMars Flight Model Drill unit inclusive proximity and Ma_MISS electronics box (Leonardo I).

- **Set of (3) Extension Rods**: Each rod is approximately 500 mm long and is designed to extend the penetration length to 2 m. They are provided with optical and electrical contacts and fiber/harness to guarantee the transmission of the electrical and optical signal to/from the drill tool to the spectrometer. They are accommodated together with the back-up drill tool on a rotary magazine through a dedicated clamping system.
- **Rotation-Translation Group**: This includes the sliding carriage motors and sensors, the gear mechanisms, the mandrel and clamps, and the Ma_Miss optical rotary joint. Its main functions are to provide the drilling torque/rotation and linear force/vertical motion as well as the automatic engage-disengage for tools and rods.
- **Drill Box Structure**: A CRFP panel-molded structure, supporting the Rotary-Translation group and providing the overall encasing for drill tools, external rods and sensors. It includes the clamping subsystem for the extension rods (rod magazine group), and the automatic engage-disengage mechanisms for the rods. Integrated onto the drill box structure are also the Ma_Miss spectrometer and the drill proximity electronics (DPE).

The drill unit is integrated on the Rover chassis front side through a 2-degrees-of-freedom positioning mechanism, an integral part of the drill system. This positioning device allows the translational and rotational movements of the drill unit. It is capable of deploying the drill from its storage/stowed position (used during launch, cruise, landing phases and during Mars surface locomotion) to its operational/drilling position, orthogonal to the terrain. Once the drilling has been performed and the sample acquired and retrieved, the positioner rotates the drill to bring it to a position suitable to allow the sample delivery to the inlet port of the Analytical Laboratory (see Figure 1.175), a complex suite of mechanisms and scientific instruments located inside the Rover body and enclosed within an Ultra-Clean volume (UCZ) where the scientific measurements will be performed.

The positioner (Figure 1.176) is attached to a jettison device that can be activated in emergency situations to disengage completely the drill unit from the Rover; for example, in case the drill string should remain blocked/seized in the terrain, endangering the Rover mobility and the continuation of the mission. The jettison device operates through a motorized screw-nut interface (brushed motor).

To achieve the challenging 2-m drilling depth requirement (particular for the EXM mission), the drill design is based upon a multi-stroke technology, using one main drill tool and three extension rods that are progressively and automatically engaged one on top of the other during the drilling. The drill penetrates the terrain by a rotary/translation action, using actuators with brushless motors. Brushed motors—specifically developed and qualified to operate in Mars conditions—are used instead for the activation of the clamping systems (lower clamp for extension rod/drill tool engage/disengage, and rod magazine clamps).

The EXM Drill is also equipped with an ad hoc IR spectrometer for the borehole exploration (Figure 1.177). The Mars Multispectral Imager for Subsurface Studies (Ma_Miss) is a miniaturized IR spectrometer, with the optical head integrated inside the drill string's primary drill tool;

FIGURE 1.175 ExoMars Drill. Sample discharge from drill tool on ALD core sample transport mechanism—breadboard tests (Leonardo I).

FIGURE 1.176 ExoMars Drill. (Left) Drill unit and Positioner QM during mechanical test (Leonardo I). (Right) Flight Positioner assembly in Leonardo I.

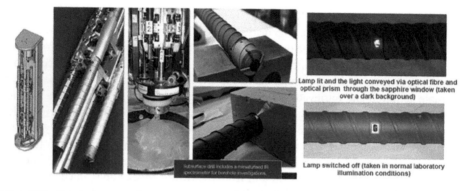

FIGURE 1.177 ExoMars Drill. Multi-rod drill unit and borehole IR spectroscopy tests.

the subsoil is observed through a sapphire window integrated at the helical profile of the drill tool. Figure 1.178 shows drilling and sampling steps during the sample acquisition process.

The drill materials and the Adopted assembly, Integration and Verification (AIV) approach is fully compatible with the class IVb planetary protection directives applicable to the ExoMars mission and with the applicable cleanliness contamination control requirements. The key features of the ExoMars drill system are in summary:

- Drilling depth max. 2 m.
- Sample size: diameter 10 mm, length 30 mm.
- Thermal environment: non-operational: −130°C/+35°C (+125°C for Sterilization/Planetary Protection). Operational: −90°C/+30°C.
- Mass: drill subsystem mass ~24.5 kg (unit + positioner + jettison) (exclusive of the Ma_Miss Instrument and control electronics).
- Mass control electronics CEU: 7.5 kg.
- Power: 70–80-W max power consumption in worst conditions.
- Vertical thrust: 10–450 N (depending on terrain characteristics).
- Torque: 1–6 Nm (upper value for hard material and widely used bits).
- Drilling speed: from 0.3 to 20 mm/min (depending on soil hardness).
- Materials: regolith/sedimentary rocks up to 110-MPa compressive strength.

The drill is designed to penetrate into a variety of soils and rocks types which are expected to be found on Mars, with the operational lifetime required by the ExoMars mission scenarios, which

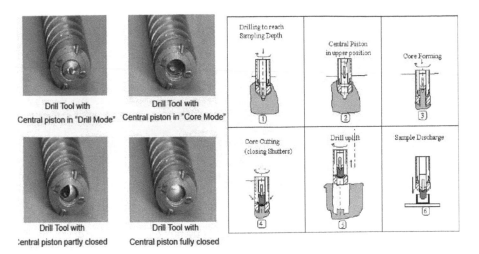

FIGURE 1.178 ExoMars Drill. Sample Acquisition Sequence.

include the completion of 6 full Experiment Cycles (EC) and at least 2 Vertical Surveys down to 2 m (with 4 subsurface sample acquisitions each survey). This means that a minimum number of 14 samples shall be acquired and delivered to the Rover Module Analytical Drawer by the drill for subsequent analysis. A Rover EC includes traveling to a selected target location, performing subsurface characterization through ground penetrating radar, drilling to a predefined depth, acquiring soil samples, and performing the in situ sample processing and analytical measurement with the Rover onboard instrumentation (IR, Raman and Mass spectrometers, Gas Chromatographer). Each EC can last up to 17–20 Sols in nominal conditions. Each drilling operation can take a number of Sols, depending on the terrain characteristics and on the possibility of leaving the tool in the ground overnight.

The drill has followed a long development, in part because of the programmatic changes that have affected the overall ExoMars program and in part because of the need to validate the technology in representative Mars conditions and on a variety of simulant soils, identified and selected with the support of the scientific community. A dedicated thermo-vacuum chamber (5 m high) has been realized (University of Padua—Italy) to perform tests in frozen soils and in the Mars CO_2 low pressure atmosphere. Different models, at the subsystem and system level, have been built to trade off technical solutions, especially for the final selection of the multi-rod automatic clamping system, for the sample acquisition chamber and for the actuators, which have to operate in Mars conditions and in compliance with stringent planetary protection and cleanliness contamination control conditions (use of specific lubricant and materials).

The following model philosophy has been adopted for development and qualification (Figures 1.179, 1.180, and 1.181):

- **Breadboards (BB)**: drill tool, drill box. Used for preliminary drilling tests, materials characterization, design validation, technical trade-offs.
- **Engineering Model (EM)**: functional validation of structure and main elements. Engineering Model of Mechanisms/Actuators and Sensors. Used to de-risk the overall assembly ahead of Qualification.
- **Qualification Model (QM)**: refurbished EM to full Qualification Standards. Used to achieve qualification (Mechanical/Thermal and Mars Environment, Functional and Life Testing Verification).
- **Flight Model (FM)**: underwent acceptance level tests and PP/CCC procedures.

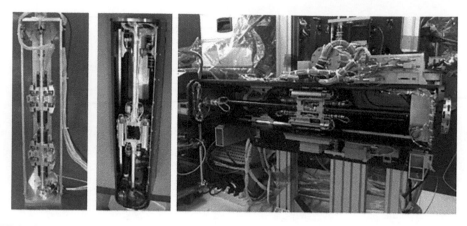

FIGURE 1.179 ExoMars Drill. Drill box breadboard (left); drill box EM (center), Drill Qualification Model at Leonardo (right).

FIGURE 1.180 ExoMars Drill. Drill tool FM details: cleaning brushes at outlet and protective cap (to be removed before operations on Mars).

FIGURE 1.181 ExoMars Drill. Drill Flight Unit in TAS-I ISO7 HC clean room.

The drilling and sample acquisition capabilities have been tested in ambient and in Mars conditions with a variety of materials, selected and procured under the supervision of the involved scientific community. These materials include different types of sedimentary rocks with hardness and compressive strength ranging from a few to 110 MPa. Tests on iced samples, i.e. samples prepared with a defined quantity of water ice in them, have also been performed, to verify drill performances and study scenarios in which the drill tool could get stuck in the terrain (e.g. in case of the drill tool remaining overnight in the terrain). Several samples have been collected in the different conditions throughout the development testing.

1.11.3   PROSPECT PACKAGE

PROSPECT is being developed by the European Space Agency (ESA) for flight on board the Russian Luna Resource mission (Luna-27), which has the overall objective to access and explore a landing site in the lunar south polar region (Sefton-Nash et al., 2018). A key science objective for PROSPECT is the detection and characterization of lunar polar volatiles. In this context, the preservation of volatiles in acquired samples up to the analysis stage is among the main challenges for PROSPECT, and specifically imposes strict constraints on the thermal impact of drilling and sample handling.

The PROSPECT package consists of two key subassemblies: the ProSEED drill, which supports the acquisition of samples from the lunar surface and subsurface, and the ProSPA instrument, which allows analysis of acquired samples by means of volatile extraction and mass spectrometry techniques. Alongside the ProSPA instrument suite, the ProSEED drill is equipped with a dedicated ProSEED Imaging System for obtaining high resolution and multispectral images of drilling operations; a Permittivity Sensor for measurement of regolith dielectric properties, and specifically remote detection of ice content; and a set of temperature sensors for close monitoring of the sample and drill thermal environment.

As part of the ESA-ROSCOSMOS cooperation on Luna-27, the PROSPECT drill will acquire samples to serve both Russian instruments and the ProSPA package, noting that the ProSEED sampling tool is designed to address the different sample needs of those instruments.

The PROSPECT development is led by Leonardo S.p.A. (Italy) who are also developing the ProSEED drill in collaboration with subcontractors, while the ProSPA instrument is under the lead of the Open University (UK). Both ProSEED and ProSPA build on existing heritage, e.g. the ExoMars Drill and the Beagle-II GAP instrument, as well as the SD2 drill and Ptolemy instrument which both flew on board the Rosetta Philae lander. Since first conceptual design and breadboarding activities started in 2015, the team has completed an extensive preliminary design process with a PDR closed in 2019, including representative testing of development models of both drill and instrument, and is embarked on the main development for a delivery of flight hardware to Russia in 2023, in advance of flight in 2025.

1.11.3.1   ProSEED Drill System Configuration and Testing

The ProSEED drill includes the main drill structure unit, a rotational positioner, and dedicated hold-down mechanisms, which are all controlled via a Control Electronic Unit (CEU) and associated software. A significant part of this architecture and technological heritage is derived from the ExoMars Drill development. In summary, the ProSEED drill consists of the following main elements (some of which are shown in Figure 1.182):

- **Positioner Rotation Joint**: This mechanism pivots the drill box along an arc including its hold-down/survival position, a range of drilling locations, the Russian sample delivery location and the ProSPA sample inlet.
- **External Hold-Down Mechanism**: This structural unit is sized to provide mechanical support to the drill structure at launch. It accommodates irreversible Frangibolt FC4 actuators for the release of the drill rotational degree of freedom after landing.

FIGURE 1.182 (Left) Integrated ProSEED development model undergoing ambient tests at Leonardo, Nerviano; (top-center) view of SIS with alignment collet in foreground, and brushed Russian sample transfer interface cylinder in background; (top right) Positioner Rotation Joint shown before integration with the drill development model; (bottom center) Russian sample auger exposed in sampling/retrieval configuration; (bottom right) ProSPA sample push-tube extended.

- **Drill Structure**: a CFRP structure supports the main drilling mechanisms, as well as the Drill Translation Joint (DTJ) which applies the main drilling thrust (up to 800 N), the imaging system, and the internal hold-down mechanism. The drill box then interfaces externally with the external hold-down mechanisms (released after landing) and the Positioner Rotation Joint.
- **Mandrel and Drill Rod**: the unit includes a mandrel and an augered aluminum tube sized to reach a sampling depth of 1 m, when deployed from its attachment position on the Luna-27 platform, with its diameter (38 mm) driven by the design of the sampling tool it accommodates. The mandrel accommodates a Maxon actuator which provides the rotation to the drill rod (nominally 60 rpm) and the needed torque (up to 11 Nm) for drilling and sampling operations. In addition, the mandrel integrates two slip rings used to provide electrical connection with the sampling tool unit and other drill-embedded sensors.
- **Sampling Tool**: The sampling tool is composed of two independent mechanisms, with corresponding actuators. A push-tube tool is accommodated at the very tip of the drill, with an internal diameter of ~3 mm and capable of retrieving cores of ~6–10 mm in length, and which acquires samples delivered directly into the ProSPA ovens. A separate sampling auger is integrated in the front end of the sampling tool, and which can be exposed from the end of the drill for sampling (~7–8 cm³) material, and retracted again before string retrieval to confine the sampled material and to prevent loss; this sampling auger acquires unconsolidated samples

which are then removed via a brushed interface for delivery to the Russian robotic arm, and subsequent transfer to the other Russian analysis instruments.

- **Internal Hold-Down Mechanism**: The structural unit sized to provide mechanical support to the drill rod and sampling tool at launch. It accommodates a melting wire-based actuator solution for low shock release after landing.
- **Imaging System**: A combined package developed by Kayser Italia, consisting of a high resolution (4 MPixel) fixed-focus camera (3DPlus, Lambda-X) and LED-based illumination unit supporting 6 spectral bands, is capable of acquiring multispectral images and up to 10-fps video of drilling operations.
- **Permittivity Sensor**: This sensor is accommodated in the rod of the ProSEED drill and measures the dielectric properties of surface and subsurface materials via the emission of an AC signal through a small electrode integrated in the drill rod surface.
- **Temperature Sensors**: In addition to engineering sensors that allow to monitor the temperatures of key mechanisms, the drill also integrates temperature sensors for monitoring the drill tip/sample temperature.

The ProSEED drill is interfaced, via the positioner rotation joint, to the Luna-27 platform. The ProSPA sample interface (so-called Solids Inlet System—IS) and the Russian sample interface are accommodated on a separate payload balcony. For this reason, and specifically considering the small size of sampling push-tube and ProSPA oven, a dedicated alignment collet is integrated into the SIS which allows for active control and alignment of the targeted oven beneath the ProSEED drill during sample transfer. The SIS, including alignment elements and the breadboard of the Russian sample transfer interface, are shown in Figure 1.183.

Searching for water ice and other volatiles in the lunar polar environment places specific requirements on PROSPECT, in particular in terms of the characteristics of the regolith which the ProSEED drill may encounter. Early testing carried out using lunar regolith simulant (NU-LHT-2M) mixed with various quantities of water ice (0%/weight up to saturation ~11%/weight), helped to define specific mechanical performance requirements of the drill and its sampling tool. Following this test phase, and taking into account system constraints, it was decided not to embark the hammer support tool which had been considered in earlier phases (see Section 1.11.3.2).

Full-scale system tests of the ProSEED design have been performed under laboratory conditions at Leonardo in Nerviano, both with materials at ambient temperature and with frozen regolith simulant containing water ice. Following unit level thermal tests at low temperature, which demonstrated functionality of key mechanisms, the integrated ProSEED drill Development Model (DM) has been extensively tested in the CISAS thermal vacuum chamber (University of Padua—Italy). This chamber, which was also used for testing of the ExoMars Drill, is capable of recreating low pressures (10^{-3} mbar) while accommodating the large-scale regolith simulant container needed for functional testing at extreme low temperatures (regolith simulant at −150°C; ProSEED interfaces at −100°C).

During early stages of testing, in which disturbances of dry regolith simulant on the surface of the sample container led to severely dusty working conditions, the ProSEED mechanisms have demonstrated a significant robustness to those dust effects, with no noticeable effects on performance. This Development Model test campaign has successfully demonstrated ProSEED functionality and performance under representative temperatures and mechanical conditions, including in particular drilling, acquisition, and successful transfer of loose dry regolith simulant, simulant containing layers of medium size inclusions, and simulant with water ice contents of ~6%/weight and ~10%/weight (see Figure 1.184). During the final phase of testing, ProSEED successfully drilled a large anorthosite rock, embedded in the base of regolith simulant containing 10%/weight water ice.

Other elements of the system already having undergone testing include the bench development model of the ProSPA instrument, the Engineering Model of the Imaging System, and the breadboards of the Permittivity Sensor and internal hold-down mechanism (see Figure 1.185).

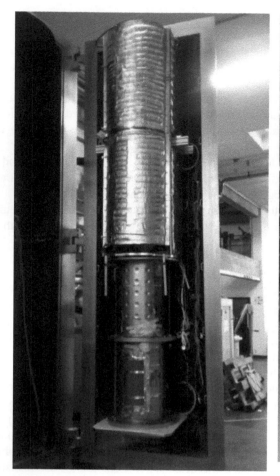 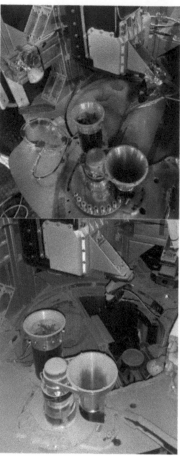

FIGURE 1.183 (Left) ProSEED drill integrated within its dedicated thermal shroud, mounted above the separately thermally controlled sample container assembly, all shown within the (open) CISAS TVAC chamber (University of Padua); (top right) post-test view of sampling area showing SIS (lower carousel with alignment collet) and Russian sample transfer interface (upper cylinder); (bottom right) post-test view of mechanical elements following effects of dust contamination during pump-down.

Specifically regarding the PROSPECT mechanisms, the project has completed an extensive test campaign of the various types of Maxon brushless actuators which are utilized across the system, concentrating on both operational and non-operational temperature compatibility, and also on robustness to mechanical environments. The next steps of the project include refurbishment of the ProSEED DM into a fully fledged Engineering Model for interface testing in Russia, before completing the flight development of the full PROSPECT package.

1.11.3.2 ProSEED Hammer Mechanism Investigations

During early phases of the PROSPECT project, it was planned to equip the Lunar Drill (or PROSPECT drill) with both rotary and percussive drives, which in principle enhance the conventional drilling process (Figure 1.186). A tool combining two types of drives will result in higher efficiency and effectiveness during penetration into the more consolidated layers of soil, which are significantly harder due to the presence of ice (Atkinson 2019). A rotary-percussive drive also enables drillability of a wider range of rocks.

FIGURE 1.184 (Left) ProSPA push-tube immediately prior to sample delivery via SIS alignment collet (shown); (center) "Russian" sample auger exposed, with sample; (right) "Russian" sample auger following sample retrieval via brushed transfer interface (cylinder on left).

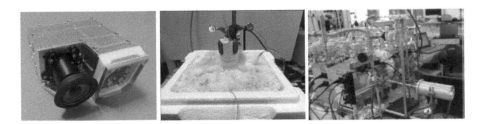

FIGURE 1.185 (Left) ProSEED Imaging System EM (camera unit on left, illumination unit on right side); (center) ProSEED Permittivity Sensor prototype during cryogenic test in glass beads/water ice mixture; (right) ProSPA BDM setup at Open University (with sample in Quartz furnace in foreground).

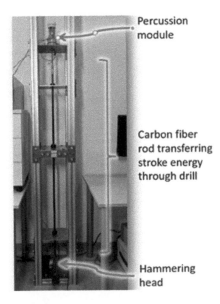

FIGURE 1.186 PROSPECT drill testbed. *Source:* Courtesy of Astronika.

Within the framework of the project, two novel solutions were developed by Astronika. The first of these solutions is based on an electromagnetic principle of operation like MUPUS PEN or CHOMIK, which ensures a highly reliable operation with high-force impulses. Essentially, in this design the efficiency of the system was significantly improved by implementation of dedicated smooth control of frequency and energy impulses, which is difficult to achieve with the use of other mechanical drives.

The second solution is propelled by a DC motor. Here, the rotational movements of the motor are converted to the linear movements of the hammer by a crank mechanism. The main advantage of this solution are much simpler principles of control as compared to the first option.

Both proposed solutions focus on obtaining the maximum efficiency and highest possible reliability.

1.11.3.2.1 Electromagnetic Actuator for Lunar Drill

In this solution the device (Figure 1.187) consists of three main subassemblies: (I) an anvil, which transfers the impact energy to the drill and soil; (II) a clutch responsible for the compensation of the radial clearance between the hammer (7) the carbon fiber rod (2), and an electromagnetic hammering mechanism (III).

The hammering mechanism uses an electromagnetic actuator technology. The current in the coil (13) creates a magnetic field mainly in the Permendur core (12), armature (8), and in the air gap

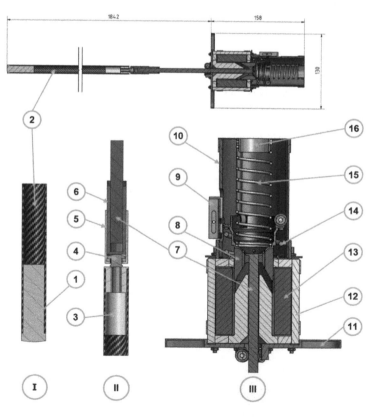

FIGURE 1.187 PROSPECT drill. Electromagnetic actuator for Lunar Drill percussion. *Source:* Courtesy of Astronika.

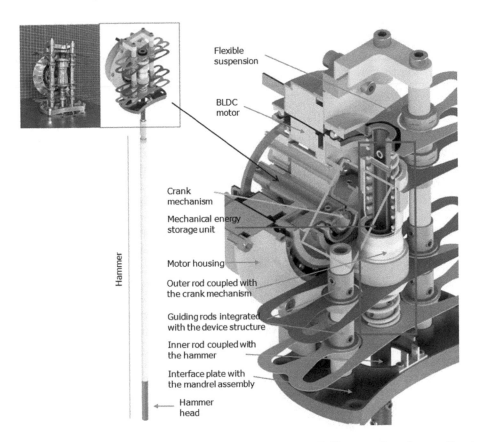

FIGURE 1.188 PROSPECT drill. Crank mechanisms for Lunar Drill percussion. *Source:* Courtesy of Astronika.

between the armature and inner core. The magnetic field generates a force in the air gap, trying to close the gap. This accelerates the hammer (7), guided by two sets of rollers (mounted on ball bearings to ensure as low friction as possible).

Furthermore, the mechanism has two springs: a return spring (14) and a drive spring (15). The return spring is responsible for the hammer's return after impact, while the drive springs' parameters determine the hammer's working frequency.

1.11.3.2.2 Crank Mechanism for Lunar Drill

The driving element in this solution is a DC motor controlled by a digital control unit and supplied with energy by the lander's batteries supported by an additional capacitor filtering power consumption peaks that occur during single stroke mode. The control unit is responsible for powering the motor at defined voltage. A crank mechanism is used to convert rotary motion into reciprocating motion. Additionally, a spring is placed between the crank-mechanism and the hammer head, which serves to accumulate the mechanical energy which is then transferred during strokes to the drilling and sampling tool, but also to protect the motor and the crank mechanism from jolts during strikes. Figure 1.188 presents a preliminary version of the hammering device driven by a DC motor, with the key subassemblies identified.

Both electromagnetic and crank mechanisms could provide a reliable source of percussive action to the Lunar Drill and had a positive effect on the drill rate of penetration (Savoia et al. 2019); however it has been decided that only rotary action would be enabled in the current iteration of the drilling system due to mass and power constraints.

1.11.4 The IceBreaker Drill

The 2008 Mars Phoenix lander used a scoop with a high-speed drill (called rasp) to penetrate less than a centimeter into the ice surface at the Martian North Pole. The next step in the exploration of the Martian subsurface is to drill at least two orders of magnitude deeper and deliver samples to a suite of instruments for analysis.

To enable this next step, we have been developing a 1-m class drill called the IceBreaker (Bergman et al. 2016; Glass et al. 2016; Paulsen et al. 2012; Zacny et al. 2010a; Zacny et al. 2013d). The drill consists of a rotary-percussive drill head, a sampling auger with a bit at the end, a Z-stage for advancing the auger into the ground, and a sampling station for moving the augered ice shavings or soil cuttings into a sample cup.

In the summer of 2010, the IceBreaker drill was tested in the University Valley (within the Beacon Valley region of the Antarctic Dry Valleys). University Valley is a good analog to the Northern Polar Regions of Mars because a layer of dry soil lies on top of either ice-cemented ground or massive ice (depending on the location within the valley). That is exactly what the Mars Phoenix mission discovered on Mars.

The first set of tests included drilling into ~20 cm of dry soil followed by 80 cm of ice-cemented ground, that is to a depth of 100 cm. The IceBreaker drill was placed outside of a science tent (our Mars-analog site), while the control system with the drill operator was inside the tent (control room on earth)—as shown in Figure 1.189. This was done in order to demonstrate remote operation of the drill. We also followed protocols used to operate the Rock Abrasion Tool on Mars Exploration Rovers and in particular, we issued only three different commands: "seek," "drill," and "sample." The drilling protocol was designed to allow for collecting samples in 10-cm intervals. The tasks for the person outside of the tent was to change out sample bags and take pictures (these tasks will be automated in the future). The exact drilling location was S77°51.891′, E160°48.029′, and the elevation was 1709 m.

FIGURE 1.189 The IceBreaker drill in the University Valley, Antarctica was used to drill 1 m in ice-cemented ground. *Source:* Courtesy Honeybee Robotics.

FIGURE 1.190 The IceBreaker drill. Samples were collected in 10-cm intervals using an autonomous sampling system (only the replacing of bags after each 10-cm interval was not automated). *Source:* Courtesy Honeybee Robotics.

During the drilling process, the bit temperature was continuously monitored by a temperature sensor embedded inside a bit. The bit temperature was one of the most important drilling telemetries since it indicated whether the subsurface ice within the soil matrix was reaching the melting point. The drilling algorithm was constructed in such a way as to either slow down or stop the drilling operation altogether if the temperature was reaching 0°C. Note that the algorithm developed for Mars drilling would limit that temperature to a temperature less than 0°C to account for possible presence of freezing point depressors within the ice-soil mixture (e.g. perchlorate).

A single test that included drilling to 1 m depth and collecting samples in 10-cm intervals took a few hours (Figure 1.190). The majority of the time was spent pulling the drill out of the hole after drilling the 10-cm interval. This was required in order to deposit the sample into a cup or a sample bag and to clean the auger surface using a passive brush to minimize cross contamination.

During the drilling process, the average power was approximately 70 W, the weight on bit (WOB) (i.e. the force the bit was pushing against the ground) was <70 N, and the penetration rate was ~1.12 m/hr. We refer to this drilling mode as 1-1-100-100 that is 1 m in 1 hr with less than 100 W of power and 100 N weight on bit. The electrical energy required to perform drilling and sampling was less than 100 Whr (the drilling energy itself was 63 Whr). The bit temperature never exceeded −5°C (the ground temperature was −19°C).

These tests have shown that drilling on Mars, in ice-cemented ground with limited power, energy, and weight on bit, and collecting samples in discrete depth intervals is possible within the given mass, power, and energy levels of a Phoenix-size lander and within the duration of a Phoenix-like mission.

Our second drilling location was massive ice with small to boulder-size rocks on top as shown in Figure 1.191. The exact location was S77°51.950′, E160°48.418′, while the elevation was 1724 m. Although the goal was to drill to 1 m only, after reaching the 1 m depth in just over an hour, we decided to proceed to a depth of 2.5 m. Drilling to 2.5 m depth took ~2.5 hr (at a penetration rate

FIGURE 1.191 The IceBreaker drill. Drilling 2.5 m in massive ice in the University Valley (the Beacon Valley region of Antarctica). *Source:* Courtesy Honeybee Robotics.

of ~1 m/hr). The average power at a depth of 2.5 m was 120 W, and the weight on bit was <100 N. Drill electrical energy to 2.5 m was <300 Whr. During the test, the maximum bit temperature was −10°C (the ground temperature was −24°C).

The ice cuttings were initially collected in 10-cm depth intervals (to a total depth of 100 cm). After 100-cm depth, the ice cuttings were collected in two depth intervals: 100–180 cm and 180–250 cm. These ice chips contained numerous ice crystals as large as 0.25 inch in size (Figure 1.192). This shows that drilling action does not necessarily crush all the ice.

The IceBreaker drill has also undergone numerous other tests in Mars chamber (Paulsen et al. 2011) as well as deployments on Devon Island in the High Arctic (Glass et al. 2011, 2014). As such, it is currently at the Technology Readiness Level (TRL) of 6.

1.11.5 TRIDENT DRILL

Since 2013, Honeybee Robotics has been contracted by NASA to develop and test a 1-m class Lunar Drill for NASA's prospecting mission to the lunar volatile-rich regions. The latest drill system called The Regolith and Ice Drill for exploration of New Terrains (TRIDENT) is a 4th-generation rotary-percussive drill that Honeybee has been refining over the last decade (Figure 1.193).

All systems share the same sampling architecture. They are all autonomous drill systems designed to capture powdered rocks and regolith samples for analysis. The drills consist of following subsystems: (1) Rotary-Percussive Drill Head, (2) Sampling Auger, (3) Brushing station, (4) Deployment stage, (5) Feed stage. The drill head has been designed with rotation and percussion decoupled. This allows the use of the more energy-intensive percussive system only when required (e.g., to penetrate harder formations). To reduce sample handling complexity, the drill auger was designed to capture cuttings and soils as opposed to cores. High sampling efficiencies are possible through a dual design of the auger (Figure 1.194). The lower section of the auger has deep and low pitch flutes. This geometry creates natural cavities ideal for retaining granular materials (cuttings and soil). The upper section of the auger has been designed to efficiently move the cuttings out of the hole.

FIGURE 1.192 The IceBreaker drill. Ice cuttings contained single ice crystals as large as 0.25 inch. *Source:* Courtesy Honeybee Robotics.

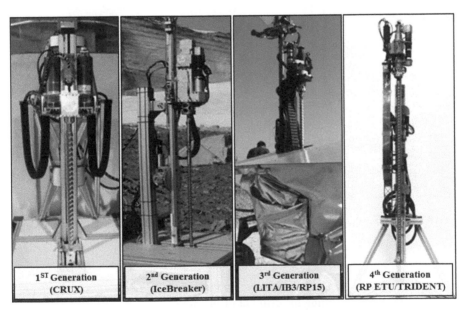

FIGURE 1.193 Evolution of Honeybee 1-m class drill systems to TRL-6. *Source:* Courtesy Honeybee Robotics.

The drill uses a "bite" sampling approach where samples are captured in ~10-cm intervals as shown in Figure 1.195. After drilling 10 cm, the auger with the sample is pulled out of the hole, and the sample is brushed off onto the ground or into 10-cc cups by a passive brush within the brushing station. This particular approach has many advantages. Drilling power can be kept to minimum since parasitic losses driven by auger convening samples from the depth to the surface are eliminated.

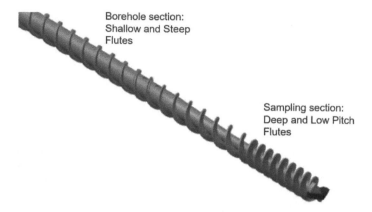

Borehole section:
Shallow and Steep
Flutes

Sampling section:
Deep and Low Pitch
Flutes

FIGURE 1.194   TRIDENT drill. Dual stage auger. *Source:* Courtesy Honeybee Robotics.

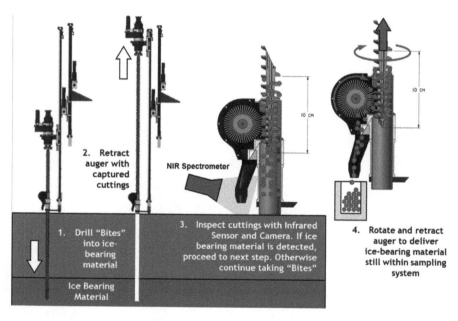

2. Retract
auger with
captured
cuttings

NIR Spectrometer

1. Drill "Bites"
into ice-
bearing
material

3. Inspect cuttings with Infrared
Sensor and Camera. If ice
bearing material is detected,
proceed to next step. Otherwise
continue taking "Bites"

4. Rotate and retract
auger to deliver
ice-bearing material
still within sampling
system

Ice Bearing
Material

FIGURE 1.195   TRIDENT drill. Implementation of "bite" sampling. *Source:* Courtesy Honeybee Robotics.

Upon retracing, the mission can take time analyzing the sample while the drill is above the hole—in its safe location. Periodically lifting the drill out of the hole allows the hole to cool down (if drilling generates a lot of heat). Since the drill bit has an integrated temperature sensor to monitor bit temperature during drilling, the same sensor could be used to capture thermal data of the hole every time the drill is lowered back into the hole. This is opportunistic science data that would contribute to the heat flow measurements on the Moon.

The deployment stage lowers and preloads the drill system on the ground. The feed stage, on the other hand, is used to advance the auger to a 1-m depth. Both stages are pulley-based to reduce system weight and vibration to the rover and improve dust tolerance.

The technology for the 1-m class drill started with the development of the TRL3 CRUX drill (Zacny et al. 2006). This was the first rotary-percussive drilling system developed to test various modes of drilling: rotary, percussive, and rotary-percussive. The CRUX drill demonstrated significant benefits of having rotation and percussion as two independent systems. In particular, it was found that in softer materials rotary drilling is ideal, while in hard formations, rotary-percussive is significantly more efficient. In some scenarios such as sticky material adhering to the auger, it was found that percussion alone could be used to clean the auger.

CRUX drill used many Commercial Off the Shelf (COTS) components which were not vacuum-rated. The IceBreaker drill was the 2nd-generation drilling system developed specifically for the Mars IceBreaker mission (McKay et al. 2013). All the components used in this drilling system were either space-rated or had a space-rating technology development path. The drill has been extensively tested in the Arctic, Antarctica, and Mars chamber at Honeybee Robotics. The system demonstrated drilling in rocks, ice-cemented ground, and ice with low power (100–200 W), low weight on bit (<100 N) and high penetration rate (1 m/hr).

The IceBreaker drill had a "Function" and the "Form" of the space drill, but not the "Fit." In particular it was not optimized for mass or volume. The 3rd-generation drill called the LITA drill (Life in the Atacama) achieved "Fit" through significant reduction of mass—the IceBreaker drill weighed 40 kg while the LITA drill weighed 10 kg (Zacny et al. 2015a). The LITA drill has been deployed from a CMU rover in Atacama and in Greenland. The LITA drill was also upgraded for the lunar environment and has undergone thermal vacuum tests at NASA GRC (Kleinhenz et al. 2015; Kleinhenz et al. 2018; Paulsen et al. 2017). During these vacuum tests, the drill captured volatile rich samples (NU-LHT-3M with 5 wt.% water) at −100°C and deposited them into cups. The drill has also been placed on a vibrating table to determine needed location for launch locks.

TRIDENT is the 4th-generation drill system specifically designed for drilling into volatile-rich regions on the Moon. The drill is baselined for the VIPER (Volatiles Investigating Polar Exploration Rover) mission. VIPER is a lunar rover planned to launch to the Moon in 2022 as part of the Commercial Lunar Payload Services (CLPS) to support the crewed Artemis Program. The rover will prospect for lunar resources in permanently shadowed areas in the lunar south pole region and in particular it will map the distribution and concentration of water ice. The mission draws significant heritage on a prior NASA rover concept called Resource Prospector. Mapping our water resources on the Moon is a next step in lunar exploration. This reconnaissance mission would pave the way for future science and In Situ Resource Utilization (ISRU) endeavors to the Moon and also Mars.

The roving platform is a powerful reconnaissance vehicle; with carefully selected instruments to achieve required goals in the shortest time possible (Figure 1.196). The Neutron Spectrometer

FIGURE 1.196 VIPER (Volatiles Investigating Polar Exploration Rover) is a planned robotic lunar rover by NASA, that will be tasked to prospect for natural lunar resources, especially water ice within a permanently shadowed region. *Source:* Courtesy of NASA.

Subsystem (NSS) is mounted at the front of the rover with a goal of searching for hydrogen-rich hot spots. Since hydrogen is an excellent proxy for water, NSS will guide the rover in search for water. Once a hydrogen hot spot has been identified, the rover will park directly above it and deploy the TRIDENT drill. The samples deposited onto the ground by the drill will be analyzed by the Near InfraRed Volatiles Spectrometer Subsystem (NIRVSS) and Mass Spectrometer (MSolo). NIRVSS is mounted so that its sensor is pointed to the location where the drill deposits the sample. The goal of NIRVSS is to characterize hydrocarbons, mineralogical context for the site, and the nature of water ice. MSolo will analyze volatiles coming off from the cuttings.

1.11.6 PLANETARY VOLATILES EXTRACTOR (PVEx)

Planetary Volatiles Extractor (PVEx) is a volatiles delivery system that can be used for both prospecting missions (provide volatiles to Gas Chromatograph Mass Spectrometer—GC/MS) as well as mining missions (capture volatiles in cold trap for processing). The heart of the PVEx is a coring auger with internal heaters (Figures 1.197, 1.198, and 1.199). The coring auger is driven into the ground by a rotary-percussive drill, TRIDENT, currently under development for VIPER missions.

To arrive at the PVEx design, we investigated many other approaches to extracting volatiles from planetary regolith (Vendiola et al. 2018, Zacny et al. 2016b). These investigations were test heavy. Numerous prototypes were developed and then tested under relevant conditions of frozen regolith doped with various water wt.% and in vacuum. The tests included NU-LHT-2M lunar highland simulant and JSC-1a lunar mare simulants. Water wt.% was varied from 2% all the way to full saturation of around 12%.

It was found that the heating of regolith is a significantly easier step than capturing sublimed volatiles. In many architectures the capture efficiency was zero—that is, all volatiles that have been sublimed, were lost to "space" (i.e. interior of a vacuum chamber), while no molecules made their way into a cold trap. In some architectures, volatile capture efficiency was in the few percent range. It became clear that for any volatile capture architecture to be effective, the path that volatiles take to the cold trap has to be physically sealed—that is molecules should have only one path to travel—to the cold trap. At one extreme scenario, one can imagine icy-regolith being placed inside a "pressure cooker" connected to a cold trap via a valve. Once regolith heats up, volatiles have nowhere else to

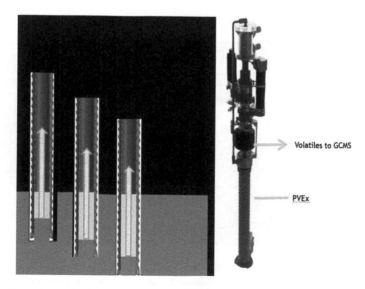

FIGURE 1.197 Operation of PVEx. *Source:* Courtesy Honeybee Robotics.

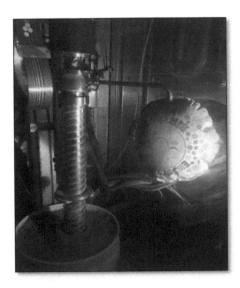

FIGURE 1.198   PVEx balloon test. Inflation of the balloon gives visual confirmation that flow occurs. *Source:* Courtesy Honeybee Robotics.

FIGURE 1.199   Details of PVEx mining system. *Source:* Courtesy Honeybee Robotics.

go but through a valve and into a cold trap where they re-condense. On another extreme, once can imagine a cold plate placed just above the regolith surface. The volatiles coming off the regolith can travel sideways and into a cold vacuum of space or up onto the cold plate. In this scenario, it was found that all volatiles pretty much "prefer" to escape into space as opposed to the cold plate. PVEx architecture works well because the coring auger seals the path the volatiles take.

Operation of the PVEx is the same as operation of the TRIDENT, with some minor additions. Once PVEx drills to a target depth and forms the regolith core, the heaters lined up on the inside of the coring auger are turned on. The conductive/radiative heat warms up the core and liberates volatiles. Volatiles then flow up the coring auger, through the swivel, and either directly into GCMS or into a cold trap where they re-condense. The benefit of an intermediate cold trap is that the volatiles flow could be metered out to the GCMS. Using this method of volatile collection ensures that far less volatiles are lost through sublimation to the vacuum of space.

PVEx is the only ISRU system that has demonstrated the end-to-end steps required to deliver volatiles to a cold trap: penetrating icy-regolith, sublimating volatiles, re-capture of volatiles on a cold finger. It has a significant potential to be both a prospecting tool as well as a mining tool—the differences are in the size of the coring auger.

1.12   10-M CLASS SAMPLERS

This section presents drilling and penetration systems in the 10-m class. This depth category opens up a range of approaches and techniques which focus on minimizing mass and volume, while trying to keep power and weight on bit within the range of the 1-m class drills. That's why this depth class has seen most innovative architectures to date. Many deeper drills are essentially scaled up 10-m class drilling systems.

1.12.1   Lunar Subsurface Explorer Robot Using Peristaltic Crawling

In recent years, planetary explorations have been earnestly conducted in order to elucidate the planetary history and discover of new resources. So far, however, past exploration missions paid little attention for subsurface investigations of the Moon or planets to discover phenomena such as moonquakes, heat, and underground conditions. Such subsurface investigations as the setting of a seismometer and the sampling of lunar soil could not only explain of the origins of the planets, but also aid in the development of planetary workstations (Oda and Kubota 2009). Recently, elucidating the lunar crustal structure has been prioritized for future lunar explorations. Hence, robots that can undertake these missions are strongly focused.

Previously, on the Earth, soil excavation was typically performed by boring machines (Soumela, Visentin, and Ylikorpi 2001). However, because the length of the drill is the same length as the target excavation depth, these machines tend to be large and heavy and the bases cannot restrain the excavating reaction force. Therefore, a small, light-weight excavation robot is expected for planetary investigations as shown in Figure 1.200.

To design a small, light-weight excavation robot for planetary investigations, the authors developed a robotics mechanism of locomotion by using the peristaltic crawling of the earthworm. The earthworm moves by peristaltic crawling (Alexander 1992). The movement is accomplished by extension and contraction of numerous body segments, as shown in Figure 1.201. At the start of peristalsis, the anterior segments are contracted and extended, propagating the contraction to the next segment. The contraction and extension mechanism progress as a wave from front-to-back segments. Peristaltic crawling motion has the following advantages: it requires less space than other locomotion methods, the large soil contact area can provide stable movement, and the excavated soil can be backward-discharged through the interior space. All of these features are incorporated into

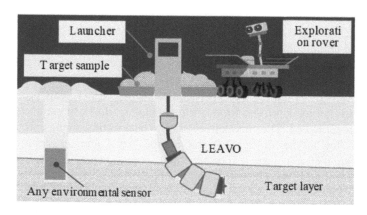

FIGURE 1.200   Overview of mission by LEAVO.

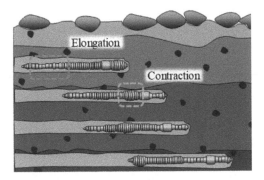

FIGURE 1.201   Earthworm locomotion.

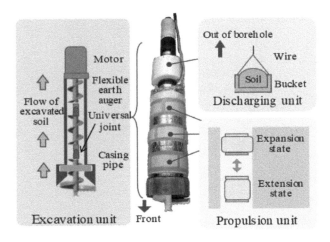

FIGURE 1.202   LEAVO: Lunar Subsurface Explorer Robot.

the proposed planetary subsurface excavation robot. The authors developed a planetary subsurface excavation robot with the newly proposed mechanism. The overview of the developed robot which is called LEAVO (Nakatake et al. 2016; Omori et al. 2011) is shown in Figure 1.202. As Figure 1.202 indicates, the robot is mainly made up of three units: a propulsion unit, an excavation unit, and a discharging unit.

LEAVO was newly developed to excavate a borehole for buried-typed environmental sensors, such as a seismometer and to collect regolith samples at a particular layer. As shown in Figure 1.203, an excavation method of the robot is described. The propulsion units are supporting the reaction torque/force of the excavation by gripping the wall of the borehole. It allows LEAVO to excavate deep underground in low-gravity space. The excavation unit mainly includes an excavation instrument, namely, a "flexible earth auger," and a casing pipe covering the auger. As shown in Figure 1.204, the flexible earth auger is divided into three parts and two universal joints, and is assembled by combining these earth auger parts by universal joints. By universal joints, the flexible auger can be passively bent.

The excavation unit excavates soil and transports it to the back of the robot. The soil in the back of the robot is discharged out of the borehole using the discharging unit.

Figure 1.205 shows the experimental setup and an overview of the experiment. The launcher supports the peristaltic crawling until the robot enters the borehole. In the real mission, the launcher is equipped with an exploration rover.

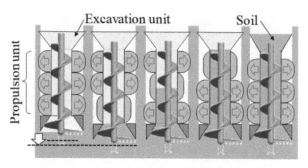

FIGURE 1.203   Excavation underground by peristaltic crawling.

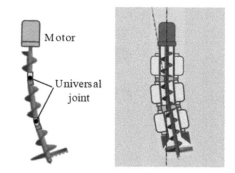

FIGURE 1.204   Methods for changing excavation direction.

The experimental procedure followed these steps:

(Step 1) The robot excavates a hole by rotating the auger while the propulsion mechanism propels the robot down using peristaltic crawling.

(Step 2) When the bucket is filled with soil, the peristaltic crawling and rotation of the auger stops.

(Step 3) The winder winds up the bucket. The soil in the bucket is then removed by a dust collector.

(Step 4) The bucket returns to the robot by the reverse rotation of the winder motor and the pulling force of a spiral spring.

(Step 5) The sequence is repeated.

In the excavation experiment, the robot successfully excavated to the maximum depth of the experimental setup (938 mm) as shown in Figure 1.205. Figure 1.206 plots the excavation depth versus time. In the phase (i), the soil began to be discharged from the discharging spout, and the discharging unit began to operate. The diameter of the borehole was gradually decreasing from the phase (i) to (ii), indicating that as the propulsion unit expanded, the dropped soil was pushed against the borehole wall. Therefore, from the phases (ii) to (iii), the robot went back about 150 mm by changing to peristaltic crawling, and excavated again from the phases (iii) to (iv). This re-excavation increased the diameter of the borehole. Finally, in the phase (iv), the excavation speed of the robot decreased because the posterior propulsion subunit could not sufficiently support the robot's excavation. Beyond the phase (v), the robot excavated to a depth of 938 mm. Since the robot reached the bottom of the soil tank, the excavation was completed at a depth of 938 mm. From this it is expected

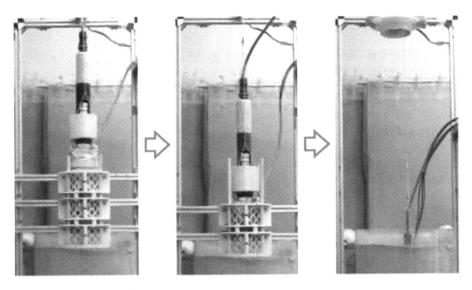

FIGURE 1.205 Situations during the excavation experiment.

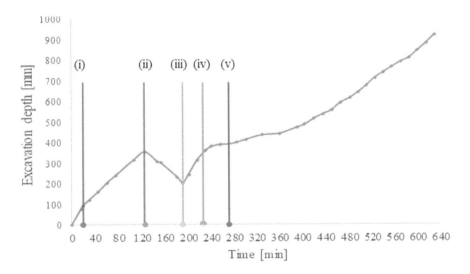

FIGURE 1.206 Excavation depth-time.

that the robot can excavate deeper because the velocity of the robot is not decreased. Using this technology, a seabed excavation robot for inspection of submarine resources has been developed (Tadami et al. 2017).

1.12.2 PNEUMATIC DRILL ON THE MOON

In evaluating options for the excavation method, it is important to understand properties of lunar regolith. Apollo 15–17 brought back regolith core samples from surface to 3-m depth (Heiken, Vaniman, and French 1991). These cores revealed that regolith deeper than ~20 cm is highly compacted, reaching >1.9 g/cc and relative densities of over 90% (Carrier 2005). Hence, the only way to penetrate it is to either remove it (e.g. using an auger drill as on Apollo) or crush it, pulverize it, and recompact it into a side wall (e.g. pile driver). However, the drill for such purposes requires significant power (450 W—based on Apollo) and the pile drivers have impact energies of 10s of Joules. Both options therefore are not suitable for small robotic missions.

The pneumatic drills (Zacny et al. 2013a) excavate soil by loosening and removing it from the hole. Essentially, the work of the auger in a conventional drill is done by gas. The soil removal is accomplished by expelling gas through a cone-shaped nozzle pushed into the soil. As it expands in vacuum, the injected gas exchanges momentum with the soil particles and carries them out of the hole. Soil transfer efficiencies can reach 1:6000 (1 g of gas lofts 6000 g of soil) because of the vacuum (Zacny et al. 2010a, b). Deflection plates on the surface divert this dusty gas away from the lander. The ejected soil follows a ballistic trajectory because of the vacuum. As such, it will not cover the lander or its instruments.

To demonstrate this approach, we used a rod with a 2.5-cm diameter cone at its end and lowered it into compacted (1.9 g/cc) JSC-1A lunar soil simulant (Figure 1.207). Initially, the rod was pushed into the soil until a specified force limit of 272 N (60 lb) was reached. Next, compressed gas was injected below the cone. During this time the insertion thrust force required to penetrate fell to zero (the soil was removed from underneath the cone)—Figure 1.208. This test showed that the rod can

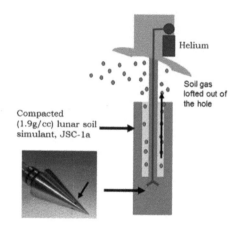

FIGURE 1.207 Pneumatic drilling approach for excavating holes in soil. *Source:* Courtesy Honeybee Robotics.

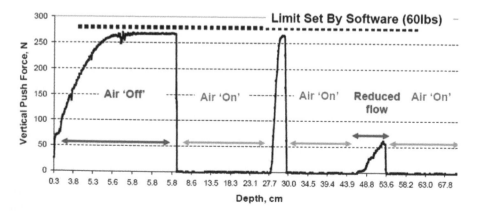

FIGURE 1.208 Vertical load applied to a penetrometer, when the gas was and was not injected. The applied load drops to zero when gas is flowing from the nozzle.

advance into highly compacted lunar analog simulant with zero force. The push force would only be needed to overcome borehole friction. The borehole, as demonstrated on Apollo 16 and 17 to 3 m in depth, will not cave in (Heiken, Vaniman, and French 1991), and friction will be minimized.

To verify that the pneumatic system can reach 2 m in compacted lunar regolith, we performed a penetration test at 1 atm with a hollow aluminum tube with a perforated cone attached to its end. It was lowered into a 3.3-m-tall soil bin filled with compacted (1.9 g/cc) NU-LHT-2M (Figure 1.209). The thrust force of 20 N was provided by dead weight. The gas flow of ~60 SLPM and ~800 kPa was directed into the rod. To reach the 3-m depth, the test was stopped every 1 m and additional 1-m sections were screwed on. The depth of 3 m was reached in ~1 min of actual pneumatic drilling. This test not only demonstrates that the penetration depth of 3 m is achievable but also the gas flow can be restarted at any depth. It should be noted that particles as large as 8 mm in diameter were lofted out of the hole.

To create a compact drilling system, the drill stem needs to be somehow stowed into a small form factor before it can be deployed—this is akin to a tape measure. We investigated numerous ways of having this sort of architecture including lenticular booms, bi-stem booms, coiled tubing, and many others. The first "space" use of such a boom were the robotic arms on Mars Viking 1 and 2 landers (Figure 1.211). The Viking boom was made from thin metal sheet. Carbon fiber reinforced plastic (CFRP) booms were applied for solar sail applications (Aguirre-Martinez et al. 1987). The pneumatic drill designed with a glass fiber STEM was tested in a vacuum chamber to 2 m (Figure 1.210). Using gas pressures/flow rates lower than prior 1-atm tests (because of vacuum) we advanced the boom to 2 m depth into the NU-LHT-2M lunar highlands simulant in ~2 min.

The Pneumatic Drill approach has been implemented into two lunar heat flow probes: LANGSETH (Lunar Apparatus for loNG-term Subsurface Exploration of the Transport of Heat) and LISTER (Lunar Instrumentation for Subsurface Thermal Exploration with Rapidity). The probes would be mounted close to the ground, either on the footpad, leg, or underneath the lander. Figure 1.212 shows an example where the probe is mounted on a footpad of a Mars InSight-size lander.

FIGURE 1.209 Pneumatic drilling. A hollow 4-m rod reached 3 m in 1.9-g/cc NU-LHT-2M simulant in 1 min. Particles (<8 mm) were blown out of the hole (see inset). *Source:* Courtesy Honeybee Robotics.

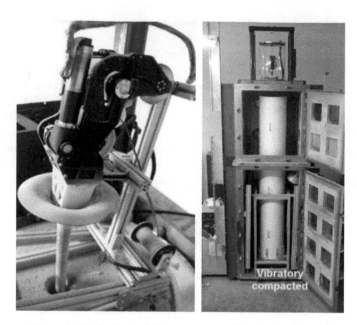

FIGURE 1.210 Pneumatic drilling. Vacuum chamber tests. The depth of 2 m was reached in 2 min. *Source:* Courtesy Honeybee Robotics.

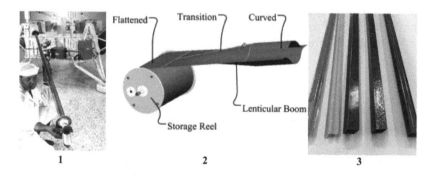

FIGURE 1.211 (1) Mars Viking boom. (2) Components of a lenticular boom. (3) Several examples of booms of different shapes that have been fabricated by Honeybee and Opterus.

The design of LANGSETH is shown in Figure 1.213. The probe is 28 cm tall, 12 cm wide, and 15 cm thick. The mass is 6 kg without electronics and 9 kg with electronics. The probe consists of four main subsystems. The Pneumatic Drill (the Boom, the Boom deployment mechanism, and the structure), the Sensing Subsystem (Cone and the Needle Probe), the Pneumatic Subsystem (tank with valves and plumbing), and Electronics. The boom consists of two halves of glass fiber bonded to a Kapton inner liner (Figure 1.214). When stored, the boom is flattened and wound onto a reel. When deployed, the tube forms a rigid cylindrical structure that can withstand significant buckling and torsional stresses. In our application, the boom will not see any torsional stresses. The buckling will also be significantly minimized because the boom will be constrained in a hole with only 10–50 cm of unsupported space between the deployment mechanism and the ground. The boom offers superior structural performance capabilities, while it doubles as wires conduit.

We conducted various trades to determine optimal gas routing. The options considered sending gas within the tube vs. using dedicated twin tubes on either side of the main boom. Test data have

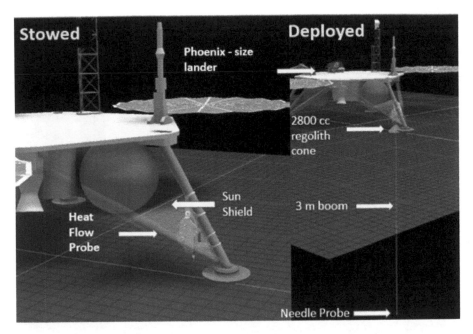

FIGURE 1.212 LANGSETH stowed (L) and deployed (R). *Source:* Courtesy Honeybee Robotics.

shown that it would be difficult to create a sealed boom and hence by default, the twin tube architecture shown in Figure 1.214 has been selected and tested. In addition, the use of twin tubes maintains a constant pressure drop within the system.

Figure 1.215 shows schematics of the boom deployment architecture. The Boom Storage Reel has the Boom and the Drive Tape (made of Kapton) spooled together. The Drive Tape is connected to the Drive Reel, powered by BLDC motor. To extend the boom, the motor pulls the Drive Tape and spools it on the Drive Reel. At the same time, the Boom is deployed and advances forward.

Excavated regolith particles will be lofted out of the hole and must be deposited nearby without affecting the experiments and activities of other payload instruments on the lander. For that reason, a deflector plate has been implemented to deflect the regolith sideways and down. Our previously developed deflector (Figure 1.216) has been successfully tested in vacuum chamber. Note the soil (even fine particles) falls down once deflected by the plate—no soil particles remain airborne. Particle settling time in the vacuum is driven by gravity (while on earth is predominately driven by atmospheric pressure) (Fuerstenau 2006).

The boom ends with a cone and a ~2-mm diameter needle probe at its tip (Figure 1.217). Compressed gas from a tank on the surface flows down the boom and into the cone, and jets through the holes at the lower end of the cone. The purpose of the cone is to thermally isolate the needle probe from the boom as well as act as a pneumatic drilling tip. Our baseline material for the cone is PTFE with a thermal conductivity of 0.14 W/mK. The needle probe has a heater wire and a platinum RTD for measuring temperature and thermal conductivity of the lunar regolith. The needle probe requires six wires—two for the heater and four for the RTD. This is required for 4-wire full bridge measurement. The design shown in Figure 1.217 is one of the 12 cone prototypes that have been modeled in Componential Fluid Dynamics and tested.

Although LANGSETH is still undergoing technology development, LISTER is scheduled to fly to the Moon in 2022 as part of Commercial Lunar Payload Services or CLPS missions (Nagihara et al. 2014; Ngo et al. 2019). LISTER is designed to penetrate at least 2 m in lunar surface and

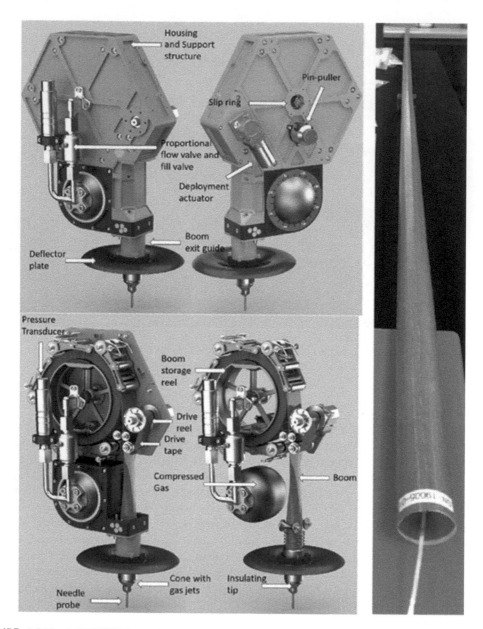

FIGURE 1.213 LANGSETH: Major subsystems (left) and Boom (right). *Source:* Courtesy Honeybee Robotics.

perform temperature and thermal conductivity measurements in Mare Crisium. These two measurements are required to determine heat flow properties of the Moon.

Figure 1.218 shows latest design and testing of the LISTER in the worst-case atmospheric pressure conditions. LISTER is using a coiled-tubing approach to extend a "drill" pipe. In this approach, a 6-mm diameter steel tubing is initially wound onto a drum. A set of drive pulleys pulls the metal tube out and at the same time deforms it to create a straight tube. This technology has been used in oil and gas for drilling deep wells; however, the method of creating a straight tube (using so-called injectors and goosenecks) is different in oil and gas applications. To deform the tube and advance it forward, a nominal power of 40 W is needed.

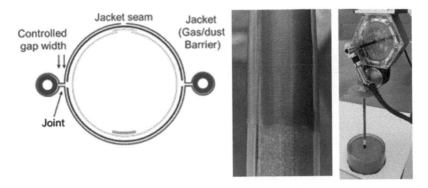

FIGURE 1.214 Boom with twin gas tubes on the side. *Source:* Courtesy Honeybee Robotics.

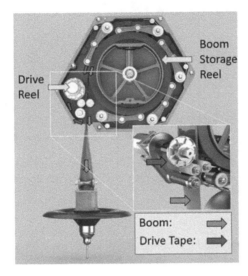

FIGURE 1.215 Boom deployment. *Source:* Courtesy Honeybee Robotics.

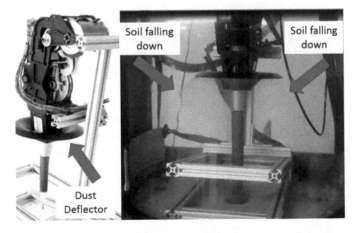

FIGURE 1.216 Dust deflector tests in vacuum chamber using an earlier model of our heat flow probe. *Source:* Courtesy Honeybee Robotics.

FIGURE 1.217   Cone and Needle Probe design. *Source:* Courtesy Honeybee Robotics.

FIGURE 1.218   LISTER heat flow probe is using a coiled-tubing design and pneumatic drilling to reach 2 m depth. (Left) LISTER mounted underneath a lander deck. (Center) CAD Render. (Right) Ambient tests in lunar simulant. *Source:* Courtesy Honeybee Robotics.

The tube is used as a compressed gas conduit. It also houses a set of wires for temperature measurements and a heater (located in a needle below the cone).

1.12.3   REDWATER DRILL

In the past decade orbital measurements revealed that a third of the Martian surface contains shallow ground ice. The Phoenix lander (67°N) uncovered both ice-cemented regolith and pure ice at the surface (Cull et al. 2010). SHARAD suggests the ice extends from 9 to 66 m depth at that location (Putzig et al. 2014). SHARAD also revealed the presence of debris-covered glaciers as well as buried ice sheets in the Arcadia Planitia (30°N to 45°N) that are up to 170 m thick and nearly pure ice. Other features suggest shallow ice deposits at Arcadia Planitia, including exposed ice within impact craters and ice-like dielectric (Byrne et al. 2009). Just recently more scarps and craters were observed that contain exposed ice below 1–2 m of regolith and extending to 100 m depth (Dundas et al. 2018). There is no doubt that, with time, we will discover more ice deposits around the planet.

The discoveries of nearly pure ice deposits in mid-latitudes on Mars make it possible for implementing two proven terrestrial technologies: Coiled Tubing (CT) for drilling and Rodriquez Well or RodWell (Haehnel, and Knuth, 2011) for reaching ice (Astrobiology goals) and water extraction (human settlement goals).

CT rigs use a continuous length of tubing (metal or composite) that is flexible enough to be wound on a reel and rigid enough to react drilling forces and torques. The tube is pushed downhole using so-called injectors (for example, a set of actuated rollers that pinch the tube and advance it down). The end of the tube has a Bottom Hole Assembly (BHA)—a motor and a drill bit for drilling

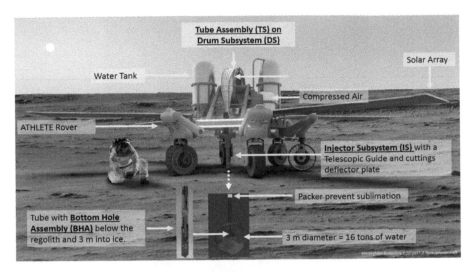

FIGURE 1.219 RedWater with all the subsystems. Major subsystems in bold/underlined. *Source:* Courtesy Honeybee Robotics.

into the subsurface. To remove chips, compressed air (or mud) is pumped down the tube. A hole is drilled by advancing coiled tubing deeper into the subsurface while blowing chips out of the way. A commercial CT rig such as RoXplorer weighs 15 tons and drills to 500 m at 1 m/min in hard rock.

RodWell is a technology where a hole is drilled in ice, subglacial ice is melted and pumped to the surface (Zacny et al. 2018a). It has been used in Antarctica's South Pole since 2002 to provide fresh water.

RedWater combines the two technologies into one: it uses the CT approach to create a hole (Figure 1.219). Once the hole is made, the coiled tubing is left in the hole and used as conduit for water extraction. The BHA contains a rotary-percussive drill subsystem, a downhole pump, and heaters. The tube houses an insulated and heated hose as well as wires for downhole motors and heaters. During drilling, compressed gas is sent downhole through the hose. The gas escapes through the annular space between the tube and borehole wall and removes chips that can be collected and analyzed for science. Upon reaching an ice layer, the drill continues for another few meters and then stops advancing forward, but the bit is continuously spinning. Heaters are turned on to melt the surrounding ice. Once ice starts to melt, the pump starts pumping a fraction of the melted water up the same hose that was used for the compressed gas, and into a storage tank on the surface via a 3-way heated valve (valve switches between the gas tank and water tank). The remaining water passes through a downhole heater and is pumped into the rotating bit for water jetting; this continuous stirring of water and injection of hot water speeds up the melting process. After melting a section of ice, the CT is reactivated to drill further and the melting process continues.

Since atmospheric pressure at Arcadia Planitia is above the triple point of water, liquid water can exist. However, it is unstable and can boil off very quickly. For this reason, it would be desirable to seal off the hole. This can be achieved via active means (a packer can expand in a hole and seal the annular space between the tube and the borehole) or passive means (water vapor would re-condense on the cold borehole wall and seal it; this in fact has been observed). The tube would have to be heated to free itself up before continuing further down, when needed.

Our pneumatic excavation tests at 7 Torr showed penetration rates of 1 m/min in regolith. The mass ratio of gas used to material removed out of the hole was 1:500. Assuming a 5-cm diameter hole (current baseline for RedWater), the required mass of gas would be 10 kg. Gas can be brought from the Earth (Mars 2020 mission brings a tank of compressed N_2 to blow dust off rocks), Helium pressurant can be used from landed systems, rocket fuel can be burned and turned into gas, and

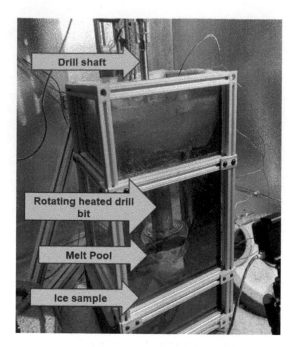

FIGURE 1.220 RedWater ice drilling and melting tests showing melt pool formation. *Source:* Courtesy Honeybee Robotics.

ISRU gasses (H_2/O_2) could also be used. Finally, a compressor could compress Martian air. MOXI on Mars 2020 has a compressor that would take 100 hr to compress 10 kg of CO_2 from 7 to 760 Torr.

To extract the required 16 tons of water per NASA requirements, a pool of approximately 3 m in diameter has to be created, which is feasible. The process of water extraction would take several weeks. Our thermal models using MathCad as well as calculations show that required heat for melting is ~1.5 kW and for keeping the water hose warm along the 25 m length is ~1 kW.

Preliminary tests have been conducted to demonstrate that drilling and melt pool formation is feasible. Figure 1.220 shows experimental setup inside a walk-in freezer at −20°C. A drill bit with internal heaters was driven into a clean block of ice by a rotary-percussive drill. Upon reaching a target depth, the drill was allowed to rotate at 120 rpm while the heaters were turned on to provide 600 W of thermal power. After approximately 2 hr, a melt pool of approximately 2 L in volume was formed. This corresponds to thermal efficiently of 20%. Additional tests will be conducted to optimize the design of the system. It is anticipated that RedWater will achieve Technology Readiness Level (TRL) of RL 6 in 2022.

1.12.4 THE ULTRA-LIGHT MOBILE DRILLING SYSTEM (UMDS)

Drilling techniques are well-understood, and widely used on Earth in underground oil and gas production. However, there is one key difference between drilling on Earth and drilling in space. For Earth, efforts are focused on optimizing production to maximize profits. In space, time spent on drilling is not so important a factor, while minimizing mass and power consumption has top priority. Therefore, designs focus on reducing, as far as possible, the amount of power that needs to be delivered to regolith or rock to allow the system to work. Another important requirement is to secure the borehole and keep the drill bit clean, via the efficient transport of cuttings.

The Ultra-Light Mobile Drilling system (UMDS) was developed by the CBK PAN and AGH University of Science and Technology in Poland (Seweryn et al. 2014a). It is designed for drilling in extreme, particularly planetary, environments. Unlike conventional drill strings that are composed

FIGURE 1.221 UMDS components. The mobile robot (blue), the support module (red), and the drill head (green).

of several pipes that are screwed together, the UMDS uses tubular boom technology as a wireline system. This approach is much more efficient in terms of mass and volume, and its main components are shown in Figure 1.221. The key parameters of the UMDS are summarized in Table 1.13.

The main component of the support module is a manipulator with 2 degrees of freedom, and the drill string. The main role of the manipulator is to set the direction for the drilling operation, as the drill head should be positioned along the gravitational vector. It is also used to collect samples and evacuate cuttings from the drilling subsystem. It is equipped with two magazines for collected cores. A total of 40 cores can be stored in separate containers to reduce the possibility of cross contamination between samples.

TABLE 1.13

Key UMDS Parameters

Parameter	Value	Unit
Weight	21	kg
Dimensions:	$740 \times 512 \times 460$	mm
Transport phase	$740 \times 512 \times 1700$	mm
Drilling phase		
Drilling depth	2	M
Drill diameter	35	mm
Maximum rover range	100	M
Maximum power consumption	120	W
Feed	1	mm/s
Rate	50	rpm
Thrust	500	N
Special features	Possible operation with unmanned aerial vehicles	

The drill string is based on tubular boom technology. A tubular boom in a stowed configuration (typically, in the form of a cylinder) takes up little space and can be deployed from a relatively stiff beam. It is composed of three main components: a deployable tube (to protect the borehole), a deployable boom (to control drill head movements and apply force), and its dedicated actuator. These three components are shown in Figure 1.222.

The specially designed linear actuator (Figure 1.223) moves the steel C-shape tubular boom (25 mm in diameter and 330° wrapping angle) and can exert up to 400 N of force in a downward direction. Such high-force values are achieved by a worm mechanism and spherical indentations in the boom, based on the leadscrew principle. The advantage of this design is that the weakest part of the boom in the load path can be omitted.

The steel boom is stored in a spool reel mechanism located at the top of the actuator. By default, it can drill up to 2 m, but this value can be easily increased with little effect on the overall mass of the system. A second tubular boom (35 mm diameter and 400° wrapping angle) is used as cladding during operations and is designed to prevent loose material from backfilling the borehole. The two-boom drill string system has been patented.

When folded, the support module is 400–720 mm wide and 250 mm high, but when unfolded it can extend up to 1290 mm. It weighs 8.5 kg (without batteries). A crucial element of the design is that samples can still be returned to the base if the support or drilling modules fail. Therefore, the support module is equipped with a mechanical interface that connects it to the rover. If either module fails, the interface is activated and the rover returns samples to the base.

In 2015, all of the UDMS's modules were manufactured and integrated (Figure 1.224). Tests were performed at both the component and system levels. Three test campaigns were performed, during which boreholes over 5 m were drilled in materials ranging from loose sand (like regolith) to 30-MPa compressive strength rock. The first campaign tested the drilling module alone. This confirmed the correct operation of the drill head, and showed that all mechanisms worked correctly, despite high contamination. Eighty cores were collected from a maximum depth of over 350 mm.

No damage was observed to the drill head and there were no serious failures. Force and torque were measured. Results obtained at 50 rpm rotational speed, and different linear feed speeds are shown in Table 1.14. The drill head and support module were tested together in the second test

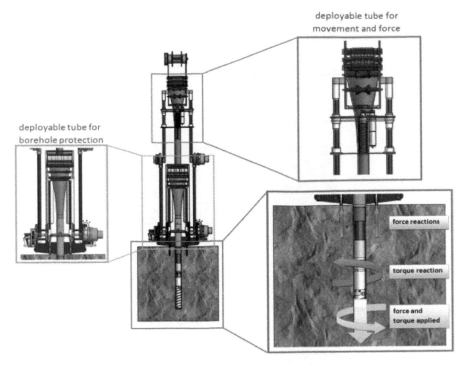

FIGURE 1.222 UMDS. Drill string components: cladding to secure the borehole (left); linear actuator to transfer force to the drill head located at the borehole bottom (top right). Torque transfer through the rotary anchor at the drill head (bottom right).

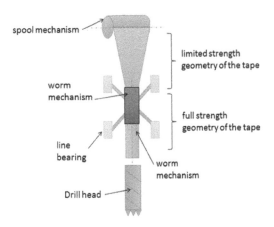

FIGURE 1.223 UMDS. Schematic view of the tubular boom actuator.

campaign (Figure 1.225). The linear actuator proved able to produce a force of over 400 N and could stabilize and guide the drill head during operations.

The third test campaign was run at a CBK facility dedicated to simulating operations on Mars (see details in Figure 1.226). All three modules (shown in Figure 1.224) were integrated and tested, together with the whole sampling process. Test results are qualitative due to the system's complexity.

The test results provided a large amount of information about the overall system, and its individual components. Specifically, they confirmed the approach of placing the drill head at the bottom of the borehole, and the decision to use tubular tapes to transfer the force to the drill head. The latter

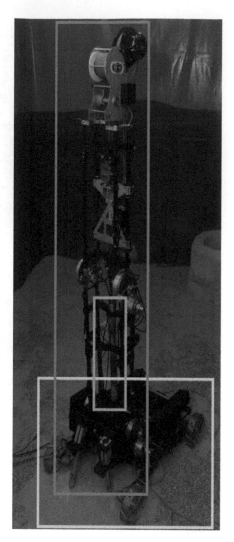

FIGURE 1.224 UMDS components: the mobile robot (green), the support module (blue), and the drilling subsystem (red).

TABLE 1.14

UMDS. Torque and Downward Force (WOB) at 50 rpm for Different Feed Rates, in Materials with Compressive Strengths 4.6, 12.5, and 30.8 MPa

Feed (mm/s)	4.6-MPa		12.5-MPa		30.8-MPa	
	Torque (Nm)	WOB (N)	Torque (Nm)	WOB (N)	Torque (Nm)	WOB (N)
0.1	0.89	7.7	1.24	59	1.92	92
0.2	1.45	7.5	1.71	79	2.74	119
0.3	1.63	13.9	2.45	108	—	—
0.4	1.72	12	2.7	119	—	—
0.5	1.88	14.6	2.93	103	—	—

FIGURE 1.225 UMDS. Drilling and support testbeds at AGH University of Science and Technology. The unfolded SM (left) holds drill head (middle). The actuator exert force on tabular boom (right).

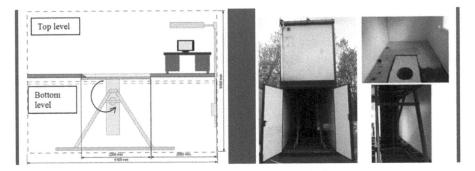

FIGURE 1.226 CBK facility where the UMDS was tested.

might prove to be very useful technology in the context of the deep drilling missions planned for Europa or Enceladus.

1.13 DEEP DRILLING

Arguably the most important goal of planetary exploration is to determine if life exists, or ever existed, on another world (Hand et al. 2017). To understand the evolution of planetary bodies and to find definitive signs of present or past life, we need to go deep into planets or the Moon's subsurface. On Mars, Europa, and Enceladus, planetary bodies with a high potential for life, one needs to penetrate deep below the surface that has been affected by damaging cosmic and solar radiation.

Conventional drill systems deployed in the oil and gas industries screw drill pipes together to form a long drill string. To get deeper, more drill pipes are added. This is a very robust approach and solves many problems related to deep drilling. These drills require high power drives, weigh a lot, and occupy significant space. These are not major issues on Earth; however, they are show-stoppers in extra-planetary settings (Dachwald et al. 2020).

The energy limitation also drives the drilling approach. Since locations of interest to Astrobiology may be covered with water ice, an obvious way to drill is to melt through ice. Melting, however, is an extremely inefficient process. Melt probes require kW-levels of power to keep penetrating and preventing the probe from freezing in a hole. Much of the heat is lost into surrounding ice, making the approach even less efficient. Another major drawback to this approach is that there are no power systems currently available that would provide multiple kW of electrical energy. Fission reactors for space exploration have been developed in the past, but their power outputs were less than 1 kW. A recent study suggests that developing a space-qualifying fission-based power system that would enable melting probes to operate, for example, would take over ten years and billions of dollars. Although, NASA-DOE work on a Kilopower-based fission power system could beat this paradigm.

1.13.1 Planetary Deep Drill (PDD)

Planetary missions are significantly constrained by limited power, volume, and mass resources. To solve the mass and volume problem while enabling deep penetration, it was decided to use a wireline drilling approach (Figure 1.227). In a wireline approach, a drill is suspended on an umbilical that provides power and data to the surface. This is the reverse of the paradigm usually used on Earth. Typical terrestrial drills are large and the drill bit is small. In a wireline approach, a drilling bit is bigger in diameter than a drill itself behind it. This is a critical requirement for this drilling approach since the drill bit needs to create space for the drill to fit into. The drill itself is a slim tube with all actuators, sensors, controllers, and instruments packaged tightly inside it.

To drill deeper, the drill first sets a set of anchors. These anchors push against the borehole wall and lock the drill from rotation or moving up or down. In a low-gravity environment, this is an extremely important consideration since now, the drill weight on bit (i.e. the force needed to be applied on the drill bit) is no longer limited by the weight of a rover or a lander. Once the anchors are set, the drill's internal Z-screw advances forward and pushes on the drill bit. A set of motors spin and hammer the drill bit at the same time. This inch-worming action could be repeated several times until the target depth is reached.

After reaching the target depth, the drill with its valuable sample is pulled back to the surface by the umbilical wound on a drum. The sample is transferred to onboard instruments for analysis, while the drill is lowered back into the hole and the entire process is repeated. To drill deeper, a longer umbilical is needed. Hence the majority of the system mass sits within the drill itself and the downstream support equipment (umbilical drum, science instruments, and so on).

This approach has some drawbacks. The major one is that it demands a stable borehole. If the borehole is unstable, it will collapse and trap the drill in the hole. There are various methods of addressing this such as expandable casing that would require the use of a bi-center bit to deal with a smaller diameter hole above the drill bit, or a system that could drill itself out. None of these are trivial solutions though. That's why this approach is best suited in ice, which coincidently is also the primary target for near-future Astrobiology missions.

Ice on Mars, Europa, and Enceladus is very cold, while the gravity in all cases is low. Both of these aspects create an extremely stable borehole that would remain open for years. On Earth, borehole closure is a real issue because of relatively warm ice and high gravity; as such holes in Antarctica are filled with ethylene glycol solutions or other low freezing point mixtures.

Honeybee Robotics, sponsored by the American Museum of Natural History in New York City and the Planetary Society, designed and fabricated Planetary Deep Drill (PDD) for penetrating 100s of meters to kilometers into Mars' ice caps and Europa's ice sheets (Zacny et al. 2013d). This prototype drill weighs approximately 40 kg and it is 4.5 m long and 6 cm in diameter. All mechanisms and electronic drivers have been integrated inside it. The drill also has a microscope with 0.5 microns per pixel resolution.

The PDD has undergone extensive testing in the Plaster City Gypsum quarry (Sharpe and Cork 1995) owned and operated by the US Gypsum Company—see Figure 1.228.

Initially one hole was drilled to a 10.5 m depth. At that depth, wet silty-clay was encountered which reduced daily penetration rates from 1.5–2 m per day to 0.5 m per day. The decision was

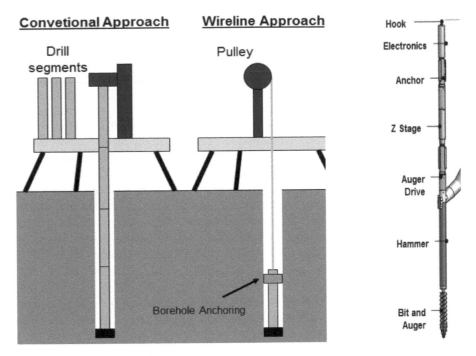

FIGURE 1.227 (Left) Conventional drill string vs. wireline drilling approach. Typical components of the wireline drill (right). *Source:* Courtesy Honeybee Robotics.

FIGURE 1.228 PDD. Drill testing at the US Gypsum quarry. Downhole microscopic imagers captured images of the borehole wall in white and UV light. *Source:* Courtesy Honeybee Robotics.

therefore made to move the drill from site 1 and establish site 2. Within a week the drill again reached 10 m depth. Drilling continued to 13.5 m depth, which is 3× the drill's length of 4.5 m.

Drilling telemetry has been extremely useful in predicting formation density. Drilling at lower power in a range of 50 W and penetration rates as fast as 2 cm/min indicated the rock is porous and in turn cuttings volume would be low. When the drilling energy increased to 250 W, and penetration rates dropped to 0.5 cm/min, this indicated the rock is very dense.

The microscopic imager worked extremely well with four LEDs (white light and UV light) as shown in Figure 1.228. Focusing has been relatively easy as well. Downhole sensors and imagers are extremely valuable for providing initial data on subsurface conditions.

1.13.2 Auto-Gopher-2 and WATSON

In order to understand the history of liquid water, water-rock interactions, and potential for a body to harbor life, subsurface samples are required for geological and astrobiological analysis. To meet these requirements a drill system to penetrate the subsurface of the explored bodies, capture samples for analysis, and present the unaltered sample to instruments for analysis is required. While it is obvious that there is a need to access the subsurface on planetary bodies, the actual act of autonomously drilling on extraterrestrial bodies is very challenging. A joint JPL and Honeybee Robotics task had the objective to develop and demonstrate a scalable drilling technology which will make deep drilling possible over the next two decades within the volume envelope, power systems, and Entry, Descent, and Landing (EDL) systems for current launch vehicles. This drill is named Auto-Gopher and its first generation was demonstrated to reach 3 m in the field (Badescu et al. 2006). It utilizes a wireline approach similar to a precursor used in Antarctica and has the potential to capture ice cores from kilometers of depth (Bar-Cohen and Zacny 2009). The technology and concept of operation has been developed in conjunction with future mission constraints including mass, power, and operation effectiveness. The drill is essentially a tube that encompasses the drill bit, mechanisms, actuators, sensors, and drive electronics. The drill is suspended at the end of a lightweight tether and in turn, penetration depth is limited only by packaging capability of the tether. This enables drilling to 10s of meters to 1000s of meters, without significant increase in system mass or complexity. The Auto-Gopher was developed in two iterations. Its first generation included external drive electronics and was demonstrated to reach 3 m in the field. The second generation included integrated electronics into the drill body and was demonstrated to reach 7.52 m deep in the field test.

Two of the limiting factors for autonomous drilling are the available mass and power. Proven power sources for landed missions are solar panels and Radioisotope Thermoelectric Generators (RTG). In order for the Auto-Gopher-2 to drill through ice, a few 100 W of power result in 10–20 min of drilling. This is very feasible with the current Mars Science Lab (MSL) power system, where it is consistent with these driving power levels.

In contrast, an ice melt probe could need up to kWs of power and this could require development of a small, space-worthy nuclear reactor. NASA and the Department of Energy (DOE) are developing technology and a system concept based on the successful Kilopower reactor that could serve this purpose, if the required launch approval can be obtained. The Auto-Gopher is periodically retracted out of the hole to extract the drill cuttings. During these up-and-down trips, the drill could engage counter-rotating reamers to keep the borehole open, if needed.

A major concern in terrestrial wireline drilling is the borehole collapse. The hole cannot only collapse above the drill, trapping the drill below the collapse point, but also the hole can get smaller through creep, and pinch the drill at a choke point. For planetary ice drilling, this is not an issue as ice creep is a function of ice temperature, pressure at depth and gravity. The gravity and ice temperatures on bodies such as Mars and Enceladus are much lower than Earth gravity and ice temperature in Antarctica is expected to keep the creep to a minimum over the mission time frame.

In 2012, the Auto-Gopher-1 wireline drill was developed and demonstrated to perform semi-autonomous coring in 40-MPa gypsum to a depth of 3 m, which was 1.5× the drill's length. Building on this promising technology, a fully autonomous Auto-Gopher-2 (AG-2) wireline system (Figure 1.229) was developed. This section of the chapter is focused on the piezoelectric actuator that produced the percussive action. After successful tests in the lab, the drill was field tested at a 40-MPa gypsum quarry (at US Gypsum Plaster City, CA) and reached 7.52 m deep reaching greater than twice the drill length as originally proposed.

1.13.2.1 The Auto-Gopher-2 (AG-2) Mechanism

The Auto-Gopher-2 (Figure 1.230) represents the latest generation in electromechanical deep drilling technology. It builds on previous successful technology demonstrations enabled by Auto-Gopher-1. The drill is a rotary percussive drill and incorporates additional capabilities over the previous generation such as embedded electronics and autonomous drilling with fault detection.

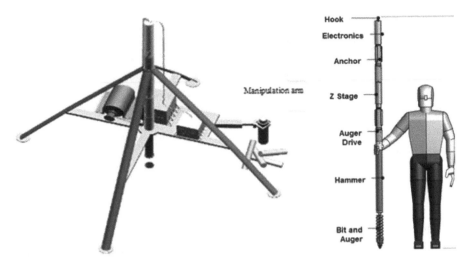

FIGURE 1.229 Notional diagram showing deployment system and drill system (left) and the Auto-Gopher-2 (right). *Source:* Courtesy Honeybee Robotics.

Section #	Section names	Description
(1)	Cable Termination	Top Connectors for E-O signals, and strength member termination
(2)	Power Converters	Front-end converters for 300-360 VDC input stepped down to various intermediate buses
(3)	Top Anchor	Anti-torque and anti-preload system for nominal drilling operation
(4)	Top Anchor & Z1Controllers	Motor controllers for Z-stage and top set of Anchors
(5)	5V Power	POL power converters outputting 5 VDC
(6)	Data Media Converters & CPU	MAGBES, and CPU for drill's controls
(7)	Camera/Sensor	Includes borehole camera and AHRS sensor
(8)	Bottom Anchor & Z2 Controllers	Motor controllers for Z-stage and Anchors
(9)	Z-stage	Provides longitudinal movement and generates WOB
(10)	Bottom Anchors	Anti-torque and anti-preload system for inch-worming
(11)	Piezo Electronics	Consists of control and power electronics for Piezo Actuator
(12)	Auger Drive	Includes a BLDC motor, controller, and gear train
(13)	Thumper & Piezo Actuator	Mechanical shock generator for Auger/Bit & An ultrasonic percussion based hammer system
(14)	Auger/Bit	Full faced bit with Auger for sample cuttings collection

FIGURE 1.230 The subsystems of the Auto-Gopher-2 Wireline Drill. *Source:* Courtesy Honeybee Robotics.

The ultrasonic hammer system employs a piezoelectric-actuated percussive mechanism for providing impacts on striker which transmits the impulses to the rock via the anvil/bit (Figure 1.231). High-impact hammering blows are realized by the principle of piezoelectric effects, where the input AC electrical signal on a piezoelectric transducer leads to a mechanical vibrating resonant system, generating large mechanical forces on the rock. The level of output mechanical forces of the ultrasonic hammer system is closely related to its vibration velocity that is proportional to the excitation frequency and displacement amplitude, and the level of displacement amplitude is proportionally related to the input voltage. Therefore, properties such as the piezoelectric coefficient, electromechanical coupling, and mechanical Q which produce a high vibration velocity are a key enabling parameters for large hammer actions.

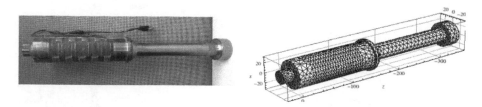

FIGURE 1.231   Auto-Gopher-2. The photo of fabricated piezoelectric actuator (left) and corresponding finite element model (right). *Source:* Courtesy of NASA JPL.

To meet the demand for high-performance hammer actions, a piezoelectric stack configuration, where a number of alternately poled piezoelectric layers are connected mechanically in series and electrically in parallel, was implemented with a combination of the optimum design of the ultrasonic horn and backing. The horn design amplifies the vibrations amplitude induced by the piezoelectric stack when it is driven at its fundamental half-wavelength axial mode through a geometric factor related to the ratio of areas and the mechanical Q. Segmenting the piezoelectric stack also amplifies the strain for a given voltage by a factor that is proportional to the number of the piezoelectric layers. The energy generated by the ultrasonic horn is then transferred to a free mass (striker), converting a high-frequency, low-amplitude excitation into a low-frequency high-impact energy hammering on the bit, producing a relatively low voltage requirement, and high-efficiency ultrasonic hammer system. The impact energy imparted to the bit is stochastic with a distribution of frequencies. Lower energy blows of the order of 0.1 J have frequencies in the hundreds of Hz range while higher energy blows of the order of 0.4 J have a frequency in the 10-Hz range.

The hammer is independent from the rotary motion and hence can be engaged when the formation becomes hard for the rotary motion to cut through or when the tungsten carbide teeth get dull. In addition, a percussive system allows the cuttings within the bailer bucket above the core to compact more and in turn occupy less volume.

1.13.2.2 Transducer Development

The piezoelectric actuator was designed with alumina disks at both ends of the piezoelectric stack to electrically insulate the piezoelectric stack electrodes from the transducer horn, backing and stress bolt allowing the ground to float. These insulators were integrated to match the interface dictated by the whole system drill design. The actuator solid model is shown in Figure 1.232.

Due to the need for higher power as a result of increasing the bit diameter, larger diameter actuators were designed and fabricated. Two piezoelectric actuators with 5.2-kHz resonant frequency were fabricated and tested; one used 5.1-mm-thick PZT rings while the other used 6.35-mm-thick PZT rings. The larger thickness PZT rings were selected to accommodate the voltage supplied by the system without the need to use an intermediate impedance matching transformer. A series of piezoelectric actuators using the 6.35-mm-thick piezoelectric rings and alumina insulating rings at both ends of the stack were analyzed, fabricated, and tested in the lab prior to integration into the drill system. Two of the actuators that were used as backup for the field trip are shown in Figure 1.233.

The geometry of the backing and horn were modified to include flat sections to produce slots which could be clamped by wrenches for increased torque in the assembly process. Epoxy was used to compensate for any surface mismatches between ceramic rings, electrodes, backing, and horn to reduce stress concentrators. In addition, 3D printed support fixtures were fabricated to maintain electrodes alignment during the stress bolt preloading. These steps have been found useful in assembling the piezoelectric actuators.

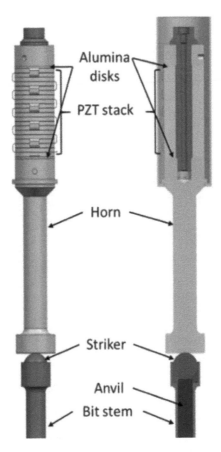

FIGURE 1.232   Auto-Gopher-2. Piezoelectric transducer design integrated into a testbed. *Source:* Courtesy of NASA JPL.

Figure 1.234 shows the measured input voltage (rms), input current (rms), vibration amplitude (peak-to-peak), impedance and input power with decreasing frequency sweeps ranging from 5.4 to 5 kHz under constant input voltage conditions (0.3, 0.5, and 0.7 V), which is amplified by 55-dB power amplifier (MODEL 1140LA). Similar trends were found, where the input current and vibration displacement of the actuator were found to be maximum, while the input voltage and impedance of the actuator become minimum at resonance (~5.2 kHz). The maximum rms displacement and vibration velocity are found to be 25.2 μm and 0.82 m/s under 67 V at 5.18 kHz.

The 5.2-kHz actuator and a testbed that was constructed allowed for the testing of the actuator performance with respect to the impact energy (Figure 1.235). The testbed includes a vertically mounted plate on which the actuator is mounted at the neutral plane flange. The testbed was updated to have the actuator mounted with the horn tip facing down to include a preload spring as it is in the drill system implementation. A solid rod is used as an analog for the bit and was mounted between two high stiffness wave springs preloaded between two plate-mounted collars. The gap between the bit replacement and the striker could be adjusted to determine an optimal value where the striker reaches resonance.

1.13.2.3   Impact between Free-Mass and Drill Bit

To study the hammering behavior of the free-mass to the bit, a Finite Element (FE) model was developed. The model is axisymmetric and includes a steel ball having 3.8 cm (1.5 inch) diameter, drill bit

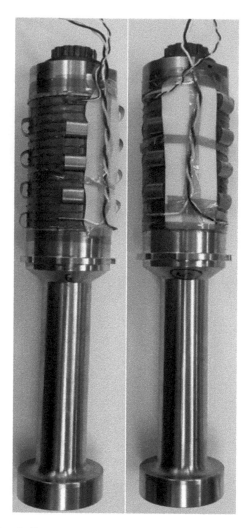

FIGURE 1.233 Auto-Gopher-2. Spare transducers fabricated with 6.35-mm (0.25″) thick PZT rings and alumina insulation rings. *Source:* Courtesy of NASA JPL.

and a block of rock as shown in Figure 1.236. The real teeth of the bit are arranged in a straight line across the diameter of the bit with a thickness of 3 mm. In the model, the teeth are assumed as being arranged in a circle of the same perimeter as the straight-line length of the cumulative bit teeth and with the same thickness. This modification makes the teeth shape compatible with the axisymmetric model and maintains the same contact area with the rock.

In the simulation, the ball hits the shaft of the bit with a speed of 3 m/s. The corresponding energy is 1 J. Figure 1.237 shows the displacements of the ball and the bit shaft. The contact time is ~125 μs and the bouncing back speed was calculated as 1.83 m/s. The force at the ball-bit and the bit-rock interfaces are presented in Figure 1.238. The maximum ball impact force reaches 16.7 kN at time of ~50 μs. The impact creates an elastic wave that propagates down into the rock. The maximum force at the bit-rock interface is up to 22.1 kN (65-MPa averaged stress) at ~280 μs. The level of stress itself may fracture rocks having medium hardness.

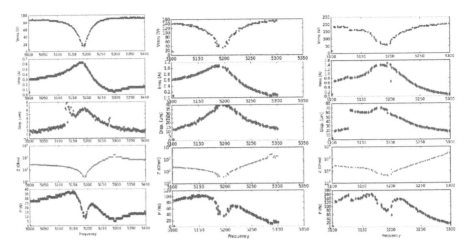

FIGURE 1.234 Auto-Gopher-2. Input voltage (rms), input current (rms), vibration displacement (peak-to-peak), impedance and input power of 5.2-kHz piezoelectric actuator as a function of frequency. *Source:* Courtesy of NASA JPL.

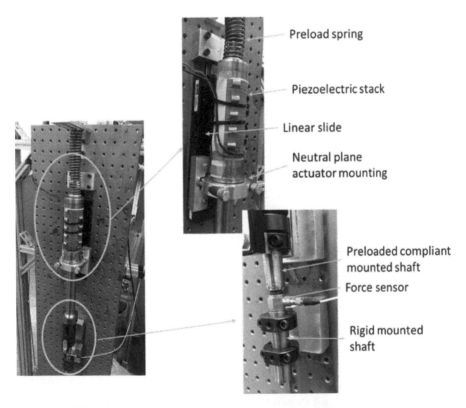

FIGURE 1.235 Auto-Gopher-2. Piezoelectric actuator testbed with transducer preload and free mass gap control features. *Source:* Courtesy of NASA JPL.

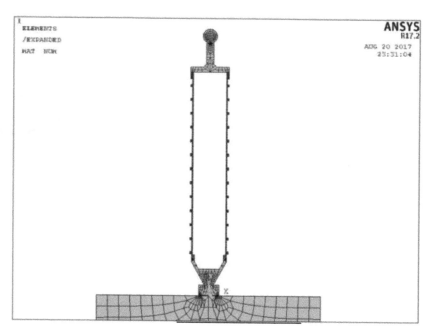

FIGURE 1.236   Auto-Gopher-2. The FE model of the drill bit impacted by a ball. *Source:* Courtesy of NASA JPL.

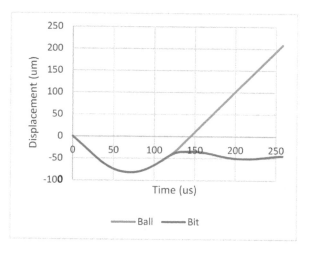

FIGURE 1.237   Auto-Gopher-2. The displacements of the ball and bit shaft after impact. *Source:* Courtesy of NASA JPL.

1.13.2.4   Drive Electronics and Control Software for Ultrasonic Hammer

The drive electronics were developed to fit inside the tube of the Auto-Gopher-2 system. The control software monitors the current and the frequency relative to the resonant device. The electronic drive development efforts were undertaken to produce an optimal combination of hardware and control software. This approach accounted for small changes in the resonance frequency caused by environmental changes in temperature, pressure, and mechanical boundary conditions. The software

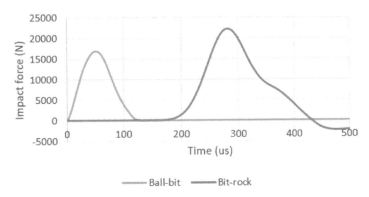

FIGURE 1.238 Auto-Gopher-2. The force at the ball-bit and the bit-rock interfaces. *Source:* Courtesy of NASA JPL.

monitors the power in real time, allowing for adjustment to the frequency to maintain the proper phase offset and current to the horn.

A decision was made to use the voltage provided by the power bus without an intermediate transformer to reduce the total volume occupied by the piezoelectric transducer drive electronics. The fixed large input voltage coming into the piezo drive electronics presented a couple of challenges to overcome. The impedance of the piezo actuator at resonance is ~34 Ohms. If driven at resonance, with a fixed voltage of 280 V, would result in 2.3 kW of power delivered to the system. In order to properly control the power delivered to the actuator, the piezoelectric transducer is driven off resonance between the resonance and anti-resonance frequency where the impedance is larger. Figure 1.239 shows a plot of the current as a function of frequency. The area of operation was chosen to be above the resonant peak for hardware safety reasons and stability of the impedance.

Driving off resonance introduces a few subtleties into the reported power measurement due to the phase angle between the current and voltage. Given that the voltage and current are out of phase, this means there are times where the voltage will be positive but the current is negative, or vice versa, which results in negative power flowing "into" the actuator. This is due to energy being stored in the mechanical system of the piezoelectric transducer and then delivered back into the drive electronics. Measuring just the piezoelectric rings current would tell us the horn velocity and apparent power but would not detail the real power delivered to the actuator/drill/rock. Therefore, two current measurements in the drive electronics are used, a low side current sense resistor where the average of the current can be used to infer the real power delivered to the actuator/drill/rock and a hall current sensor for monitoring the piezoelectric rings RMS current to control horn-tip velocity.

Looking to the future, implementing an efficient buck voltage converter to adjust the input voltage would be very beneficial. Using this design approach, the actuator could operate at resonance and the apparent power would equal real power delivered, and no current would flow back into the drive electronics.

A communication overview schematic is shown in Figure 1.240 for the Auto-Gopher-2 command and control. Successful communication from the JPL client to the embedded piezoelectric microcontroller using the TCP/IP to RS485 interface was accomplished. The JPL client is a GIMP Toolkit Graphical User Interface (GUI) utilizing a gopher library that was written in Python to allow easy portability and platform independence. The Python library uses the ZeroMQ messaging protocol to communicate with the Auto-Gopher-2 server located in the drill using TCP/IP commands. The gopher server located on the AM3358 platform inside the drill communicates over two ports including a subscriber-based model on one port, and a request and reply model on the second port. The publisher broadcasts the event records of commands received, Modbus communication, and other diagnostic information to clients that have subscribed to the server. This allows status to be

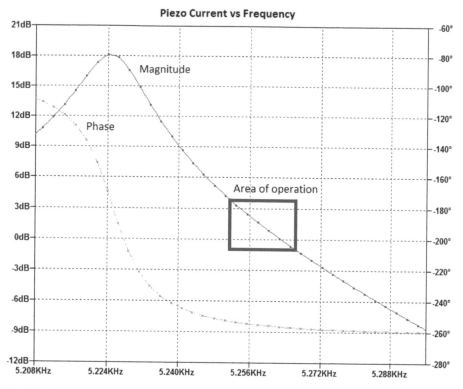

FIGURE 1.239   Auto-Gopher-2. Plot of piezo current vs. frequency. *Source:* Courtesy of NASA JPL.

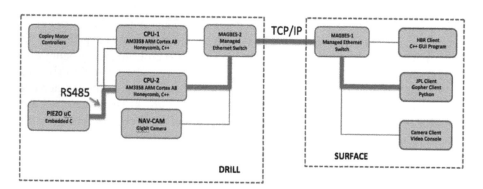

FIGURE 1.240   Auto-Gopher-2. Communication Overview, successful communication from surface client to drill piezoelectric driver. *Source:* Courtesy of NASA JPL.

broadcast across multiple devices, where the failure of one client does not take down the remaining subscribers. The request and reply portion of the server receives commands from a client, executes the command, and replies with the result of that command.

1.13.2.5   Lab Tests: Verification of Auto-Gopher-2 Performance

Laboratory tests of the integrated drill were conducted in-house on a testbed at Honeybee facilities by following the same steps as during system-level testing of the Auto-Gopher-1 drill. Additional

TABLE 1.15

Auto-Gopher-2. Summary of Drilling Performance for Rotary-only vs. Rotary-hammer Mode of Operation in Two Different Strength Rocks

Target Material	Rock Hardness (MPa)	Hammer (Piezo) Setting	Set Torque (Nm)	Set WOB (N)	Avg. ROP (mm/s)	Avg. Power (W)	Specific Energy (Wh/m)
Indiana Limestone	45	OFF	30	700	0.159	408.5	711.5
Indiana Limestone	45	100%, 1A	30	700	0.214	473.4	613.5
Cordova Crème	25	OFF	30	700	0.654	381.5	162.1
Cordova Crème	25	100%, 1A	30	700	0.729	438.2	167.1
Cordova Crème	25	1s ON/1s OFF, 1A	30	700	0.688	397.9	160.7

testing was performed to characterize the drill performance in two modes of operation: rotary only vs. rotary-hammer. The performance was characterized by drilling 10-cm-deep holes in the homogenous limestone blocks. Two limestone varieties were used: a 25-MPa compressive strength Cordova Crème Limestone and a harder 45-MPa Indiana Limestone. The tests were performed in a cold chamber at Honeybee where ambient temperature was approximately −20°C and relative humidity was measured to about 70–80%. The cold chamber was used for these tests since the deployment structure was already set up there.

Drilling performance in terms of WOB, ROP, and power in both rock types are summarized in Table 1.15. As expected, the average ROP is faster for the softer Cordova Crème Limestone specimen by more than three times (~0.7 mm/s compared to ~0.2 mm/s) irrespective of the mode of operation.

Drilling with rotary-hammer mode (piezo-actuator turned on at 100% DC) in 45-MPa Indiana Limestone was about 30% faster than rotary only (0.21 vs. 0.16 mm/s). This resulted in a decrease in the specific energy measured in Whr/m for the rotary-hammer mode in compared to rotary only (614 vs. 712 Whr/m) despite an overall higher power consumption. Note that the piezo actuator required 68 W of total power when run at 100% duty cycle. The power however does not include the inefficiency of the drill itself such as electrical losses (e.g. power converters, motor drive, cable) and mechanical losses (e.g. friction in gearbox). The efficiency of the entire drill is estimated to be about 70%.

The tests showed that the rotary-hammer mode of drilling is more efficient when drilling in the harder 45-MPa Indiana Limestone with regard to ROP and the specific energy required. When drilling in the softer 25-MPa Cordova Crème however, there was only a slight improvement in the ROP and no advantage in specific energy with rotary-hammer compared to rotary-only.

1.13.2.6 Field Test: Validation of the Auto-Gopher-2 Performance

To validate the deep drilling capabilities, the plan was to test the drill and reach of depth of 5 m or twice the drill's length, whichever is greater. The tests were very successful and the drill reached 7.52 m deep during the field campaign (Figure 1.241). This depth was sufficient to demonstrate the deep drilling capabilities of the Auto-Gopher-2 system. The field tests took place at the US Gypsum Company gypsum quarry (Plaster City, CA). The Plaster City site was also the site of the Auto-Gopher-1 tests. The US Gypsum Corp. performed Unconfined Compressive Stress tests on three gypsum cores and measured the strength to be 38 MPa ± 2 MPa, which is in the range of ice at cryogenic temperature. The reasons for field testing include:

1. Costs—It is expensive to set up a 4–6-m column of rock and brace it to prevent collapse during earthquake (California rules) and add scaffolding for placing a >2-m drill on top. It is less expensive to rent a truck, pack up the gear and drive 3 hr to the field site.

FIGURE 1.241   The field test of the Auto-Gopher-2. *Source:* Courtesy Honeybee Robotics.

2. Realistic interactions—It is impossible to introduce geological uncertainty in a lab. Field trips offer geological uncertainty and, in turn, it is a better location for testing robustness of the drill to changing conditions. For example, in the laboratory tests all rock blocks are without cracks or voids. In the field, a large rock crack or void presents the risk of creating pebbles large enough to jam the drill bit. Designing the drill control software to monitor for sudden increases in the bit drive torque helps mitigate these possible problems.
3. Realistic consequences—There is an inherit risk to conducting tests in a field that makes decision making more conservative and in turn more like it was a real mission. If a drill gets stuck in the field, it will take considerable effort to pull it out. In a lab, if a drill is stuck, it will be relatively easy to break the rock and free the drill.

During the field trip the logging of the piezo-actuator and the MOSFET temperature was accomplished using a custom design GUI control software. The results showed: piezo temp started at ~25°C and increased on average about 8°C per drill run and cooled about 4°C during cuttings removal.

While under the controlled conditions of the lab tests, where it was easy to get reliable measurement of the ROP with and without piezo actuation, when drilling the gypsum formation in the field it was not feasible to make reliable measurements. The difficulties resulted from the fact that the gypsum formation exhibits significant variability along the depth path caused by its inhomogeneity and the presence of variety of physical inclusions including clay, air pockets, and embedded rocks. Therefore, no conclusive comparison of the drilling performance has been obtained when applying rotary vs. rotary-hammer during field testing.

All of the high-level s/w operations were tested and modified in real time for optimal performance including inch-worming out of the borehole and the binding fault recovery procedure which automatically detected high torque on a few occasions including when the auger had overfilled with cuttings and while drilling through a clay layer. The gypsum formation exhibited a lot of variability where other embedded materials were found such as clays and silty clays, and even voids caused by subterranean water (gypsum is soluble) were present. The voids caused loose rocks to fall in, which caused spikes in auger torque. It could have also made anchoring difficult had we drilled deeper.

While the transducer performed well in initial testing, its performance decreased in time and investigation is underway to determine the reason for this behavior. The drive electronics and control and communication software for driving the transducer and exchanging information with the rest of the drill system were developed and integrated in the Honeybee Robotics developed drill system. The task was successfully completed after the demonstration of drilling 7.52 m deep in gypsum during field testing.

During this development, no efforts were made to optimize the mass and dimensions of the drill. One of the key objectives has been to develop an integrated system (including the drive electronics) within the drill and, as reported, this objective has been successfully met. Future work will include further development of critical subsystems such as cutting, mobility, sensing, percussion, electronics, and autonomy.

1.13.2.7 WATSON Drill Tests

The WATSON (Wireline Analysis Tool for Subsurface Observation of Northern-ice-sheets) drill was a modified Auto-Gopher-2 drill (Bhartia et al. 2019; Mellerowicz et al. 2019). The modifications included replacement of a piezo-driven hammering system with a more conventional cam-spring systems, as well as a high-TRL deep UV fluorescence/Raman instrument as an analog SHERLOC, the Mars 2020 deep-UV fluorescence and Raman spectrometer.

WATSON represents one of the first steps in paradigm changing of the space exploration: instead of bringing a sample to an instrument (conventional paradigm), WATSON brings an instrument to a sample. The sensing systems uses motorized stages to raster the laser across cm-scale regions of the interior surface of the borehole, obtaining fluorescence spectral maps with a 100-μm spatial resolution and a spectral range from 265 to 440 nm which are highly sensitive to many organic compounds, including microbes (Bhartia et al. 2008, 2010; Salas et al. 2015). Interrogation into the ice wall with a laser allows for a non-destructive in situ measurement that preserves the spatial distribution of material within the ice (Eshelman et al. 2019). A high-fidelity prototype was developed and initially tested in a lab in rocks and ice (including cryogenic ice). The drill mass was 70 kg, drill length 4.1 m and drill diameter 10 cm. The average drilling power in most formations (including ice) was less than 500 W.

During the summer of 2019, WATSON was deployed at a site located near Summit Station at an elevation of 3200 m above sea level in the vast interior of Greenland (Figure 1.242). WATSON successfully demonstrated robotic drilling to 111 m and mapped the spatial distribution of organics and microbes in the borehole while acquiring 50-cm-long cores used later for correlation of the acquired instrument data (Malaska et al. 2020).

WATSON can be easily optimized to meet the power and mass/volume constraints of a lander or rover similar in size as NASA's Curiosity Rover which was launched in 2011 (Figure 1.243). Future astrobiological missions to the icy regions of planetary bodies such as the Mars poles, Enceladus, and/or Europa may benefit from a similar architecture capable of in situ organic mapping and life detection within hundreds of meters beneath the surface.

1.13.3 Search for Life Using Submersible Heated (SLUSH) Drill

Europa is a primary target in the search for past or present life because it is potentially geologically active and likely possesses a deep global ocean in contact with a rocky core underneath its outer ice

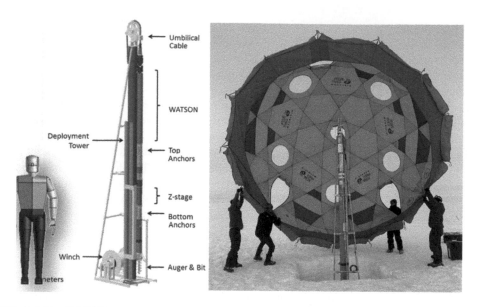

FIGURE 1.242 WATSON prototype reached 111 m depth in Greenland's Summit. *Source:* Courtesy Honeybee Robotics.

FIGURE 1.243 WATSON can be deployed from Curiosity/Perseverance class rover. *Source:* Courtesy Honeybee Robotics.

shell. Galileo spacecraft observations and theoretical models predict that the ice shell is 3–30 km thick and overlays an ocean ~100 km deep (Hand et al. 2017; Howell and Pappalardo 2020).

To reach the subsurface ocean where life may be most prevalent, a probe would need to penetrate the ice shell while moving the excavated material aft. This can be achieved by melting the material (thermal penetration) and cutting the material (mechanical penetration). Mechanical systems break the icy material efficiently but transport ice chips inefficiently. Thermal systems have an effective chip removal approach but a power-intensive ice-melting step. The Search for Life Using Submersible Heated (SLUSH) drill is a hybrid thermo-mechanical drill probe system that combines the most efficient aspects of these two techniques (Zacny et al. 2018b).

Figure 1.244 shows the comparison of the three approaches in 240-K ice using the same bit with integrated heater. From the penetration and power standpoint, the melt probe was slowest and required the most power, while mechanical drilling was the fastest and required the least power (mechanical drill energy did not include the energy of transporting chips). Importantly, slushing approach had higher performance than pure melting (penetration rate was ~1 order of magnitude

FIGURE 1.244 Melting, slushing, and drilling (w/o chips transport). *Source:* Courtesy Honeybee Robotics.

greater). It should be noted that it is very unlikely the drill will need to cut ≤100-K ice, except for the initial hole starting, because the reactor will continuously warm up the ice in front of the bit.

SLUSH has a heated drill bit in front, antitorque cutters on the side, and several tether bays on top (Figure 1.245). Critical subsystems are inside a pressure vessel. SLUSH utilizes a mechanical drill to break the ice and a reactor to partially melt the fragments, enabling the efficient transport of material behind the probe. The resulting slush behaves like liquid despite being partially frozen, significantly reducing the power required for melting the full volume of ice. Further, because the mechanical approach generates higher penetration rates than melting, SLUSH can reach the ocean in a much shorter time than a pure melt probe. Once SLUSH passes through the top cryogenic ice and penetrates deeper into warmer ice, it can use a purely thermal approach to melt through this warmer ice without the need for mechanical cutting.

SLUSH incorporates the Kilopower reactor for both thermal and electrical needs. The fission reactor can be turned on/off and is self-moderating, significantly simplifying thermal management. The probe is physically connected to a surface lander by a communications tether, housed in several spool bays that are left behind in the ice once the spool is depleted. This allows each tether section to be purpose-designed. For example, the top section, which may see 150-kPa shear stresses on a diurnal cycle, will be reinforced with Kevlar and or Vectran. Leaving the spools behind also shortens the probe length as it descends, making penetration more efficient.

While Kevlar/Vectran reinforcement and the refrozen channel left behind by the probe may provide protection from the diurnal stress environment, if the tether does break, broken sections could become tune antennas to form a "Tunable Tether." In-line RF communications nodes would sense the broken tether and adjust from wired to wireless mode for communication. A mixture of in-line RF and acoustic communication nodes may potentially comprise the backup communications link by incorporating frequency adjustable transceivers combined with transducers into each spool section at frequency-selected distances.

SLUSH is notionally a 5-m long, 57-cm diameter probe, with a diameter being driven predominately by the size of the Kilopower reactor. If the reactor can be scaled down, this will directly

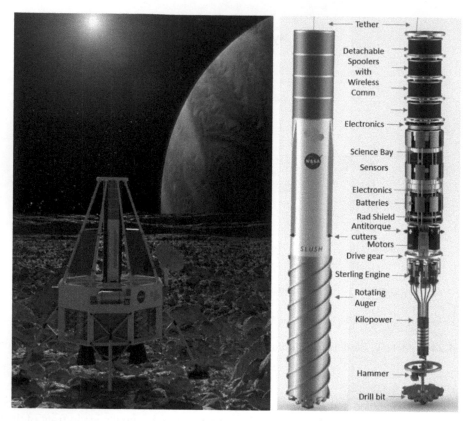

FIGURE 1.245 Conceptual design of SLUSH (Zacny et al. 2018b). *Source:* Courtesy Honeybee Robotics.

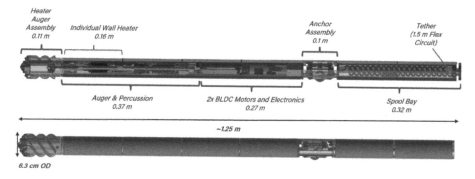

FIGURE 1.246 SLUSH prototype to be tested in August 2020. Major components are shown. *Source:* Courtesy Honeybee Robotics.

reduce the probe diameter. Since the excavation area is a function of Diameter^2, a small reduction in diameter has significant (^2) reduction in the area, drilling power, and improvement in penetration rate.

Currently SLUSH is being prototyped and tested with an anticipated Technology Readiness Level or TRL of 4 by 2021. Figure 1.246 shows scaled SLUSH prototype with its major subsystems. The diameter is 6.3 cm and length is 1.25 m. The probe includes heated drill bit, external wall heaters

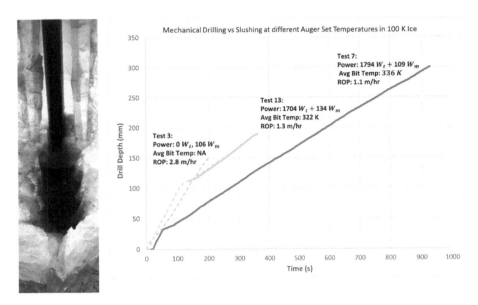

FIGURE 1.247 SLUSH prototype tests. (Left) SLUSH penetrated 3× its length in cryogenic (100 K) ice. (Right) Penetration data. Wt = Thermal Power; Wm = Mechanical Power (i.e. drilling power); ROP = rate of penetration. *Source:* Courtesy Honeybee Robotics.

that are individually activated and controlled, decoupled rotary (auger) and percussion subsystems driven by BLDC actuators, antitorque (anchor) assembly and flooded tether bay with its 1.5-m flex cable. The probe will be tested to ~2.5 m depth in warm (250 K) and cryogenic ice (100 K). The arm ice testing will occur in a 5-m-tall walk-in freezer while 100-K ice tests will be done in vacuum chamber. The goal of the vacuum chamber tests is to demonstrate the "start" problem in vacuum (i.e. ice would sublime and not melt due to vacuum conditions).

The latest tests with a scaled-down SLUSH prototype shown in Figure 1.247 demonstrated that SLUSH approach is feasible. Figure 1.247 shows a graph for three tests that have been performed with a heated drill bit. In cases where the bit was not heated, the thermal power, Wt was zero. Wm refers to mechanical or drilling power. Test #3 show purely mechanical drilling. The penetration rate was very fast but as soon as the drill bit penetrated its length into ice, the cuttings clearing stopped and the bit chocked. Test #13 shows an example when, after the mechanical drilling resulted in the bit chocking, heaters were used to melt ice to form slush. The penetration continued at somewhat slower speed. Test #7 shows slushing (mechanical drilling and partial melting)—this test did not see as high of the penetration rate as mechanical drilling but the penetration rate achieved a steady state, demonstrating the slushing for greater depths is feasible. In this particular case, the bit penetrated 3× its length with slush refreezing behind the bit.

A potential reason for slower penetration rate using slushing approach as mechanical approach could be due to the fact that slushing warms up ice ahead of a drill bit and as such, changes its mechanical properties (cryogenic ice is brittle, while warm ice is ductile). Brittle ice is ideal for rotary-percussive drilling (i.e. baseline for SLUSH drill), while warm and ductile ice is normally drilled using purely rotary drilling. Hence by changing method of drilling from rotary-percussive to purely rotary, higher penetration rates may be achievable. SLUSH, however, should include percussion in cases where the probe encounters harder salt deposits.

Figure 1.248 shows performance data of SLUSH in purely rotary-percussive mode (with heaters turned off) in 100-K ice and in 250-K ice. Note that ice ductile to brittle transition occurs at around 165-K ice—i.e., below 165 K the ice is brittle and above 200 K the ice is ductile. Data

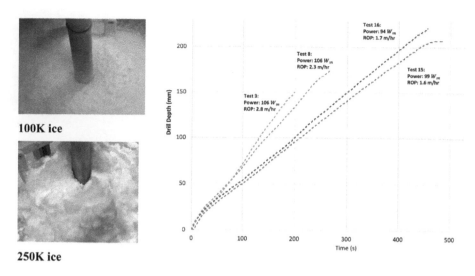

100K ice

250K ice

FIGURE 1.248 SLUSH—mechanical drilling tests. Rotary-percussive drilling in cold ice is faster than in warm ice. *Source:* Courtesy Honeybee Robotics.

shows that the penetration rate in cold brittle ice is higher (2.3–2.8 m/hr) than in warmer ductile ice (1.6–1.7 m/hr), even though the drilling power was nominally the same (106 W in 100-K ice and 96–99 K in 250-K ice). Slushing is therefore essential to the probe performance as it solves the cuttings clearing problem.

1.13.4 CRYOBOT FOR THE EXPLORATION OF OCEAN WORLDS

Icy moon oceans beckon with ingredients that potentially may harbor extant life. Beginning with the Galileo and Cassini missions, measurements have revealed the presence of global oceans under the icy crust of several moons of Jupiter and Saturn as well as others. Among those moons, Europa and Enceladus have their ocean in contact with the rocky core, providing an environment similar to the conditions existing on the terrestrial seafloor where life has developed at hydrothermal vents (Hand et al. 2017; Lunine 2017; Thomas et al. 2016). Accessing these oceans presents considerable difficulty due to a number of issues including the depth and composition of the icy crust, the time needed to travel through the crust, the power needed to propel a probe, communication of scientific and engineering data though the ice and back to Earth, entry and mobility in the ocean and autonomous operations for the life of the mission. The JPL Cryobot architecture is being developed to provide a feasible system design for descending through a crust of Europa to facilitate the detection of the evidence of life in its ocean. The Cryobot architecture consists of a Cryobot head, power system, thermal management system, hazard detection and navigation system and communications components. Also, under development, in key support of this architecture, is a validated Cryobot Descent Simulator that will be a fundamental design tool for the Cryobot. Based on a design principle to feasibly integrate redundant capabilities that will mitigate unknown environmental risks, this architecture is conceptualized with the form, fit, and function that can be infused into a flight mission that will access the ocean. The work outlined in this section provides a technology path that can reach the beginning of mission design in a decade. This includes developing a trade space of power sources and Cryobot size that minimizes mission duration. Reducing descent time is essential for both system reliability and reducing probability of failure. The development also outlines a field campaign in warm terrestrial ice, complementing lab cryo-ice experiments. A field campaign is being planned for an arctic location to provide validation of the Cryobot Descent Simulator in conditions similar to the warm deep crust of Europa. The described technology will also be available for use on other ocean worlds with similar ice crust like Enceladus or with extensions to bodies with ice beneath regolith such as Mars.

1.13.4.1 Cryobot System Architecture

The Cryobot architecture is based on designs that have a viable path-to-flight and due to the anticipated mission scope, are developed to meet the requirements of a NASA flagship project. The key drivers for this class of mission are reasonable science return and acceptable mission risk. These key drivers have informed the development of the work outlined in (Figure 1.249).

1.13.4.1.1 Requirements and Assumptions

A set of key requirements and assumptions have been defined as follows: minimum of 15-km descent in less than three Earth years with graceful degradation for thicker ice. A system mass target of a do-not-exceed value of 500 kg, inclusive of payload is targeted; Europa Clipper reconnaissance data is assumed for safe landing requirements; planetary protection is expected to be a significant path-to-flight driver but is considered out of scope for the current technology effort. However, we will leverage previous funded science sampling cleanliness effort mentioned later; the Jupiter-Europa radiation environment will drive surface system architecture and the cryogenic temperature, vacuum, ice salinity, dirty ice with insoluble particulates are assumed; within the crust we assume embedded hazards including cracks, active faults, voids, obstacles, slush, and lakes (shown in Figure 1.249).

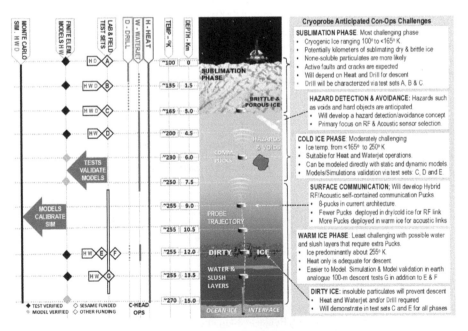

FIGURE 1.249 (Left) Descent Simulator modes, calibrated by finite element models, validated by lab and field data. (Middle) Cryobot mode depends on physical conditions and temperature profile (from Manga 2009). (Right) Anticipated ConOps informs system architecture and sim design. The letters A–G indicate various tests that will be performed. *Source:* Courtesy of NASA JPL.

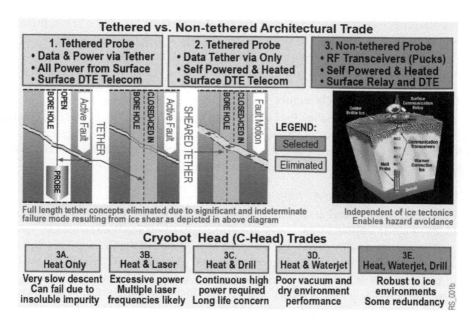

FIGURE 1.250   Non-tethered Cryobot architecture was selected with 8 puck transceivers. The nose is equipped with mechanical drill and water jet for increasing descent speed. *Source:* Courtesy of NASA JPL.

1.13.4.1.2   The Cryobot Architecture

The proposed architectural trade space for the Cryobot is shown in Figure 1.250. The original development dates to the work of Zimmerman (2000), Zimmerman et al. (2001). The selected non-tethered Cryobot architecture 3E in Figure 1.250 is robust to differential ice motion across cracks and faults. The architecture under development is shown in Figure 1.251. The Cryobot head heat, water jet, and drill capabilities optimize performance for a broad range of ice thermo-mechanical environments. Drilling can speed descent in sublimating configurations and cut through dirty ice and

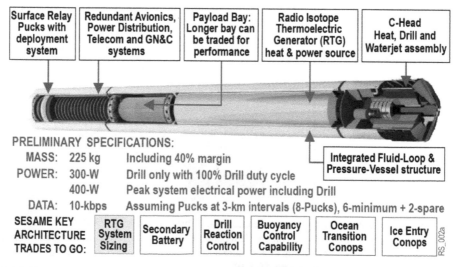

FIGURE 1.251   Cryobot preliminary architecture, resource specs, and key trades. *Source:* Courtesy of NASA JPL.

hard obstacles. Water jetting is a very efficient excavation technique in a submerged environment. It significantly increases descent speed and harvests otherwise wasted energy from the sides of the probe and directs it to the head.

A validated modeling effort including uncertainty quantification is being developed to inform the acceptable balance between probe diameter, length, and thermal energy density. Additional technology trades and technology concept definition to develop a recommendation to thermal sources and thermal management that will meet the unique requirements of a Europa ice probe are also underway. In addition to the Cryobot head, power, and thermal technologies, additional technology concept definition is planned to address long-lead enabling systems. Two key enabling technologies identified for the Cryobot are telecom transceivers and hazard detection and avoidance. The latter is required to avoid potential voids and hard layers during Europa descent. A summary of proposed Cryobot modeling, testing, and ConOps is shown in Figure 1.249.

1.13.4.2 Heat Source Considerations

The overall heat density of the Cryobot is the driving design consideration for overall time to descend to the ocean. It is also a key driver for defining the overall probe mass, length, and width. Similarly, a number of thermal control engineering systems couple directly to the heat source and further drive the volume. Keeping system volume low is important but there are fundamental limitations for both engineering and science that will drive the total volume and hence require either more thermal power or necessitate a longer time-to-ocean. Given the small number of years that the probe is expected to operate in the ice, an efficient overall design is critical.

A radioisotope power system is used, first, to power thermal melting and, second, to generate electrical power to store and drive other systems. Note this is different from radioisotope space systems to first produce electrical power, with the majority of heat production generally being unused. The general-purpose heat source (GPHS) offers a starting point for a design but is nominally too low in heat density for the architecture to close with a reasonable time-to-ocean of as little as 2–3 years. It does allow ocean access with descent times a few times longer. Advanced concepts such as microspheres or Kilopower can do considerably better but are currently immature, requiring development, or suffer large mass penalties that may render a Cryobot design infeasible. Repackaging of GPHS is highly promising, getting two-thirds of the improvements of microspheres but with significant heritage. This work requires development based on existing systems.

1.13.4.3 Cryobot Descent Simulator

The proposed Cryobot Descent Simulator provides a proof of concept that the Cryobot system can attain the Europan ocean in the allocated time. This is a critical element of path-to-flight, requiring assessment of performance against uncertain environmental conditions, and extrapolation of empirical and numerical results to a mission-scale simulation. As an analog, entry descent and landing (EDL) for Mars rovers is developed through a rigorously validated Monte Carlo simulation, achieving high *a priori* confidence of success despite an uncertain atmosphere (Kornfeld et al. 2014). To achieve an equivalent program for Europa, the Cryobot Descent Simulator brings together computational fluid dynamics and finite element modeling and detailed parametric analysis of probe performance. The simulation is validated with empirical data collected in warm-ice lab facilities, cryogenic test facilities, and in the field.

A detailed analytic model (extending the work of Aamot, Ulamec et al. 2007) provides a benchmark for trade studies, elucidating dependencies on probe dimensions, available power, and physical properties of the ice. However, taken alone, analytic models are insufficient for providing the desired degree of predicative confidence in Cryobot descent speeds. For example, while the Cryobot will mostly operate continuously on Europa, laboratory test data includes transients that need to be calibrated before applying the results to the calculation of descent speed.

Finite element and other 2- and 3-dimensional mesh-based methods will rectify many of these deficiencies, including modeling subtle heat sinks and transients, and the fluid dynamics of water

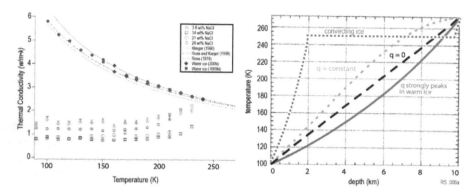

FIGURE 1.252 (Left) Thermal conductivity-salinity dependence, as measured in the lab (Carey et al. 2018). (Right) Models of ice temperature as a function of depth (Chyba and Ostro 1998). These introduce variance into a model of probe performance.

jetting and flow around the probe (Matheou and Dimotakis 2016). All models require, as inputs, the ice physical properties as a function of depth, including temperature, thermal conductivity, and heat capacity. These in turn depend on global properties such as ocean-surface energy exchange and salinity. Properties of planetary cryogenic ice that characterize physical properties and their interplay with Europa's structure are shown in Figure 1.252

1.13.4.3.1 Simulation and Uncertainty Quantification

The real-world values of the ice physical properties have large uncertainties and the presence of dust layers and voids in the ice is unknown. Environmental conditions can significantly impact the proof of concept for a given design, but neither analytic nor finite element models readily incorporate physical variances into their results. Furthermore, no model incorporates drilling efficacy, which must be determined by empirical tests. To remedy these issues, the Cryobot Descent Simulator will utilize an uncertainty quantification approach to integrate the several model components.

The program to achieve a validated Monte Carlo simulation is illustrated on the left side of Figure 1.249. Parametric equations form the core of the simulation, with several free parameters to describe behavior that is not easily included analytically (for example, see Schuller 2017, 2019). The finite element computations fix the additional parameters at discrete environmental points, and empirical data validates the finite element code and provides any unmodeled terms. The now-calibrated parametric equations can be interpolated to a continuum of behavior, from the Europan surface down to the ocean.

While the finite element code may take days to execute for a single set of parameters, the previously mentioned formulation enables rapid execution of many simulations. For each simulation, the environmental conditions will be varied based on the best scientific understanding. For a given Cryobot design, the aggregate simulations provide statistical metrics of performance, establishing proof of concept that the mission objectives can be met in the allocated time. The fast code speed also allows rapid trade studies between Cryobot designs.

1.13.4.3.2 Cryobot Testing

In order to anchor the numerical models, a Cryobot test program is underway. Test sets that cover a range of representative temperatures, ice salinity, and dust layering, enabling characterization and validation of the simulator for different portions of the probe's descent and for each of the operational modes, are being performed. Sublimation is anticipated to dominate the energy budget early in the descent, so additional testing is carried out under vacuum for the coldest ice. Testing in the field and in a dedicated ice tower enables sustained steady-state operation in a large body of ice, providing critical validation data that can't be acquired through smaller-scale laboratory tests. This data will be collected and used to back out transient and boundary condition artifacts from the model. Furthermore, field testing will provide an end-to-end test case for the Descent Simulator.

1.13.4.4 The Cryobot Head

The Cryobot head will use three different excavation approaches to propel descent in the ice: mechanical drilling, water jetting, and thermal melting (Figure 1.253). These could be used separately or in any combination. By integrating these three excavation options, the Cryobot has the ability to work in any combination of materials within the Europan ice shell (cryo and warm ice, salts, sediments, etc.). The mechanical drill is ideally suited for drilling hard materials such as cryogenic ice, salts, or rocks. As such, this approach would be used in the upper portion of the Europa crust and whenever the Cryobot encounters sediments or rocks. Water jetting works well in ice (any temperature) and ice with some sediments. As such, this system could be used after penetrating cryogenic ice with a mechanical drill or to assist mechanical drilling. Thermal melting works best in warm and relatively clean ice. As such, this approach is best suited for lower sections of the Europan ice shell. By using these three approaches, the overall life requirement on each is significantly reduced. In addition, in some instances, the systems could be redundant (Zacny et al. 2018b).

The Cryobot head (Figure 1.254) will consist of a drilling system (drill bit, motors, and percussive system), water jetting system (pump and motor), and melting system (simulated with resistive heaters).

Although each of the components is a proven technology and frequently used in terrestrial applications, the requirements driven by the Europan environment and Cryobot accommodation are new. This development is prototyping full-scale hardware and testing in warm ice and a cryogenic ice facility. Key to the current work is development of a drill bit that needs to perform several functions: cut cryogenic ice as well as salts and sediments in warmer ice with limited power and weight on bit; work in conjunction with water jetting; and, double as a melt probe head. The systems will be tested

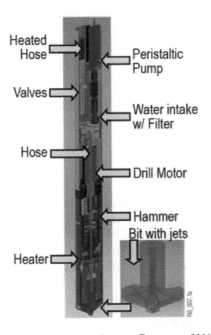

FIGURE 1.253 A baseline design for the Cryobot. *Source:* Courtesy of NASA JPL.

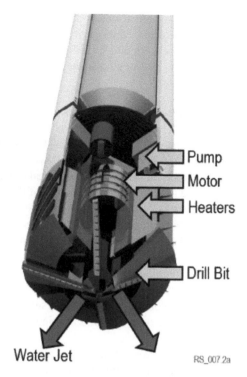

FIGURE 1.254   Notional design of Cryobot head. The design follows Zacny et al. (2018b).

under relevant conditions and will include a feed stage for deploying the Cryobot head into Europa analog ice. It will also measure the penetration rate, weight on bit, and drilling torque. Ground support equipment will use slip rings for power and data transfer and swivel for water transfer to the Cryobot head. Temperature sensors and pressure sensors will be included. The Cryobot head will also include a rotary-percussive drill head to provide rotation and percussion to the bit—this functionality will only be used in the initial stage of the development. Ultimately, the Cryobot head will include these 2 degrees of freedom.

1.13.4.4.1   Advanced Cryobot Head and Further Characterization

Once the drill bit has been finalized in the initial design, we will use the design data to size the motor, percussive system, pump, and heaters to design and fabricate an advanced Cryobot head. In particular, this portion of the project will:

- Update thermal models of performance and design. Perform final down select and placement of heaters.
- Update fluid models based on test data. Perform pump trade study that guides selection, design, and fabrication of the optimal pump.
- Perform motor selection and percussive mechanism selection that guides design and fabrication of the drill bit and its rotary-percussive system.
- Perform Cryobot head system-level design, fabrication and integration of all the components.
- Test under relevant conditions.
- Characterize performance, power required, reaction torques, and weight-on-bit sensitivity.

1.13.4.5   Probe Laboratory and Field Testing for Simulator Validation

The project will use and extend developments of the University of Washington's Applied Physics Laboratory group in performing measurements of ice probes both in the lab and then in the field

(Winebrenner and Elam 2020). The existing probe maintains vertical travel autonomously by means of pendulum steering using a flange at its upper end. This probe has been tested in Greenland in 2013 and 2014 to depths up to 400 m and descent speeds up to 6.5 m/hr, using 2.1–4.5 kW electrical power supplied at the surface by gasoline-powered generators. Most recently, this probe has been modified to incorporate jetting to suspend sediment in dirty ice, control of the melt-channel using antifreeze to enable probe recovery, and meltwater sampling for geobiology and geochemistry. This includes the implementing and extending methods for probe-cleaning and sample-acquisition (Dachwald et al. 2014), and to quantify controls on forward contamination.

This existing ice melt probe is being deployed initially as a scaled-down model of the previously described Cryobot. It has all the relevant functionality required for the warm ice phase. The key functionalities for this phase are heating and water jetting. Modifications will be made to the existing probe to acquire data on probe skin temperature and jetting, sediment, and meltwater parameters (e.g., salinity) in descent through dirty ice, in support of the model validation. Also, this probe will be used in the Madison, Wisconsin ice-tower facility with the data analysis results used to validate scaled probe models and to benchmark its performance in comparison to the full Cryobot performance in similar and controlled environments.

Next, the modified probe employing the Cryobot water jet design parameters will be tested in arctic ice where ice depths commensurate with the needs of model validation are available. This allows direct validation of the simulator for over long distances. The experiment will build on the lab tests to tune the melt heat, data acquisition, water jet flow rates and descend as much as 100 m in the ice sheet.

1.13.4.6 Key Technology Systems

During this project, a set of enabling, long-lead technologies will be developed to a conceptual level to inform the Cryobot architecture and ConOps interfaces. Three key subsystems have been identified.

1.13.4.6.1 Thermal Management System

The energy source will require thermal management to distribute the thermal energy, operate in the extreme ranges of the environment, and maintain all systems at flight-allowable temperatures. To solve this challenge, the project will examine novel thermal-structural system designs enabled by advanced manufacturing technologies (for example Figure 1.255). Our baseline design will consist of an adaptive, smart algorithm-controlled network of multiple-diameter ammonia loop heat pipes embedded in the structure with valves to direct and manage heat pipe fluid flow (Park et al. 2006). These have recently been demonstrated in the fabrication and operation of multifunctional parts and mechanically pumped two-phase thermal control and conformal large flat heat pipe (Furst 2018).

1.13.4.6.2 Ice Communication via Hybrid RF/Acoustic Transceivers

Current research on Europan ice properties (Barr et al. 2007) that effect signal transmission in the ice shell suggest that the upper brittle crust (perhaps 2.5 km down to 7.5 km) will exhibit grain sizes of 1 mm to 10 cm which effect signal attenuation across grain gap boundaries. From about 7.5–15 km, grain sizes are expected to increase from 1–10s of centimeters but, due to the higher pressures and temperatures, it is expected that the porosity decreases, causing boundary healing thus providing a denser transmission medium.

Modeling for the RF portion (Bryant 2002) has determined that signal attenuation (dB/km) is minimal in the cold/small-grain and crevassed region (only 4.3 dB/km at 100 MHz) of the shell down to 7.5 km. But in warm ice or water the attenuation increases by an order of magnitude and is exacerbated by the presence of salts.) For this project, research will focus on the actual transceiver design of critical components like the patch array antenna (e.g., a radiation-hardened ceramic), refinement of power requirements, and data packet transmission. Similarly, the transceiver acoustic

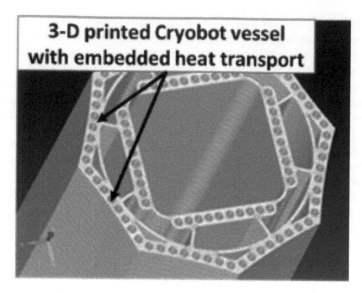

FIGURE 1.255 Cryobot. Potential thermal pressure vessel options. *Source:* Courtesy of NASA JPL.

element provides redundancy in the lower region of the ice sheet, employing low dB loss piezo transducers effective in transmitting/receiving acoustic signals in Europa's environment (Haider et al. 2018).

1.13.4.6.3   Navigation/Hazard Avoidance and Control

The Cryobot control subsystem architecture will draw on the heritage sensor and control system designed for the *Cryo-Hydro Ice Robotic Penetrating System* (CHIRPS) developed by JPL and tested in a dirty glacier on the island of Svalbard, Norway in 2001 (Zimmerman et al. 2001). That architecture coupled an inclinometer for tracking and correction of tilt, with a 2-axis accelerometer for control of rotation and descent rate, to facilitate steering. Other state sensors monitored temperature, pressure, and water jet flow rate. For this project, that heritage architecture will be expanded and matured, drawing on advancements in electronic component technologies (Bolotin 2018). This work will focus on identifying and designing advanced miniature radar (microstrip RF (Jin and Towfic 2018)) and acoustic (piezo) ranging sensors for navigation, hazard identification and avoidance, and water detection.

1.13.4.7   Summary

The Cryobot project is developing a promising cryogenic ice penetration system capable of facilitating the detection of life in the Europan ocean, with a baseline design that descends 15 km in three years and with mass <500 kg. An architecture is being developed with form fit and function that allows it to be engineered into a future flight system. This work is also identifying key technologies that represent the greatest technical risk to the overall Cryobot. In addition to developing the prototype Cryobot head, the high-fidelity Cryobot Descent Simulator is being developed and validated to reduce key technology risks of the architecture. These systems will be validated through lab and field measurements. The design is built to be capable of long duration operations in the extreme Europan environment and is based on a design principle to integrate redundant capabilities to mitigate the unknown environmental risks. The project is using the best scientific understanding of the ice environment for input to the Descent Simulator and lab ice. During the project, additional enabling technologies needed for the Cryobot will be examined, identifying specific components needing further development.

1.14 CONCLUSIONS

Exploration normally follows a typical path of trying to go after the so-called "low-hanging fruit" questions. As such, exploration started with observing sky using telescopes, launching flybys and then orbiters, landers, and rovers. As science questions have been addressed, new science questions were posed. We reached a point where to answer the next science questions, we need to go below the surface and either analyze samples in situ or bring samples back. This drives development of new, low-power and low-mass technologies.

This chapter summarized some of the technologies that have been developed over the past decade for the purpose of planetary exploration, whether it's search for life or identification of mineral resources for the purpose of space mining (In Situ Resource Utilization or ISRU). The examples given provide a glimpse into technologies and approaches that are being developed and certainly do not represent all the technologies that are being worked on.

The reader is encouraged to seek additional information by examining references at the end of this chapter, as well as papers by the co-authors.

ACKNOWLEDGMENTS

Some of the research reported in this chapter was conducted at the Jet Propulsion Laboratory (JPL), California Institute of Technology, and Honeybee Robotics under various contracts with the National Aeronautics and Space Administration (NASA). The Auto-Gopher task has been funded by the NASA MatISSE (Maturation of Instruments for Solar System Exploration) program under the Technical Manager Dr. James R. Gaier, Program Officer, Instrument Development Programs, Planetary Science Division, NASA Headquarters.

Some of the research reported in this chapter was supported by the Polish National Centre for Research and Development project no. PBS1/A2/0/2012 and by European Space Agency (ESA) project no. 4000112603/12/NL/CBi.

The design, building of, and research into the HP³ has been supported by the German Aerospace Center DLR, by NASA JPL, the ÖAW, and the Polish Academy of Science.

Reference herein to any specific commercial product, process, or service by trade name, trademark, manufacturer, or otherwise does not constitute or imply its endorsement by the United States Government or the Jet Propulsion Laboratory, California Institute of Technology.

The information presented about future space mission concepts and related nuclear space power systems is pre-decisional and is provided for planning and discussion purposes only.

The authors would like to thank Mircea Badescu; Yoseph Bar-Cohen; Stewart Sherrit; Jet Propulsion Laboratory/California Institute of Technology, Pasadena, CA; Rohit Bhartia, Optical Informatics, LLC, Altadena, CA; and Dara Sabahi from Honeybee Robotics, Altadena, CA, for reviewing this chapter and providing valuable technical comments and suggestions. The authors would like to thank many other employees of NASA, Honeybee Robotics and other agencies and institutions that have contributed to this work.

REFERENCES

Abbey, W., Anderson, R., Beegle, L. et al. (2019). A look back: The drilling campaign of the Curiosity rover during the Mars Science Laboratory's Prime Mission. *Icarus* 319: 1–13.

Adachi, M. and Kawamoto, H. (2017). Electrostatic sampler for large regolith particles on asteroids. *Journal of Aerospace Engineering* 30: 04016098-1–04016098-9.

Adachi, M., Maezono, H., and Kawamoto, H. (2016). Sampling of regolith on asteroids using electrostatic force. *Journal of Aerospace Engineering* 29: 04015081-1–04015081-9.

Adachi, M., Hamazawa, K., Mimuro, Y., and Kawamoto, H. (2017). Vibration transport system for lunar and Martian regolith using dielectric elastomer actuator. *Journal of Electrostatics* 89: 88–98.

Adachi, M., Obata, R., Kawamoto, H., Wakabayashi, S., and Hoshino, T. (2018). Magnetic sampler for regolith particles on asteroids. *Journal of Aerospace Engineering* 31: 04017095-1–0401709512.

Aguirre-Martinez, M.A., Bureo-Dacal, R., Del Campo, F., and Fuentes, M. (1987). The CTM family of masts and the CTM Engineering Model. In: *3rd European Space Mechanisms & Tribology Symposium*, Madrid, Spain.

Alexander, R.M. (1992). *Exploring Biomechanics, Animals in Motion*, W. H. Freeman and Co., New York, pp. 77–79.

Atkinson, J. (2019). Cryogenic Penetration and Relaxation Behavior of Dry and Icy Lunar Regolith Simulants. PhD thesis, Colorado School of Mines, Golden, CO

Atkinson, J., and K. Zacny. (2018). Mechanical Properties of Icy Lunar Regolith: Application to ISRU on the Moon and Mars, *ASCE Earth and Space 2018 Conference*, Cleveland, OH, April 10–12, 2018.

Backes, P. et al. (2014). Sampling system concepts for a touch and go architecture comet surface sample return mission. In: *AIAA Space 2014*, August 4–7, 2014. San Diego, CA.

Backes, P. et al. (2017). BiBlade sampling tool balidation for comet surface environments. In: *IEEE Aerospace Conference*, Big Sky, MT, March 4–11, 2017.

Backes, P. et al. (2020). The dual-rasp sampling system for an Enceladus Lander. In: *IEEE Aerospace Conference*, Big Sky, MT, March 5–10, 2020.

Badescu, M., Sherrit, S., Olorunsola, A.K., Aldrich, J., Bao, X., Bar-Cohen, Y., Chang, Z., Doran, P.T., Kenig, F., Fritsen, C., Murray, A., McKay, C.P., Peterson, T., Du, S., and Tao, S. (2006). Ultrasonic/sonic gopher for subsurface ice and brine sampling: Analysis and fabrication challenges, and testing results. In: *Proceedings of the SPIE 13th Annual Symposium on Smart Structures and Materials*, San Diego, CA, SPIE Vol. 6171-07, February 26–March 2, 2006.

Badescu, M., Sherrit, S., Jones, J., and Hall, J. (2009). Self-contained harpoon and sample handling device for a remote platform. In: *Sensors and Smart Structures Technologies for Civil, Mechanical, and Aerospace Systems 2009*, International Society for Optics and Photonics, San Diego, CA, March 2009, Vol. 7292, p. 72921I.

Badescu, M. et al. (2013). Dynamic Acquisition and Retrieval Tool (DART) for comet sample return. In: *IEEE Aero Conference*, Big Sky, MT, March 1–7, 2013.

Ball, A.J. et al. (2007). *Planetary Landers and Entry Probes*, 1st edition, Cambridge University Press, Cambridge.

Banaszkiewicz, M., Seweryn, K., and Wawrzaszek, R. (2007). Thermal conductivity determination of cometary and asteroid materials. *Advances in Space Research* 40: 226–237.

Banik, J. and Murphey, T. (2010). Performance validation of the triangular rollable and collapsible mast. In: *24th Annual AIAA/USU Conference on Small Satellites*, Logan UT, August 9–12, 2010.

Bar-Cohen, Y. and Zacny, K. (Eds.). (2009). *Drilling in Extreme Environments—Penetration and Sampling on Earth and Other Planets*, Wiley-VCH, Hoboken, NJ, ISBN-10: 3527408525, ISBN-13: 9783527408528, 827 pages.

Barmin and Shevchenko (1983). Soil scooping mechanism for the Venera 13 and 14 unmanned interplanetary spacecraft. Translated from *Kosmicheskie Issledovaniya* (Cosmic Research) 21(2): 171–175, March–April.

Barnouin-Jha et al. (2004). Sampling a planetary surface with a pyrotechnic rock chipper. In: *Proceedings of the IEEE Aerospace Conference*, Big Sky, MT.

Barr, A. et al. (2007). Convection in ice shells and mantles with self-consistent grain size. *Journal of Geophysical Research* 112: E02012.

Beegle, L.W., Wilson, M.G., Abilleira, F., Jordan, J.F., and Wilson, G.R. (2007). A concept for NASA's Mars 2016 Astrobiology Field Laboratory. *Astrobiology* 7(4): 545–577.

Bergman, D., Glass, B., Stucky, T., Zacny, K., Paulsen, G., and McKay, C. (2016). Autonomous structural health monitoring techniques for the icebreaker drill. In: *ASCE Earth and Space Conference*, Orlando, FL, April 11–15, 2016.

Bhartia, R., Hug, W., Salas, E., Reid, R., Sijapati, K., Tsapin, A., Abbey, W., Nealson, K., Lane, A., and Conrad, P. (2008). Classification of organic and biological materials with deep ultraviolet excitation. *Applied Spectroscopy* 62(10): 1070–1077.

Bhartia, R., Salas, E., Hug, W., Reid, R., Lane, A., Edwards, K., and Nealson, K. (2010). Label-free bacterial imaging with deep-UV-laser-induced native fluorescence. *Applied and Environmental Microbiology* 76(21): 7231–7237.

Bhartia et al. (2019). Wireline analysis tool for subsurface observation of Northern-ice-sheets. In: *AGU Fall Meeting*, San Francisco, CA, December 9–13, 2019.

Blake, D., Vaniman, D., Achilles, C. et al. (2012). Characterization and calibration of the CheMin Mineralogical Instrument on Mars Science Laboratory. *Space Science Reviews* 170: 341–399. doi:10.1007/s11214-012-9905-1

Bolotin, G. Cold survivable distributed motor controller 2017 [cited 2018 September]; Available from: https://www.lpi.usra.edu/opag/meetings/sep2017/posters/Bolotin-2.pdf

Bonitz, R. et al. (2009). The Phoenix Mars Lander robotic arm. In: *2009 IEEE Aerospace Conference*, Big Sky, MT, March 7–14, 2009.

Bonitz, R. (2012). The brush wheel sampler—A sampling device for small-body touch-and-go missions, IEEE Aerospace Conference, March 3–10, 2012, Big Sky, MT.

Bryant, S. (2002). Ice-embedded transceivers for Europa cryobot communications. In: *Proceedings, IEEE Aerospace Conference*.

Byrne et al. (2009). Distribution of mid-latitude ground ice on Mars from new impact craters. *Science* 325: 1674–1676.

Carey, E.M., Choukroun, T.V.M., Zhong, Cohen, B., Barmatz, M., Zimmerman, W., and Cwik, T. (2018). Thermal properties of hydrated salts with implications for icy satellites. 49th LPSC, March 19–23, 2018, The Woodlands, TX.

Carrier, D. (2005). The four things you need to know about the geotechnical properties of lunar soil. Lunar Geotechnical Institute, September 2005.

Choukroun et al. (2018). Sintering of fine-grained porous water ice: Preliminary investigation of microstructure and strength evolution. In: *IEEE Aerospace Conference*, Big Sky, MT, March 5–11, 2018.

Chu, P., Wilson, J., Davis, K., Shirishi, L., and Burke, K. (2008). Icy soil acquisition device for the 2007 Phoenix Mars Lander. In: *39th Aerospace Mechanisms Symposium*, Huntsville, AL, May 2008.

Chu, P., Spring, J., and Zacny, K. (2014). ROPEC—Rotary Percussive Coring drill for Mars sample return. In: *Proceedings of the 42nd Aerospace Mechanisms Symposium*, NASA Goddard Space Flight Center, Washington, DC, May 14–16, 2014, pp. 195–210.

Chu, P., Indyk, S., Zacny, K., and James, W. (2014a). A comet surface sample return probe [#1536]. In: *45th Lunar and Planetary Science Conference*, The Woodlands, TX, March 17–21, 2014.

Chyba, C. F. and Ostro, S.J. (1998). Radar detectability of a subsurface ocean on Europa. *Icarus* 134: 292–302.

Clark et al. (2016). TAGSAM: A gas-driven system for collecting samples from solar system bodies. In: *2016 IEEE Aerospace Conference*, Big Sky, MT, March 5–12, 2016.

Clow, G. and Koci, B. (2002). A fast mechanical-access drill for polar glaciology, paleoclimatology, geology, tectonics and biology. *Memoirs of National Institute of Polar Research*, Spec Issue 56: 1–30.

Costa, J. et al. (2019). Development and testing of a sample cup for laser-based instruments. In: *Abscicon*, Seattle, WA.

Cull et al. (2010). Compositions of subsurface ices at the Mars Phoenix landing site. *Geophysical Research Letter* 37: L24203.

Dachwald, B., Mikucki, J., Tulaczk, S., Digel, I., Espe, C., Feldmann, M., Franke, G., Kowalski, J., and Xu, C. (2014). IceMole: A maneuverable probe for clean in situ analysis and sampling of subsurface ice and subglacial aquatic ecosystems. *Annals of Glaciology* 55(65): 14–22.

Dachwald, B., Ulamec, S., Postberg, F., Sohl, F., de Vera, J.P., Waldmann, C., Lorenz, R.D., Zacny, K., Hellard, H., Biele, J., and Rettberg, P. (2020). Key technologies and instrumentation for subsurface exploration of ocean worlds. *Space Science Reviews* 216: 83. https://doi.org/10.1007/s11214-020-00707-5

Davis, K. et al. (2012). Mars science laboratory's dust removal tool. In: *Proceedings of the Aerospace Mechanisms Symposium*, May 16–18, 2012, Pasadena, CA.

Delage, P., Karakostas, E., Belmokhtar, M., Lognonné, P. et al. (2017). An Investigation of the Mechanical Properties of Some Martian Regolith Simulants with Respect to the Surface Properties at the InSight Mission Landing Site, Space Science Reviews doi:10.1007/s11214-017-0339-7.

Dundas et al. (2018). Exposed subsurface ice sheets in the Martian mid-latitudes. *Science* 359: 199–201.

Eimer, B.C. and Taylor, L.A. (2007). Lunar regolith, soil, and dust mass mover on the Moon. In: *Lunar and Planetary Science Conference*, League City, TX.

Eshelman, E., Malaska, M., Manatt, K., Doloboff, I., Wanger, G., Willis, M., Abbey, W., Beegle, L., Priscu, J., and Bhartia, R. (2019). WATSON: In situ organic detection in subsurface ice using deep-UV fluorescence spectroscopy. *Astrobiology* 19(6): 771–784.

Farias et al. (2019). Thermal design and validation of Mars 2020 Gas Dust Removal Tool (gDRT). In: *49th International Conference on Environmental Systems*, ICES-2019-249, Boston, MA, July 7–11, 2019.

Finzi et al. (2007). SD2—How to sample a comet. *Space Science Reviews* 128: 281–299.

Flagan, R.C. and Seinfeld, J.H. (1988). Chapter 7: Removal of particles from gas streams. In: *Fundamentals of Air Pollution Engineering*, Prentice-Hall, Inc., Englewood Cliffs, NJ, ISBN: 0-13-332537-7. http://resolver.caltech.edu/CaltechBOOK:1988.001

Flowerdew, M., Tyrrell, S., Riley, T.R., Whitehouse, M.J. et al. (2012). Distinguishing East and West Antarctic sediment sources using the Pb isotope composition of detrital K-feldspar. *Chemical Geology* 293: 88–102.

Fuerstenau, S.D. (2006). Solar heating of suspended particles and the dynamics of Martian dust devils. *Geophysical Research Letter* 33: L19S03. doi:10.1029/2006GL026798

Furst, B. (2018). An additively manufactured evaporator with integrated porous structures for two-phase thermal control. In: *48th International Conference on Environmental Systems*.

Glass, G., McKay, C., Thompson, S., and Zacny, K. (2011). Automated Mars drilling for "IceBreaker". In: *IEEE Aerospace Conference*, Big Sky, MT, March 5–12, 2011.

Glass, B., Dave, A., Lee, P., and Zacny, K. (2014). Testing of mars-prototype drills at an analog site. In: *14th ASCE International Conference on Engineering, Science, Construction and Operations in Challenging Environments*, St. Louis, MO, October 27–29, 2014.

Glass, B., Bergman, D., Yaggi, B., Dave, A., and Zacny, K. (2016). Icebreaker-3 drill integration and testing at two Mars-analog sites. In: *ASCE Earth and Space Conference*, Orlando, FL, April 11–15, 2016.

Goetz, W. et al. (2007). The nature of Martian airborne dust. Indication of long-lasting dry periods on the surface of Mars. LPI Contributions. 1353 3104.

Golombek, M., Kipp, D., Warner, N. et al. (2017). Selection of the InSight Landing Site, Space Science Reviews 211: 5. doi:10.1007/s11214-016-0321-9

Golombek, M., Grott, M., Kargl, G., Andrade, J. et al. (2018). *Space Science Reviews*. doi:10.1007/s11214-018-0512-7

Goodge, J.W. and Severinghaus, J.P. (2016). Rapid access ice drill: A new tool for exploration of the deep Antarctic ice sheets and subglacial geology. *Journal of Glaciology* 62(36): 1049–1054.

Gorevan, S.P., Myrick, T., Davis, K., Chau, J.J., Bartlett, P., Mukherjee, S., Anderson, R., Squyres, S.W., Arvidson, R.E., Madsen, M.B., Bertelsen, P., Goetz, W., Binau, C.S., Richter, L., and Hoehe, L. (2003). Rock abrasion tool: Mars exploration Rover mission. *Journal of Geophysical Research* 108: 8068. doi:10.1029/2003JE002061

Gouache, T., Gao, Y., Gourinat, Y., and Coste, P. (2010). Wood wasp inspired space and earth drill. In: *Biomimetics Learning from Nature*, InTech, Rijeka, Croatia, pp. 467–486.

Green, A, Zacny, K., Pestana, J., Lieu, D., and Mueller, R. (2012). Percussive excavation and its nullifying effect on the influence of soil relative density. In: *ASCE Earth and Space 2012*, Pasadena, CA, April 16–18, 2012.

Green, A, Zacny, K., Pestana, J., Lieu, D., and Mueller, R. (2013). Investigating the effects of percussion on excavation forces. *Journal of Aerospace Engineering* 26(1): 87–96, doi:10.1061/(ASCE)AS.1943-5525.0000216

Gromov, V.V. et al. (1997). The mobile Penetrometer, a "Mole" for sub-surface soil investigation. In: *Proceedings 7th European Space Mechanisms & Tribology Symposium*, ESTEC, Noordwijk, The Netherlands, pp. 151–156.

Grott, M., Spohn, T., Knollenberg, J. et al. (2019). Calibration of the heat flow and physical properties package (HP3) for the InSight Mars Mission. *Earth and Space Science* 6(12): 2556–2574. doi:10.1029/2019EA000670

Grygorczuk, J., Banaskiewicz, M., Seweryn, K., and Spohn, T. (2007). *Journal of Telecommunications and Information Technology* 1: 50.

Grygorczuk, J. et al. (2011). Advanced mechanisms and tribological tests of the hammering sampling device CHOMIK. In: *Proceedings of the 14th European Space Mechanisms and Tribology Symposium*, Constance, Germany.

Grygorczuk, J. et al. (2013). High energy and efficiency penetrator—HEEP. In: *Proceedings 15th European Space Mechanisms and Tribology Symposium*.

Grygorczuk, J. et al. (2015). Mole penetrator driven by an electromagnetic direct drive (EMOLE). In: *Proceedings 16th European Space Mechanisms and Tribology Symposium*.

Grygorczuk, J. et al. (2016a). Hammering mechanism for HP3 experiment (InSight). In: *43rd Aerospace Mechanisms Symposium Papers*, pp. 415–428.

Grygorczuk, J., Wiśniewski, Ł., Dobrowolski, M., and Kędziora, B. (2016b). Electromagnetic Drive and Method of Production Thereof. United States of America, Patent No. US 2016/0067855A1.

Grygorczuk, J. et al. (2016c). A multi-sectioning, reconfigurable electromagnetic hammering propulsion for mole penetrators. In: *43rd Aerospace Mechanisms Symposium Papers*, pp. 171–176.

Grygorczuk, J. et al. (2018). Method for producing a torque, preferably for drilling rigs and the driving device for pulsed production of the torque. Poland, Patent No. PL 229850.

Haehnel, R.B. and Knuth, M.A. (2011). Potable Water Supply Feasibility Study for Summit Station, Greenland, CRREL TR-11-4.

Haider, M.F. et al. (2018). Piezoelectric wafer active sensors under gamma radiation exposure. In: *SPIE*, Denver, CO.

Hand, K.P., Murray, A.E., Garvin, J.B. et al. (2017). Europa Lander Study 2016 Report, JPL D-97667, Task Order NNN16DO11T, Europa Lander Mission Pre-Phase A, NASA.

Hasegawa, K. (2019). Japan's Hayabusa2 probe makes 'perfect' touchdown on asteroid, accessed June 30, 2020, https://phys.org/news/2019-07-japan-hayabusa2-probe-touchdown-asteroid.html

Heiken, G.H., Vaniman, D.T., and French, B.M. (Eds.). (1991). *Lunar Sourcebook: A User's Guide to the Moon*, Cambridge University Press, Cambridge.

Helmick, D., McCloskey, S., Okon, A., Carsten, J., Kim, W., and Leger, C. (2013). Mars science laboratory algorithms and flight software for autonomously drilling rocks. *Journal of Field Robotics* 30(6): 847–874. doi:10.1002/rob.21475

Holmberg, N.A., Faust, R.P., and Holt, H.M. (1980). Viking '75 Spacecraft Design and Test Summary: Volume 1—Lander Design. Langley Research Center, NASA Reference Publication 1027, November 1980.

Howell, S.M. and Pappalardo, R.T. (2020). NASA's Europa Clipper—A mission to a potentially habitable ocean world. *NatureCommunications* 11: 1311. https://doi.org/10.1038/s41467-020-15160-9

IDDO. (2015). Agile sub-ice geological drill system design review. U.S. Ice Drilling Program.

Jandura, L. (2010). Mars science laboratory sample acquisition, sample processing and handling: Subsystem design and test challenges. In: *Proceedings of the 40th Aerospace Mechanisms Symposium*, May 12–14, 2010 Cocoa Beach, Florida.

Jens, E., Tarantino, P., Preudhomme, M., Hinchman, R., Nakazono, B., and Vaughan, D. (2017). Precision cleaning samples for science analysis using a gas-based dust removal tool. In: *Aerospace Conference, 2017 IEEE*, Big Sky, MT, pp. 1–12.

Jens, E., Nakazono, B., Brockie, I., Vaughan, D., and Klatte, M. (2018). Design, development and qualification of a gas-based durst removal tool for Mars exploration missions. In: *2018 IEEE Aerospace Conference*, Big Sky, MT, March 2018.

Jin, C. and Towfic, Z. (2018). SESAME radar options study: Direction finding using X-band micro-strip antenna patch array. JPL.

Karras, J. et al. (2017). Pop-up Mars rover with textile-enhanced rigid-flex PCB body. In: *2017 IEEE International Conference on Robotics and Automation (ICRA)*, Singapore.

Kato, H., Satou, Y., Yoshikawa, K., Otsuki, M., and Sawada, H. (2020). Subsurface sampling robot for time-limited asteroid exploration. In: *Proceedings of IEEE/RSJ International Conference on Intelligent Robots and Systems (IROS)*, Las Vegas, NV, October 2020.

Kawakatsu et al. (2019). Mission design of Martian Moons eXploration (MMX). In: *70th International Astronautical Congress (IAC)*, Washington, DC, October 21–25, 2019.

Kawamoto, H. (2014). Sampling of small regolith particles from asteroids utilizing alternative electrostatic field and electrostatic traveling wave. *Journal of Aerospace Engineering* 27: 631–635.

Kawamoto, H. and Shirai, K. (2012). Electrostatic transport of lunar soil for in-situ resource utilization. *Journal of Aerospace Engineering* 25: 132–138.

Kawamoto H. and Yoshida, N. (2018). Electrostatic sampling and transport of ice for in-situ resource utilization. *Journal of Aerospace Engineering* 31: 04018044-1–04018044-6.

Kawamoto, H., Shigeta, A., and Adachi, M. (2016a). Utilizing electrostatic force and mechanical vibration to obtain regolith sample from the Moon and Mars. *Journal of Aerospace Engineering* 29: 04015031-1–04015031-6.

Kawamoto, H., Kato, M., and Adachi, M. (2016b). Electrostatic transport of regolith particles for sample return mission from asteroids. *Journal of Electrostatics* 84: 42–47.

Kleinhenz, J., Paulsen, G., Zacny, K., and Smith, J. (2015). Impact of drilling operations on lunar volatiles capture: Thermal vacuum tests. In: *AIAA SciTech*, 2015-1177, Kissimmee, FL, January 5–9, 2015.

Kleinhenz, J., Smith, J., Roush, T., Colaprete, A., Zacny, K., Paulsen, G., Wang, A., and Paz, A. (2018). Volatiles loss from water bearing regolith simulant at lunar environments. In: *ASCE Earth and Space 2018 Conference*, Cleveland, OH, April 10–12, 2018.

Knapmeyer, M. et al. (2018). Structure and elastic parameters of the near surface of Abydos site on comet 67P/Churyumov-Gerasimenko, as obtained by SESAME/CASSE listening to the MUPUS insertion phase. *Icarus* 310: 165–193.

Kornfeld, R.P. et al. (2014). Verification and validation of the Mars Science Laboratory/Curiosity Rover entry, descent, and landing system. *Journal of Spacecraft and Rockets* 51(4): 1251–1269.

Kozlov, O.E. and Kozlova, T.O. (2014). Manipulators of the Phobos-Grunt project and Lunar projects. In: Sąsiadek, J. (Ed.), *Aerospace Robotics II*, Springer-Verlag, Cham.

Krömer, O., Scharringhausen, M., Fittock, M. et al. (2019). Design details of the HP3 mole onboard the InSight mission. *Acta Astronautica* 164: 152–167. doi:10.1016/j.actaastro.2019.06.031

Lichtenheldt, R., Schäfer, B., Krömer, O., van Zoest, T. (2014a). In: Sung-Soo, K. and Hwan, C.J. (Eds.), *3rd International Conference on Multibody System Dynamics*, ISBN: 978-89-950027-7-3.

Lichtenheldt, R., Schäfer, B., and Krömer, O. (2014b). Shaping the future by engineering. In: *58th Ilmenau Scientific Colloquium IWK*, URN (Paper): urn:nbn:de:gbv:ilm1-2014iwk-155:2.

Liu, Y. et al. (2007). Unique properties of lunar impact glass: Nanophase metallic Fe synthesis. *American Mineralogist* 92: 1420–1427.

Lorenz, R. et al. (2018). Pneumatic sample acquisition and transfer for 'Ocean Worlds' landers. In: *15th International Planetary Probe Workshop*, Boulder, CO.

Lunine, J.I. (2017). Ocean worlds exploration. *Acta Astronautica* 131: 123–130.

Mahaffy, P.R., Webster, C.R., Cabane, M. et al. (2012). The sample analysis at Mars investigation and instrument suite. *Space Science Reviews* 170: 401–478. doi:10.1007/s11214-012-9879-z

Malaska, M.J., Bhartia, R., Manatt, K.S., Priscu, J.C., Abbey, W.J., Mellerowicz, B., Palmowski, J., Paulsen, G.L. Zacny, K., Eshelman, E.J., and D'Andrilli, J. (2020). Subsurface in situ detection of microbes and diverse organic matter hotspots in the Greenland ice sheet. *Astrobiology* 20(12).

Manga, F.N.M. (2009). Geodynamics of Europa's icy shell. In: Robert, W.B.M., Pappalardo, T., and Khurana, K. (Eds.), *Europa*, University of Arizona Press, pp. 381–394.

Mantovani, J.G. and Townsend, I. (2013). Planetary regolith delivery systems for ISRU. *Journal of Aerospace Engineering* 26: 169–175.

Matheou, G. and Dimotakis, P.E. (2016). Scalar excursions in large-eddy simulations. *Journal of Computational Physics* 327(C): 97–120.

McKay, C.P., Stoker, C.R., Glass, B.J., Davé, A.I., Davila, A.F., Heldmann, J.L., Marinova, M.M., Fairen, A.G., Quinn, R.C., Zacny, K.A., Paulsen, G., Smith, P.H., Parro, V., Andersen, D.T., Hecht, M.H., Lacelle, D., and Pollard, W.H. (2013). The Icebreaker Life Mission to Mars: A search for biomolecular evidence for life. *Astrobiology* 13(4): 2013.

Melamed, Y., Kiselev, A., Gelfgat, M., Dreesen, D., and Blacic, J. (2000). Hydraulic hammer drilling technology: Developments and capabilities. *Journal of Energy Resources Technology* 122: 1–8.

Mellerowicz et al. (2019). Development of a deep drilling probe with integrated deep-UV fluorescence & Raman spectrometer for Mars. In: *AGU Fall Meeting*, San Francisco, CA, December 9–13, 2019.

Miyamoto, H. et al. (2019). Phobos environment model and regolith simulant for MMX mission. In: *49th Lunar and Planetary Science Conference* (Abstract. No. 1882), The Woodlands, TX, March 19–23, 2018.

Molaro, J.L. et al. (2019). The microstructural evolution of water ice in the solar system through sintering. *Journal of Geophysical Research (Planets)* 124: 243–277.

Moreland et al. (2018). Full-scale dynamic touch-and-go validation of the BiBlade comet surface sample chain. In: *IEEE Aerospace Conference*, Big Sky, MT, March 3–10, 2018.

Mulvaney, R., Bremner, S., Tait, A., and Audley, N. (2002). A medium-depth ice core drill. In: *Fifth International Workshop on Ice Drilling Technologies*, Tokyo, Japan.

Nagihara, S., Zacny, K., Hedlund, M., and Taylor, P.T. (2014). Compact, modular heat flow probe for the lunar geophysical network mission. In: *International Workshop on Instrumentation for Planetary Missions*, p. 1101.

Nagihara, S., Kiefer, W.S., Taylor, P.T., Williams, D.R., and Nakamura, Y. (2018). Examination of the long-term subsurface warming observed at the Apollo 15 and 17 sites utilizing the newly restored heat flow experiment data from 1975 to 1977. *Journal of Geophysical Research: Planets* 123: 1125–1139. https://doi.org/10.1029/2018JE005579

Nakatake, T., Konno, M., Mizushina, A., Yamada, Y., Nakamura, T., and Kubota, T. (2016). Soil circulating system for a lunar subsurface explorer robot using a peristaltic crawling mechanism. *Advanced Intelligent Mechatronics (AIM)* 2016: 407–412.

National Research Council. (2011a). *Future Scientific Opportunities in Antarctica and the Southern Ocean*, National Academies Press, Washington, DC.

National Research Council. (2011b). *Vision and Voyages for Planetary Science in the Decade 2013–2022*, National Academies Press, Washington, DC, 422 pages.

Nesnas, I.A., Matthews, J., Abad-Manterola, P., Burdick, J.W., Edlund, J., Morrison, J., Peters, R., Tanner, M., Miyake, R., and Solish, B. (2012). Axel and DuAxel rovers for the sustainable exploration of extreme terrains. *Journal of Field Robotics* 29(4): 533–685, July.

Ngo, P., Nagihara, S., Sanigepalli, V., Sanasarian, L., and Zacny, K. (2019). Heat flow probe for short-duration Lander Missions under NASA's Commercial Lunar Payload Service Program. In: *AGU Fall Meeting*, San Francisco, CA, December 9–13, 2019.

Oda, M. and Kubota, T. (2009). Roadmap for space development and exploration in Japan. *Journal of the Robotics Society of Japan* 27(5): 2–9.

Okon, A. (2010). Mars science laboratory drill *Proceedings of the 40th Aerospace Mechanisms Symposium*. May 12–14, 2010, Cocoa Beach, Florida.

Omori, H., Murakami, T., Nagai, H., Nakamura, T., and Kubota, T. (2011). Planetary subsurface explorer robot with propulsion units for peristaltic crawling. In: *Proceedings of the IEEE International Conference on Robotics and Automation*, pp. 649–654.

Park, C., Vallury, A., and Perez, J. (2006). Advanced hybrid cooling loop technology for high performance thermal management. In: *4th International Energy Conversion Engineering Conference*, San Diego, CA.

Paśko, P., Seweryn, K., Kłak, M., Teper, W., Visentin, G., and Żyliński, B. (2017). Novel sampling tool for low gravity planetary bodies. In: *Proceedings of the ASTRA Conference*, ESA ESTEC, The Netherlands.

Paulsen, G., Zacny, K., McKay, C., Shiraishi, L., Kriechbaum, K., Glass, B., Szczesiak, M., Santoro, C., Craft, J., Malla, R.B., and Maksymuk, M. (2010). Rotary-percussive deep drill for planetary applications. In: *ASCE Earth and Space 2010*, Honolulu, HI, March 15–17, 2010.

Paulsen, G., Zacny, K., Szczesiak, M., Santoro, C., Mellerowicz, B., Craft, J., McKay, C., Glass, B., Davila, A., and Marinova, M. (2011). Testing of a 1-meter Mars IceBreaker drill in a 3.5-meter vacuum chamber and in an Antarctic Mars analog site. In: *AIAA SPACE 2011 Conference & Exposition*, Long Beach, CA, September 26–29, 2011.

Paulsen, G., Szczesiak, M., Maksymuk, M., Santoro, C., and Zacny, K. (2012). SONIC drilling for space exploration. In: *ASCE Earth and Space*, Pasadena, CA, April 15–18, 2012.

Paulsen, G., Zacny, K., Kim, D., Mank, Z., Wang, A., Thomas, T., Hyman, C., Mellerowicz, B., Yaggi, B., Fitzgerald, Z., Ridilla, A., Atkinson, J., Quinn, J., Smith, J., and Kleinhenz, J. (2017). Development and testing of the lunar resource prospector drill, LEAG, October 10–12, 2017.

Peters, G.H., Carey, E.M., Anderson, R.C., Abbey, W.J., Kinnett, R., Watkins, J.A., ... Vasavada, A.R. (2018). Uniaxial compressive strengths of rocks drilled at Gale crater, Mars. *Geophysical Research Letters* 45: 108–116. https://doi.org/10.1002/2017GL075965

Purves, L. and Nuth, J.A. (2017). Development and testing of Harpoon-based approaches for collecting comet samples. https://ssed.gsfc.nasa.gov/harpoon/SAS_Paper-V1.pdf

Putzig et al. (2014). SHARAD soundings and surface roughness at past, present, and proposed landing sites on Mars: Reflections at Phoenix may be attributable to deep ground ice. *Journal of Geophysical Research: Planets* 119: 1936–1949.

Rehnmark, F., Cloninger, E., Hyman, C., Bailey, J., Traeden, N., Zacny, K., Kriechbaum, K., Melko, J., Wilcox, B., Hall, J., and Sherrill, K. (2017). High temperature actuator and sampling drill for Venus exploration. In: *ESMATS*, Hatfield, UK.

Rehnmark, F. et al. (2018a). Surface and subsurface sampling drills for life detection on Ocean Worlds. In: *15th International Planetary Probe Workshop*, Boulder, CO.

Rehnmark, F., Cloninger, E., Hyman, C., Bailey, J., Traeden, N., Zacny, K., Kriechbaum, K., Melko, J., Wilcox, B., Hall, J., and Sherrill, K. (2018b). Environmental chamber testing of the VISAGE rock sampling drill for Venus exploration. In: *44th Aerospace Mechanisms Symposium*, NASA/CP-2018-219887, Cleveland, OH, May 16–18.

Richter, L., Coste, P., Gromov, V.V., and Greszik, A. (2004). The mole with sampling mechanism (MSM)—Technology Development and Payload of Beagle 2 Mars Lander, *Proceedings of the 8th ESA Workshop on Advanced Space Technologies for Robotics and Automation*, ESTEC, Noordwijk, The Netherlands, pp. I-11–I-14.

Rickman, H. et al. (2014). CHOMIK: A multi-method approach for studying phobos. *Solar System Research* 48: 279–286.

Rix, J., Mulvaney, R., Hong, J., and Ashurst, D. (2018). Development of the British Antarctic survey rapid access isotope drill. *Journal of Glaciology* 65: 288–298.

Rybus, T. and Seweryn, K. (2016). Planar air-bearing microgravity simulators: Review of applications, existing solutions and design parameters. *Acta Astronautica* 120: 239–259.

Salas, E., Bhartia, R., Anderson, L., Hug, W., Reid, R., Iturrino, G., and Edwards, K. (2015). In situ detection of microbial life in the deep biosphere in igneous ocean crust. *Frontiers in Microbiology* 6: 1260. https://dx.doi.org/10.3389/fmicb.2015.01260

Sanders et al. (2011). Comparison of lunar and Mars in situ resource utilization for future robotic and human missions. In: *AIAA Aero Science*.

Savoia, M. et al. (2019). PROSPECT: key aspects of drilling and collecting samples at moon south pole for Luna Resource Mission. s.l. In: *European Lunar Symposium*.

Sefton-Nash, E., Carpenter, J.D., Fisackerly, R., and Trautner, R. (2018). ESA'S PROSPECT package for exploration of lunar resources. In: *49th Lunar and Planetary Science Conference*, Woodlands, TX.

Seger, R. and Gillespie, V. (1974). The Viking surface sampler, retrieved on January 27, 2013, http://ntrs.nasa.gov/archive/nasa/casi.ntrs.nasa.gov/19740003574_1974003574.pdf

Seweryn, K. (2016). The new concept of sampling device driven by rotary hammering actions. *IEEE/ASME Transactions on Mechatronic Systems* 21(5): 2477–2489.

Seweryn, K., Bednarz, S., Buratowski, T., Chmaj, G., Ciszewski, M., Gallina, A., Gonet, A., Grassmann, K., Kuciński, T., Lisowski, J., Paśko, P., Rutkowski, K., Teper, W., Uhl, T., Wawrzaszek, R., and Zwierzyński, A.J. (2014a). Novel concept and design of ultralight mobile drilling system dedicated for planetary environment. In: *Proceedings of 14th ASCE Earth and Space Conference*, St Louis, MO.

Seweryn, K., Skocki, K., Banaszkiewicz, M., Grygorczuk, J., Kolano, M., Kuciński, T., Mazurek, J., Morawski, M., Białek, A., Rickman, H., and Wawrzaszek, R. (2014b). Determining the geotechnical properties of planetary regolith using low velocity penetrometers. *Planetary and Space Sciences* 99: 70–83.

Seweryn, K. et al. (2014c). Autonomous regolith sampling on the planetary bodies—Test results from the common operation of the CHOMIK penetrator and manipulator arm. In: *Science and Challenges of Lunar Sample Return Workshop*, Noordwijk, the Netherlands. https://www.lpi.usra.edu/lunar/strategies/WorkshopOutcomesRecommendations033114.pdf

Seweryn, K., Grygorczuk, J., Wawrzaszek, R., Banaszkiewicz, M., Rybus, T., and Wiśniewski, Ł. (2014d). Low velocity penetrators (LVP) driven by hammering action—Definition of the principle of operation based on numerical models and experimental tests. *Acta Astronautica* 99: 303–317.

Sharpe, R. and Cork, G. (1995). Geology and mining of the Miocene Fish Creek gypsum in Imperial County, California. In: Tabillio (Ed.), *29th Forum on the Geology of Industrial Minerals: Proceedings*, California Department of Conservation.

Skonieczny, K. (2013). Lightweight Robotic Excavation. PhD thesis, Carnegie Mellon University, Pittsburgh, PA.

Smith, W. (2018). Low risk technique for sample acquisition from remote and hazardous sites on a comet. In: *IPM 2018*.

Smith, I.B. et al. (2020). The Holy Grail: A roadmap for unlocking the climate record stored within Mars' polar layered deposits. *Planetary and Space Science* 184: 104841.

Soumela, J., Visentin, G., and Ylikorpi, T. (2001). Robotic deep driller for exobiology. In: *Proceedings of the 6th International Symposium on Artificial Intelligence and Robotics & Automation in Space*.

Sparta, J. et al. (2018). Development of a pneumatic sample transport system for ocean worlds. In: *15th International Planetary Probe Workshop*, Boulder, CO.

Sparta, J. et al. (2019). Sampling the ocean worlds: Drilling and pneumatic transfer. In: *Abscicon*, Seattle, WA.

Spector, P., Stone, J., Pollard, D., Hillebrand, T., Lewis, C., and Gombiner, J. (2018). West Antarctic sites for subglacial drilling to test for past ice-sheet collapse. *The Cryosphere* 12: 2741–2757.

Spohn, T., Ball, A., Seiferlin, K. et al. (2001). A heat flow and physical properties package for the surface of Mercury. *Planetary and Space Science* 49: 1571.

Spohn, T. et al. (2007). MUPUS—A thermal and mechanical properties probe for the Rosetta Lander Philae. *Space Science Reviews* 128: 339–362.

Spohn, T. et al. (2015). Thermal and mechanical properties of the near-surface layers of comet 67P/Churyumov-Gerasimenko. *Science* 349: aab0464.

Spohn, T. et al. (2018). The heat flow and physical properties package (HP3) for the InSight mission. *Space Science Review* 214: 96.

Spohn, T., Smrekar, S.E., Grott, M., Hudson, T.L. et al. (2019). The heat flow and physical properties package HP3 on InSight: Status and first results. *American Geophysical Union, Fall Meeting*, Abstract DI42A-03.

Spring, J., Zacny, K., Betts, B., Chu, P., Chu, P., Ford, S., Luczek, K., Peekema, A., Traeden, N., Garcia, R., and Heidenberger, I. (2019). PlanetVac Xodiac: Lander foot pad integrated planetary sampling system. In: *IEEE Aerospace Conference*, Big Sky, MT, March 2–9, 2019.

Squyres, S. (2018). CAESAR. In: *18th Meeting of the NASA Small Bodies Assessment Group*, NASA Ames Research Center, January 17–18, 2018.

Sullivan et al. (1994). Pneumatic conveying of materials at partial gravity. *Journal of Aerospace Engineering* 7(2): 199–208, April.

Sunshine, D. Mars Science Laboratory CHIMRA: A device for processing powdered Martian samples. Proceedings of the 40th Aerospace Mechanisms Symposium, Cocoa Beach, Florida. May 12–14, 2010.

Tachibana et al. (2013). The sampling system of Hayabusa2: Improvements from the Hayabusa sampler. In: *44th Lunar and Planetary Science Conference*.

Tadami, N., Nagai, M., Nakatake, T., and Fujiwara, A. (2017). Curved excavation by a sub-seafloor excavation robot. In: *Proceedings of the 2017 IEEE/RSJ International Conference on Intelligent Robots and Systems*, Vancouver, BC, pp. 4950–4955.

Talalay, P. (2013). Subglacial till and bedrock drilling. *Cold Regions Science and Technology* 86: 142–166.

Talalay, P.G. et al. (2014). Recoverable autonomous sonde for environmental exploration of Antarctic subglacial lakes: General concept. *Annals of Glaciology* 55(65): 23–30. doi:10.3189/2014AoG65A200

Thomas, P.C. et al. (2016). Enceladus's measured physical libration requires a global subsurface ocean. *Icarus* 264: 37–47.

Toda et al. (2016). FiSI: Fiberscope sample imaging system for robotic comet surface sample return missions. In: *IEEE Aerospace Conference*, Big Sky, MT, March 5–12, 2016.

Turtle, E.P., Barnes, J.W., Trainer, M.G., Lorenz, R.D. (2017a). Dragonfly: In Situ Exploration of Titan's Organic Chemistry and Habitability, abstract #P53D-2667, American Geophysical Union, Fall Meeting 2017, San Francisco, CA.

Turtle, E.P., Barnes, J.W., Trainer, M.G., Lorenz, R.D., Hibbard, K.E., Adams, D., Bedini, P., Langelaan, J.W., and Zacny, K. (2017b). DRAGONFLY: Exploring Titan's pre-biotic chemistry and habitability. Lunar and Planetary Science Conference, Mar 20–24, 2017, The Woodlands, TX.

Ulamec, S. et al. (2007). Access to glacial and subglacial environments in the Solar System by melting probe technology. *Reviews in Environmental Science and Biotechnology* 6: 71–94. doi:10.1007/s11157-006-9108-x

Vendiola, V., Zacny, K., Morrison, P., Wang, A., Yaggi, B., Hattori, A., and Paz, A. (2018). Testing of the Planetary Volatiles Extractor (PVEx). In: *ASCE Earth and Space 2018 Conference*, Cleveland, OH, April 10–12, 2018.

Vrettos, C., Becker, A., Merz, K., and Witte, L. (2014). In: Gertsch, L.S., and Malla, R.B. (Eds.), *Earth & Space: Engineering for Extreme Environments*, ASCE Library, Reston, p. 10.

Wang, J., Cao, P., and Talalay, P. G. (2015). Comparison and analysis of subglacial bedrock core drilling technology in the polar regions. *Polar Science* 9: 208–220.

Winebrenner, D.P. and Elam, W.T. (2020). A thermal melt probe system for extensive, low-cost instrument deployment within and beneath ice sheets. *Annals of Glaciology*, in preparation.

Wippermann, T. Hudson, T.L., Spohn, T. et al. (2020). Penetration and performance testing of the HP³ Mole for the InSight Mars mission. *Planetary and Space Science* 81: 104780. doi:10.1016/j.pss.2019.104780 ISSN 0032-0633

Yano et al. (2002). Asteroidal surface sampling by the MUSES-C spacecraft. *Asteroids, Comets, Meteors* 500: 103–106.

Zacny, K. and Spring, J. (2013). Modeling tool and vibratory scoop for excavation, e-NTR #: 1369942214, Report Date: May 30, 2013.

Zacny, K. et al. (2004). Lunar soil extraction using flow of gas. In: *(RASC-AL) Conference*, Cocoa Beach, FL, April 28–May 1, 2004.

Zacny, K., Glaser, D., Bartlett, P., Davis, K., and Wilson, J. (2006). Test results of core drilling in simulated ice-bound lunar regolith for the subsurface access system of the Construction & Resource Utilization eXplorer (CRUX) Project. In: *10th International Conference on Engineering, Construction, and Operations in Challenging Environments, Earth & Space 2006 Conference*, League City, TX, March 5–8, 2006.

Zacny, K., Bar-Cohen, Y., Brennan, M., Briggs, G., Cooper, G., Davis, K., Dolgin, B., Glaser, D., Glass, B., Gorevan, S., Guerrero, J., McKay, C., Paulsen, G., Stanley, S., and Stoker, C. (2008a). Drilling systems for extraterrestrial subsurface exploration. *Astrobiology Journal*, 8(3): 665–706, doi:10.1089/ast.2007.0179

Zacny, K. et al. (2008b). Pneumatic excavator and regolith transport system for lunar ISRU and construction. In: *AIAA Space 2008 Conference and Exposition*, San Diego, CA.

Zacny, K., Mueller, R., Craft, J., Wilson, J., and Chu, P. (2009). Percussive digging approach to lunar excavation and mining. In: *Annual Meeting of LEAG and SRR*, Houston, TX, November 16–19, 2009.

Zacny, K., Paulsen, G., and Glass, B. (2010a). Field testing of planetary drill in the Arctic. In: *AIAA Space 2010*, AIAA-2010-8701, Anaheim, CA, August 31–September 2, 2010.

Zacny, K., Craft, J., Hedlund, M., Chu, P., Galloway, G., and Mueller, R. (2010b). Investigating the efficiency of pneumatic transfer of JSC-1A lunar regolith simulant in vacuum and lunar gravity during parabolic flights. In: *AIAA Space 2010*, AIAA-2010-8702, Anaheim, CA, August 31–September 2, 2010.

Zacny, K., Mueller, R., Paulsen, G., Chu, P., and Craft, J. (2012a). The ultimate lunar prospecting rover utilizing a drill, pneumatic and percussive excavator, and the gas jet trencher. In: *AIAA Space 2012*, Pasadena, CA, September 11–13, 2012.

Zacny, K., Paulsen, G. Chu, P. Avanesyan, A., Craft, J., and Szwarc, T. (2012b). Mars drill for the Mars sample return mission with a brushing and abrading bit, regolith and powder bit, core PreView bit and a coring bit. In: *IEEE Aerospace Conference*, Big Sky, MT, March 4–10, 2012.

Zacny, K. et al. (2013a). Chapter 10: Asteroids: Anchoring and sample acquisition approaches in support of science, exploration, and in situ resource utilization. In: Badescu, V. (Ed.), *Asteroids: Prospective Energy and Material Resources*, Springer, Berlin.

Zacny, K., Paulsen, G., Chu, P., Hedlund, M., Spring, J., Osborne, L., Matthews, J., Zarzhitsky, D., Nesnas, I.A., Szwarc, T., and Indyk, S.l. (2013b). Axel rover NanoDrill and PowderDrill: Acquisition of cores, regolith and powder from steep walls. In: *2013 IEEE Aerospace Conference*, Big Sky, MT, March 2–9, 2013.

Zacny, K., Paulsen, G., McKay, C.P., Glass, B., Dave, A.', Davila, A., Marinova, M., Mellerowicz, B., Heldmann, J., Stoker, C., Cabrol, N., Hedlund, M., and Craft, J. (2013c). Reaching 1 m deep on Mars: The Icebreaker drill. *Astrobiology* 13: 1166–1198, doi:10.1089/ast.2013.1038

Zacny, K. et al. (2013d). Wireline deep drill for exploration of Mars, Europa, and Enceladus. In: *Aerospace Conference, 2013 IEEE*, Big Sky, MT.

Zacny, K., Nagihara, S., Hedlund, M., Paulsen, G., Shasho, J., Mumm, E., Kumar, N., Szwarc, T., Chu, P., Craft, J., Taylor, P., and Milam, M. (2013e). Pneumatic and percussive penetration approaches for heat flow probe emplacement on robotic lunar missions. *Earth, Moon, and Planets* 111: 47–77.

Zacny, K., Chu, P., Davis, K., Paulsen, G., and Craft, J. (2014a). Mars 2020 sample acquisition and caching technologies and architectures. In: *IEEE Aerospace Conference*, Big Sky MT, March 3–7, 2014.

Zacny, K., Mueller, R.P., Ebert, T., Dupuis, M., Mumm, E., Neal, D., Spring, J., Paulsen, G., Chu, P., Mellerowicz, B., and Hedlund, M. (2014b). MicroDrill sample acquisition system for small class exploration spacecrafts. In: *14th ASCE International Conference on Engineering, Science, Construction and Operations in Challenging Environments*, St. Louis, MO, October 27–29, 2014.

Zacny, K., Spring, J., Paulsen, G., Ford, S., Chu, P., Kondos, S. (2015a). Chapter 8: Pneumatic drilling and excavation in support of Venus science and exploration. In: Badescu, V. and Zacny, K. (Eds.), *Inner Solar System: Prospective Energy and Material Resources*, Springer, Cham.

Zacny, K., Paulsen, G., Yaggi, B., Wettergreen, D., and Cabrol, N.A. (2015b). Life in the Atacama (LITA) drill and sample delivery system: Results from the field campaigns [#7020]. In: *AbSciCon*, Chicago, IL, June 15–19, 2015.

Zacny, K. et al. (2016a). Chapter 10: Drilling and breaking ice. In: Bar-Cohen, Y. (Ed.), *Low Temperature Materials and Mechanisms*, CRC Press, Boca Raton, FL.

Zacny, K., Luczek, K., Paz, A., and Hedlund, M. (2016b). Planetary Volatiles Extractor (PVEx) for In Situ Resource Utilization (ISRU). In: *ASCE Earth and Space Conference*, Orlando, FL, April 11–15, 2016.

Zacny, K., Indyk, S., Spring, J., Hyman, C., Chu, P., Thomas, T., Vendiola, V., Fitzgerald, Z., Lu, K., Luczek, K., Mueller, K., Paulsen, G., Ridilla, A., and Chow, K. (2017a). Air Dust Removal Tool (AirDRT). Lunar and Planetary Science Conference Mar 20–24, 2017, The Woodlands, TX.

Zacny, K., Rehnmark, F., Hall, J., Cloninger, E., Hyman, C., Kriechbaum, K., Melko, J., Rabinovitch, J., Wilcox, B., Lambert, J., Mumm, E., Paulsen, G., Vendiola, V., Chow, K., and Traeden, N. (2017b). Development of Venus drill. In: *IEEE Aerospace Conference*, Big Sky, MT, March 4–11, 2017.

Zacny, K., van Susante, P., Putzig, T., Hecht, M., and Sabahi, D. (2018a). RedWater: Extraction of water from Mars ice deposits. In: *Space Resources Roundtable*, Colorado School of Mines, Golden, CO, June 12–14, 2018.

Zacny, K. et al. (2018b). SLUSH: Europa hybrid deep drill. In: *IEEE Aerospace Conference*, Big Sky, MT, March 1–5, 2018.

Zacny, K. et al. (2018c). Development of a deep drill system with integrated deep UV/Raman spectrometer for Mars and Europa. In: *AIAA Space Conference*, Orlando, FL.

Zacny, K., Lorenz, R., Rehnmark, F., Costa, J., Sparta, J., Sanigepalli, V., Yen, B., Yu, D., Bailey, J., Bergman, D., and Hovik, W. (2019). Application of pneumatics in delivering samples to instruments on planetary missions. In *IEEE Aerospace Conference*, Big Sky, MT, March 2–9, 2019.

Zacny, K., Thomas, L., Paulsen, G., van Dyne, D., Lam, S., Williams, H., Sabahi, D., Ng, P., Chu, P., Spring, J., Satou, Y., Kato, H., Sawada, H., Usui, T., Fujimoto, M., Mueller, R., Zolensky, M., Statler, T., and Dudzinski, L. (2020). Pneumatic Sampler (P-Sampler) for the Martian Moons eXploration (MMX) mission. In: *IEEE Aerospace Conference*, Big Sky, MT, March 7–14, 2020.

Zimmerman, W. (2000). Extreme electronics for in situ robotic/sensing systems. In: *Aerospace Conference Proceedings, 2000 IEEE*, Vol. 7.

Zimmerman, W., Bonitz, R., and Feldman, J. (2001). Cryobot: An ice penetrating robotic vehicle for Mars and Europa. In: *2001 IEEE Aerospace Conference Proceedings* (Cat. No.01TH8542).

2 Novel Methods for Deep Ice Access on Planetary Bodies

William Stone, Vickie Siegel, Bart Hogan, Kristof Richmond, Corey Hackley, John Harman, Chris Flesher, Alberto Lopez, Scott Lelievre, Krista Myers, and Nathan Wright

Stone Aerospace, Del Valle, TX

CONTENTS

2.1 INTRODUCTION

Under NASA funding several new ice penetration technologies were developed at Stone Aerospace that represent significant departures from traditional approaches. The concepts discussed in this chapter are self-contained devices, known as cryobots, that are intended to be deployed by a planetary lander. Heretofore, terrestrial ice penetration systems have fallen into three major classes: mechanical drills (Vago et al. 2006, Weiss et al. 2008, Hsu 2010, Vasiliev et al. 2011, Zacny et al. 2013, Chu et al. 2014, Goodge and Severinghaus 2016), hot water drills (Humphrey and Echelmeyer 1990, Thorsteinsson et al. 2008, Benson et al. 2014, Rack 2016), and passive melt probes (Philberth 1962, Zimmerman 2000, Ulamec et al. 2007, Kaufmann et al. 2009, Biele et al. 2011, Lorenz 2012, Winebrenner et al. 2013, Dachwald et al. 2014, Talalay et al. 2014, Wirtz and Hildebrandt 2016). We first describe new test results from a fourth approach, Direct Laser Penetration (DLP), using a specific wavelength laser as the power source. The second half of this chapter will focus on current work in planetary mission closed-cycle hot water drilling technology (CCHWD) that will down-convert to a less effective, but still functional, thermal passive probe for the purpose of addressing the "Starting Problem" (Stone et al. 2018) on planetary bodies with no atmosphere. Each of these approaches fills a specific niche in planetary mission ice penetration. The direct laser approach is well suited for lightweight, short range penetrations (10–100 m depth) from a lander with the power source being based on the lander; the CCHWD approach is better suited for very deep penetrations (10–30 km) where the power source, presumed to be a 50–100-kW fission reactor, would be carried by the cryobot descent vehicle.

2.2 DIRECT LASER PENETRATION (DLP)

The VALKYRIE project (Stone et al. 2014, 2018) investigated the idea of powering a CCHWD cryobot using a vehicle-deployed fiber optic thread that was connected to a surface-based (or lander-based) high power fiber laser (Stone and Hogan 2015, 2016). The optical power was channeled to an internal beam dump where it heated the inside surface of a heat exchanger, which in turn heated meltwater being drawn in from the nose of the vehicle. Successful field use of this concept took place on the Matanuska glacier in Alaska in both 2014 and 2015 (Stone et al. 2018) and demonstrated penetration rates that closely followed early theoretical predictions based on Aamot theory (Aamot 1967). It was during this time that one of the authors (Hogan) became aware of a curious dichotomy in the energy absorption spectra of light in liquid water and ice. At approximately 1070 nm (near-IR) light is preferentially absorbed in a volume of ice ahead of the probe with only a small amount being absorbed in the liquid water film below the nose of the probe. The VALKYRIE probe was only using the optical beam for heat conversion: the beam was captured internally in a beam dump.

An alternative vehicle architecture was developed in which the laser beam exiting the transmission fiber was expanded, collimated, and then re-focused using a string of antireflective coated lenses to re-project the beam out of the nose of the cryobot. Under this scenario, ice ahead of the vehicle absorbs the optical energy directly, melts, and the probe descends under its own weight. The approach is remarkably efficient because the energy is coupled directly into the ice with negligible sidewall heating loss. In short, it becomes possible to put the energy right where it is needed and not waste any heat where it is not needed. This is not true of wavelengths outside of the anomalous 1070 nm wavelength; the use of a CO_2 long wavelength laser at high-power levels can flash the water, leading to a potential overpressure pulse that may damage the vehicle (Sakurai et al. 2015).

An initial test of the DLP concept (Figure 2.1) was conducted in May 2016 at Stone Aerospace using 5 kW of laser light from an IPG Photonics YLS-5000 fiber laser. In this test a 3-cm diameter tube with scored depth graduations held the end optics (Stone et al. 2018) and the entire package descended into the ice when the laser was fired. The ice blocks used for these tests were relatively "warm" ($-10°C$) such that the melt hole remained filled with liquid water during the test after passage of the probe. Eventually the beam punched a hole in the bottom of the ice block and the water surrounding the probe drained and the test was terminated (Figure 2.2). The melt hole, after removal of the probe, provides a compelling visual verification of how the beam energy was being absorbed into the ice at points well below the probe. A wide range of power levels and focal point distances from the front of the probe were investigated. From these tests thermal models were validated and used to predict the performance of such an ice penetrator in the Ross Ice Shelf (Antarctica) for various probe diameters and laser power levels (Figure 2.3).

The initiative to develop a lightweight Europa lander mission (Hand et al. 2017) led to further research into concepts for mid-range (10–100 m depth) ice penetration using a low-power board-level laser mounted on the lander and a miniature probe that used the DLP principle (Stone 2016, Hogan and Stone 2018). Figure 2.4 shows a close-up solid model cutaway of the ARCHIMEDES DLP probe. Importantly, for a planetary probe the vehicle must be able to free fall into the ice with no rigid supporting elements connecting it to the lander. This means that the vehicle itself has to carry and deploy the fiber umbilical as it descends. ARCHIMEDES combines optical power for ice penetration and a dual fiber fluorescence spectrometer for biomarker detection and measures 3 cm in diameter. Figure 2.5 shows an artist's concept of an ARCHIMEDES-class probe beginning ice penetration on Europa.

In 2017, operating under NASA COLDTech program funding, an experimental program was initiated to test the ARCHIMEDES DLP concept under Europa surface conditions. A series of test probes (Figure 2.6) were designed to carry up to 100 m of optical power fiber (Figure 2.7). All test vehicles were 3.2 cm in diameter and of varying length as will be explained later. The housing material was copper, primarily for heat transfer but also for negative buoyant mass for gravity-driven descent into the ice.

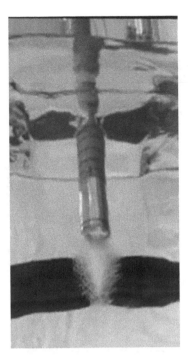

FIGURE 2.1 Video frame of the first Direct Laser Probe test conducted at Stone Aerospace in the summer of 2016. The near infrared beam is visible to the CCD video camera chip but not to the naked eye. The length of the beam penetration can be changed optically to move the focal point deeper into the ice. Ice "crazing" can be seen adjacent the beam.

The power fiber (OFS HCXtreme 365-22) consists of a 365-micron diameter fused silica core; 400-micron cladding; 430-micron hard-coating diameter; and 550-micron buffer diameter (overall OD, blue coating). The fiber is terminated by fusion splicing to a 1 mm diameter cylindrical fused silica end cap (OFS HCXtreme 940-22) with a polished exit surface with antireflective coating and a numerical aperture of 0.22. This was sealed into a copper carrier for heat dissipation into the vehicle and also provided a convenient means for establishing a waterproof o-ring sealing surface so that the interior optics cavity of the vehicle remained watertight.

Two versions of the ARCHIMEDES probe were developed. For the "short" probe (Figure 2.8), the vehicle measured 3.2 cm in diameter and 6.8 cm in length and had a 12.7 mm length internal beam dump. For the "long" probe (Figure 2.8) the vehicle measured 3.2 cm in diameter and 12.5 cm in length and had a 70 mm length internal beam dump. While the numerical aperture of the fused fiber endcap is fixed, the lens optics could be varied to achieve different effects. Initially we investigated three types of lens systems that produced the following beam patterns:

- **Divergent**: The beam expands from the optic system. This was the least efficient since as the beam diverges, power is deposited in the ice column outside the diameter of the probe.
- **Convergent**: The beam is focused at a point inside the ice some distance below the nose of the probe. This provided the best performance in general. The degree of convergence (focal length) is based on the power density (W/cm^2); the lower the power density the more convergent, the higher the power density the less convergent until you reach a collimated lens (parallel rays) at which point it becomes divergent and less efficient.
- **Collimated**: Best for very high power, as it spreads out the beam to limit power density (water does not boil as much) but also is not divergent, which is inefficient as described previously.

FIGURE 2.2   Direct Laser Probe melt hole in temperate ice (−10°C) following test and subsequent draining of the hole. The tapered hole below the cylindrical section shows the effect of focusing the beam ahead of the probe.

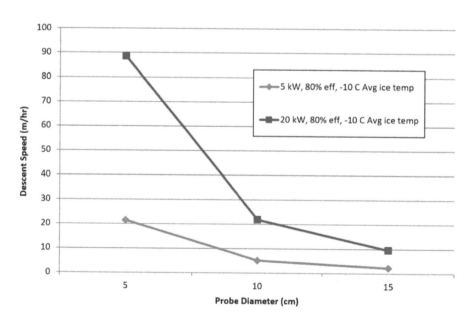

FIGURE 2.3   Predicted performance plots for Direct Laser Probe vehicles of various diameters at varying optical power levels for the Ross Ice Shelf, Antarctica.

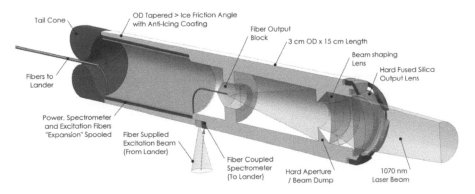

FIGURE 2.4   Cut-away solid model of the ARCHIMEDES Direct Laser Probe showing key features.

FIGURE 2.5   Artist's conception of ARCHIMEDES being deployed on the Europa Lander.

FIGURE 2.6   Planetary "short" Direct Laser Probe, 3.2 cm diameter × 6.8 cm length, test article used in the ARCHIMEDES project.

FIGURE 2.7   Close-up of the optical power fiber used for the ARCHIMEDES tests.

The precise angle (focal point distance) for best efficiency depends on probe diameter, laser power, and ice temperature. The initial DLP tests had focal depths of 6–10 vehicle diameters below the nose of the probe.

For the ARCHIMEDES tests under Europa surface conditions there was an additional factor to consider that is well known in the design of thermal passive probes (Harman et al. 2018). In very cold ice a portion of the energy input needed to descend goes to the nose of the probe to heat the ice and allow descent to begin. But as the vehicle descends, water behind the vehicle begins to refreeze. Unless the vehicle descends at an extremely fast rate, the trailing edge (stern) of the vehicle will freeze into the ice wall. Power must therefore be diverted along the length of the vehicle to heat the shell and prevent freeze in. Any more than the absolute minimum energy diversion needed to prevent the tail of the probe freezing into the ice is parasitic.

The short ARCHIMEDES probe (Figure 2.8, left image) was designed to channel 100% of the laser power into the ice beneath the probe, with no power going into sidewall heating. The long probe, on the other hand (Figure 2.8, right image), used optics and a deeper beam dump cavity such that 54% of the beam power was transferred into the copper shell. The effects of this are discussed later. Each ARCHIMEDES probe in the initial testing phase was suspended from a small diameter "launch rod" (Figure 2.8) that was used to guide the probe vertically into the ice at the start of the test.

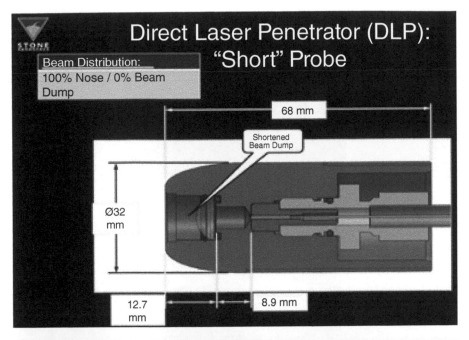

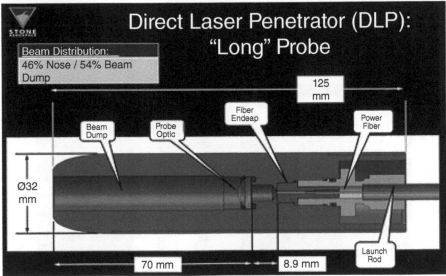

FIGURE 2.8   ARCHIMEDES "short" and "long" test probe geometries.

A parallel ground truth test series was developed for comparison with the DLP laser probe. This was provided in the form of a passive thermal probe with the same diameter as the DLP probes (Figures 2.9 and 2.10). Instead of the optical power chain used in DLP, the passive probe contained an electrical resistance cartridge heater inside its hollow shell. The heater was surrounded by a thermally conductive fluid to effectively transfer heat to the surrounding shell and probe nose.

In order to test ARCHIMEDES in a Europa-analog environment we designed and built a facility called *Europa Tower* (Figure 2.11) at the Stone Aerospace lab in Austin, Texas. *Europa Tower* is a tri-wall cryogenic vacuum chamber: the volume between the outer and first inner shell is evacuated to a hard vacuum. The volume between the middle and inner shell is filled with liquid nitrogen,

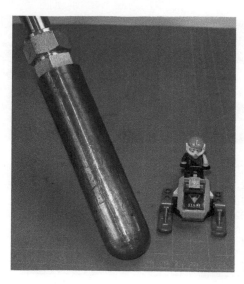

FIGURE 2.9 Close-up image of the resistively heated, passive ("hot penny") melt probe used as ground truth for the ARCHIMEDES DLP tests under Europa surface conditions.

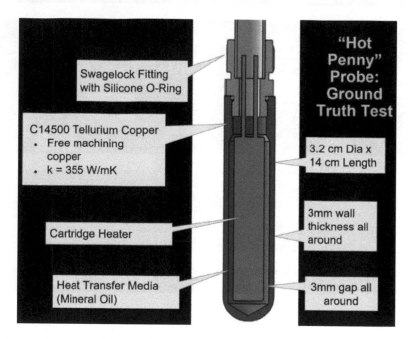

FIGURE 2.10 Section view of the passive probe used for ARCHIMEDES test validation. Passive ("hot penny") melt probe used as ground truth for DLP tests under Europa surface conditions.

while the interior of the chamber contains a large cylinder of frozen water measuring 0.75 m in diameter and 2.11 m in height. This allowed us to assess the probe's performance under environmental conditions similar to those on Europa's surface. The chamber maintains ice at a temperature of 83 K and holds a vacuum of 10^{-3} Torr, below the triple point of water, despite the generation of water vapor by the probe during a penetration test (vacuum pumps were designed to maintain this criteria).

FIGURE 2.11 The *Europa Tower* cryovac chamber developed at Stone Aerospace for simulating the surface conditions of Europa. Unlike all prior thermal-vac chambers, Europa Tower was designed around a cylinder of contained ice, 0.75 m diameter × 3 m tall, held at 83 K temperature and hard vacuum for the purpose of testing cryobot descent initiation from a lander.

Measured variables inside the chamber during a typical ARCHIMEDES test are shown in Figure 2.12. A few points are worth discussion. As energy is pumped into the probe (top red line graph) the chamber vacuum is gradually reduced from the sublimated ice that is released. As power is reduced, the vacuum pumps catch up. The spikes in reduced vacuum correlate to events when a significant amount of material was ejected from the hole. This only occurred with the laser probe tests and has been attributed to the beam creating a slush pocket beneath the nose. When the pocket becomes wide enough, the probe falls into it, displacing the slush out in a burst. This was only strictly true for the short probe where 100% of the beam energy was focused into the ice. The material ejected from the hole during these incidents appeared more like small graupel than the micro-fine dust appearance of condensed sublimate. The wall temperature of the ice cylinder remained near 83 K during the tests but thermocouples near the cylinder centerline (dashed line in the bottom graph, Figure 2.12) measured by sensor A-1 in Figure 2.13 show the ice temperature rising as the probe passes. Such sensors can be used to track vertical position in the otherwise opaque ice during a test.

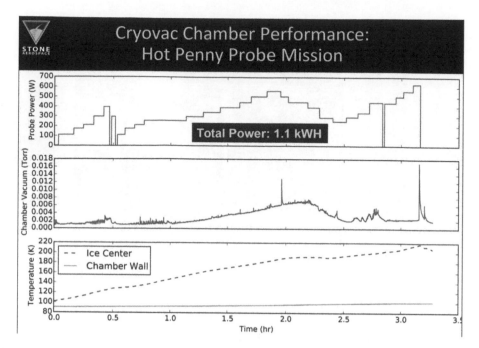

FIGURE 2.12 Typical performance of the Europa Tower facility during an actual cryobot test. When viewing these data keep in mind that the cryobot is injecting heat into the system and ice is being sublimated to gas. The ice center temperature is obtained from temperature sensor #2 while the Chamber Wall temperatures is obtained from sensor #5 (see Figure 2.13).

2.2.1 PASSIVE PROBE RESULTS

Previous tests of passive thermal probes in warmer ice under vacuum (Ulamec et al. 2007, Biele et al. 2011) suggested that there may be a problem with radiative heat loss precluding entry of the probe. Our tests (Figure 2.14) showed wide conical crater generation initially, similar to Biele et al. (2011). However, our probe, working in much colder ice, continued downward at an essentially linear rate (a velocity of 0.09 m/hour at 200 W power level and a progress metric of 24 W-hours/cm). Numerical finite element thermal modeling agreed well with the steady state downward motion of the probe. The power for this test series was limited to 200 W by the resistive cartridge heater that was used.

The temporary stalled descent in the middle of the Figure 2.15 was not related to vehicle performance, but rather due to the support rod binding at its vacuum pass-through into the chamber. Once this was freed the test continued. The test was terminated when the sublimate rising from the probe eventually sealed the hole and bound the support guide tube. In later testing (January 2020) the power tether was spooled from the probe itself as it descended and the test was terminated at the full depth of the chamber. Perhaps the most important single fact to arise from this testing series was that a passive thermal probe, in clean ice, will initiate descent under a cold start at the surface of Europa and will continue to descend as long as it has sufficient power.

2.2.2 ARCHIMEDES: LONG AND SHORT PROBE TESTING

Figure 2.16 shows a time lapse sequence of images taken inside the Europa Tower chamber with a 3.2 cm diameter × 12.5 cm ("long") laser-powered ARCHIMEDES probe descending into the ice from a cold start. Figure 2.16a and b, respectively, show the initiation of optical power delivery and

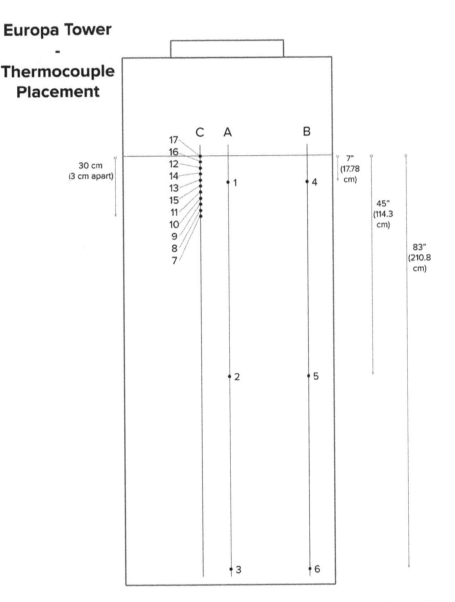

Europa Tower - Thermocouple Placement

FIGURE 2.13 Locations of in-ice temperature sensors used for the Europa Tower tests of the ARCHIMEDES DLP vehicle.

the eventual progress of heating from the nose of the vehicle backwards, stripping condensate frost along the way. Figure 2.16c shows the probe half descended into the ice surface and the beginning of a frost cone forming around the vehicle where sublimate is freezing out. In Figure 2.16d the probe is fully embedded in the ice and a cone of ejecta material is seen around the rear of the probe. In Figure 2.16e the probe has descended approximate 2 cm further before the hole closes shut and freezes onto the tail support tube, freezing the probe in place. This provides direct evidence that hole closure due to sublimate refreezing on the wall of the borehole not only is a real phenomenon, but that it happens rather rapidly. This is good news, for it suggests that a passive thermal probe can be converted to a more efficient CCHWD drill within a relatively short distance below the surface where, when the hole is closed off, the heat can generate vapor pressure allowing formation of liquid water needed to operate the drill.

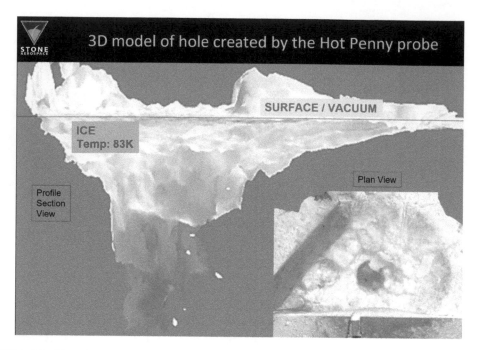

FIGURE 2.14 Top view and cross section of the penetration hole produced by the passive thermal ("hot penny") probe.

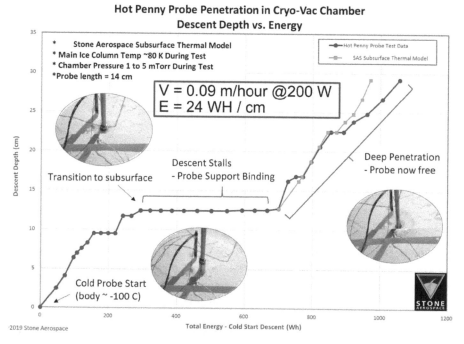

FIGURE 2.15 Passive thermal ("hot penny") probe penetration performance under Europa surface conditions.

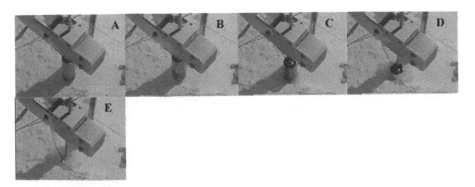

FIGURE 2.16 Key stages of the descent of the "long" DLP ARCHIMEDES probe under Europa surface conditions: (a) probe nose heat up; (b) full length of probe shedding final ice shell; (c) half vehicle length into the ice surface; (d) full vehicle length into the ice; and (e) termination of test when support rod is frozen into the ice from condensed sublimate that closes the hole.

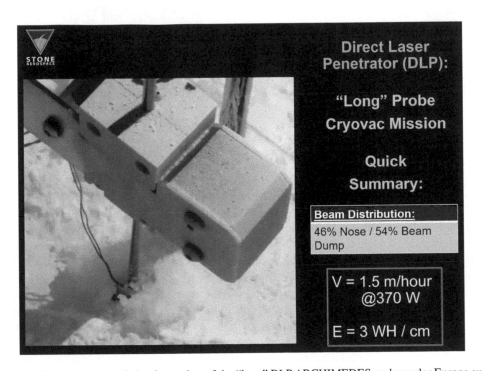

FIGURE 2.17 Summary statistics for testing of the "long" DLP ARCHIMEDES probe under Europa surface conditions.

Figure 2.17 shows an image of the termination of the long probe ARCHIMEDES test along the performance metrics: a steady-state descent velocity of 1.5 m/hour was achieved at a power level of 370 W, corresponding to an energy expense of 3 W-hours per cm of depth. Figure 2.18 shows the termination of the short probe ARCHIMEDES test. Unlike the long probe, which intentionally diverted heat into the rear part of the probe to prevent hull refreeze and binding, the short probe was designed such that 100% of the beam power went directly into the ice below the nose. The rear end of the probe froze into the ice (not the support tube) at the point where it was just starting to disappear below the surface. The penetration efficiency for the short DLP probe was 2.5 W-hours/cm of

FIGURE 2.18 Summary statistics for testing of the "short" DLP ARCHIMEDES probe under Europa surface conditions.

depth while the recorded velocity was 13 m/hour. Some discussion is in order here. For a lander that is using battery power, every watt-hour of energy used is precious, so the metric of watt-hours/cm of penetration is the proper metric for comparison as opposed to speed, which can be misleading. The long DLP probe diverted over half of its power to sidewall heating, and in that sense, it was a hybrid vehicle, half DLP and half passive probe. The short probe, in sending all of its power optically into the ice below it, took some time to heat a volume of ice below it, at which point it turned to slush and the vehicle essentially fell into the hole, with the slush burping out, assisted by the vacuum. At that point the sublimate closed the hole and froze the support rod into place. The long probe, because it was dumping energy into the surrounding ice, went deeper before the sublimate closed the hole and froze the support rod in place. Both probes successfully went subsurface. Even equipped with a power tether spooler on the vehicle (no support rod), these tests confirm that a portion of the thermal power will need to be diverted to the sidewalls of the vehicle to sustain descent.

Figure 2.19 provides a performance comparison between the long ARCHIMEDES DLP probe and the traditional electrical resistance heated passive probe. Both tests were eventually halted due to sublimate refreeze trapping the test support rods (in later tests this limitation was removed since the vehicle paid out its own power tether as it descended). We have found that the metric of watt-hours per cm of descent provides a useful means of comparison as it is independent of the absolute power level used. The clear implication from this figure is that the DLP probe was 8× more efficient than a passive thermal probe under the Europan surface conditions of this test. However, equally important is that both probe technologies succeeded in descending into the ice under Europan surface conditions.

Figure 2.20 shows a comparison between the long and short DLP probe test articles, again plotting absolute descent depth as a function of the total energy. The mechanism of descent appears to be different, with the long probe (which had heated sidewalls) proceeded on a continuous descent—faster but similar to the passive probe—while the short DLP probe took time to heat a larger volume

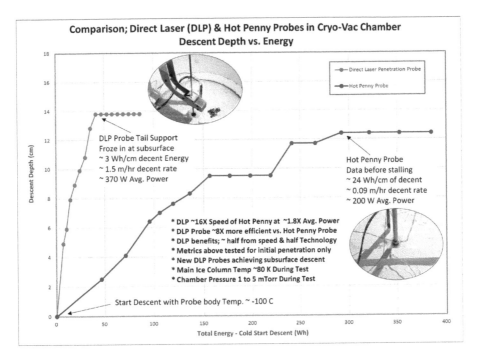

FIGURE 2.19 Comparison performance of "long" DLP ARCHIMEDES probe (blue) and "hot penny" passive probe (red) under Europa surface conditions. Critical metrics are watt-hours per centimeter of descent and maximum descent speed (meters per hour).

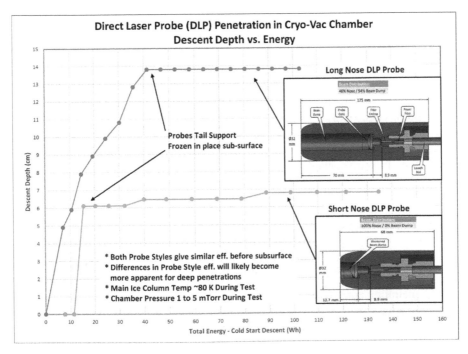

FIGURE 2.20 Comparison performance of "long" DLP ARCHIMEDES probe (blue) and "short" DLP ARCHIMEDES probe (green) under Europa surface conditions. Critical metrics are watt-hours per centimeter of descent and maximum descent speed (meters per hour). *Source:* Courtesy of Stone et al. (2018).

of ice beneath the nose and once that turned to slush it fell rapidly. Similar pulsed descent behavior was observed in the VALKYRIE CCHWD vehicle during field tests in Alaska.

2.2.3 DISCUSSION: DLP

DLP may present a unique solution to the Starting Problem for cryobots attempting to initiate travel into the ice shell of Europa or other Ocean Worlds. The Starting Problem is where the cryobot must begin descent at the lander in hard vacuum at approximately 90–100 K temperature. The lack of atmospheric pressure precludes the formation of liquid water. Therefore, energy transferred into the ice will cause it to sublimate, not melt. Fortunately, as shown in the tests described earlier, DLP has no problems operating in vacuum. The beam will initially cause a hole to form with the ice sublimating to space, but after a short descent distance the sublimated ice will refreeze on the upper sidewalls of the hole and will eventually close it off, allowing vapor pressure to build. Once that happens the DLP probe will proceed downward in a fashion similar to how it would behave on Earth (e.g. Figure 2.1).

For planetary (e.g. lander or rover) missions to the Martian polar caps or to Ocean Worlds, there appears to be a niche that DLP fills well: for short to intermediate penetrations (10–100 m depths) where the power source remains on the surface and the DLP-powered vehicle descends, paying out its power tether as it goes. The reason for this limitation, and why we are not recommending DLP for Ocean Worlds access through tens of kilometers of ice, is because of the methods by which the power is generated and transmitted. For a battery-powered lander, conversion from electrical energy used by the laser source to optical energy in the laser beam output is slightly over 50% efficient with current technology. There will be further power losses through transmission through the fiber (Stone et al. 2018) that amount to approximately 12% attenuation per kilometer. In 2010 tests were conducted relating to high power optical energy transmission through silica fibers (Stone et al. 2014, 2018). In particular, we successfully transmitted 11 kW of optical power through 1.1 km of such fiber, validating the loss due to Rayleigh scattering in the fiber. Such attenuation must be taken into account in mission planning, but it is minimal for short penetrations.

Because the power to move a cryobot downwards in an ice column is proportional to the square of the vehicle diameter, there is strong motivation to maintain the vehicle diameter to the minimum possible in consideration of a fixed lander power supply (i.e. battery) on planetary missions. Preliminary test results for fiber-spooled vehicles as part of the ARCHIMEDES project show that up to 100 m of power fiber can be carried by a 3.2 cm diameter probe without significant power loss due to fiber bending. In fact, silica fiber toughness and bend tolerance actually improve at lower temperatures (Stone et al. 2018).

The compact and relatively simple DLP design means that the probe can be readily sterilized prior to deployment. Recent tests (Schuler et al. 2019) have shown that pre-sterilized cryobots can be designed to limit forward contamination as the vehicle descends into the ice column, thus satisfying planetary protection protocols without complex procedures.

We have only tested DLP in pure water ice at this point. It is possible that impurities may pose problems. Planning is currently underway to perform tests of DLP performance in briny ice as part of the PROMETHEUS project.

2.2.4 DLP APPLICATIONS IN POLAR RESEARCH ON EARTH

Direct Laser Penetration could be used in terrestrial glaciated environments as a clean-access tool to investigate thick ice sheets, ice shelves, and subglacial water bodies. Theoretically, the approach can be scaled to work through more than 4 km of ice (e.g., the full thickness of the Antarctic ice sheet) at very high-power levels. At high power (100 kW) DLP could also be used to reach deep subglacial lakes in very short times, theoretically 16 h for 4 km with a 10 cm diameter vehicle under Antarctic ice temperature profiles. Additionally, it could be used to transit through ice shelves and install instruments in subglacial water bodies or to rapidly install fiber sensor arrays.

The reason for this difference in capability, vis-à-vis planetary mission limitation, is because, in Antarctica, large diesel generators and much larger lasers can be used on the surface, carried on tractor-towed sleds. Importantly, the direct laser penetrator would have a much smaller logistics footprint than a typical deep ice mechanical or hot water drilling rig. The field DLP equipment would mainly consist of a laser source unit (currently approximately 1 m × 1 m × 2 m), and a 50-kW generator. Consequently, used in the field, DLP would be easily deployable and offer extremely rapid descent rates.

2.3   A PLANETARY CLOSED-CYCLE HOT WATER DRILL: THOR

On Earth, in polar regions, hot water drills have proven to currently be the most efficient means of penetrating deep ice (Benson et al. 2014, Rack 2016) and they are robust in the face of substantial insoluble debris (Thorsteinsson et al. 2008). The laser-powered VALKYRIE probe (Stone et al. 2014, 2018) was an effort to take the best aspects of hot water drills and create an efficient closed-cycle hot water drill (CCHWD) that could be encapsulated into a self-contained cryobot. The power levels used—5 kW—were sufficient to prove out, in field tests, theoretical models of descent rates as well as to validate the significant advantage of CCHWD over a passive melt probe, based on the previously described metric of watt-hours per centimeter of descent. For planetary penetration of ultra-cold ice, a much larger power source is needed to ensure downward progress. For a viable flight vehicle—in the 25–35 cm diameter range to enable inclusion of ice penetration machinery, power plant, science payload, and communications links—a power source on the order of 50 kW is required (Stone et al. 2018). The successfully hot-fired "Kilopower" micro fission reactor developed by NASA Glenn (NASA 2018) provides a nominal 43 kW of thermal power and 1 kW of electrical power and is a template for a viable flight cryobot power source. Using this as a design requirement, Stone Aerospace designed and fabricated a full scale CCHWD cryobot called THOR.

The goal of the THOR (Thermal High-voltage Ocean-penetrator Research platform) project is to field-test a full-scale flight-compatible prototype for a new class of exploration vehicle which combines elements of an ice-penetrator (cryobot) and an ocean-profiler (sonde) for future Ocean Worlds exploration. The full THOR vehicle has been designed as a closed-cycle hot water drill (CCHWD) cryobot able to penetrate a minimum of 500 m (limited only by time considerations and depth limits imposed by accessible glaciers in the United States) of ice and able to transition to an actively spooled sonde upon breakthrough into a subglacial lake. We refer to this architecture as an Ice-Ocean Penetrator (IOP). The vehicle will actively spool a tether, similar to that initially designed for the SPINDLE project (Figure 2.21), providing a vertical control capability enabling controlled breakthrough into the water body, active profiling of the lake chemistry, and re-spooling and recovery to the surface.

Initial field testing of the CCHWD section of THOR (Figure 2.22), is scheduled for summer of 2021 on the Matanuska Glacier in Alaska. While Ocean World applications will require nuclear power, testing of prototype ice penetrators on Earth needs to be accomplished with an alternate power source. For THOR we developed a high-voltage electrical power transmission system as shown in Figures 2.23, 2.24, and 2.25. The pallet-mounted power system shown in Figure 2.25, along with the 50-kW single phase AC generator, is designed to be helicopter transportable. The full THOR Phase II vehicle will carry a sampling subsystem as well as spatial, environmental, and life-detection sensors to characterize the ice and water it transits through. This platform and sensor suite will then be used to develop and test strategies and behaviors for autonomous exploration and sampling as the vehicle descends through an ice column and into and through a subglacial water body below.

2.4   THE THOR CCHWD

Figure 2.26 shows an internal schematic of the THOR CCHWD elements. At the nose of the THOR CCHWD vehicle (Figure 2.27) the hot water jet has two important functions. First, it increases the

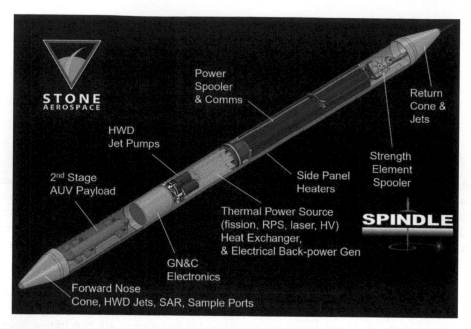

FIGURE 2.21 Computer solid model of the SPINDLE bi-directional cryobot. The vehicle was designed to descend to an Antarctic subglacial lake, release a small autonomous underwater vehicle, and, at the conclusion of the mission, melt its way back to the surface using reverse hot water jets and an onboard servo-controlled winch. The vehicle was never built, but some of its subsystems led to breakthrough designs for CCHWD cryobots.

heat transferred to the ice by increasing the surface area of the water pocket melted out in front of the vehicle and providing turbulent mixing within it. Secondly, the jet helps to break up sediment and other debris which can block standard passive thermal probes. The wall of the nose section also incorporates a heat exchanger loop which can be activated in place of the active jet when there is insufficient water in the borehole—such as at the start of penetration or for restarts in the event of a loss of water.

Behind the nose sits the vehicle pumping bay, containing the CCHWD heat transfer components which convert the high-voltage electrical power in the tether into hot water to be pumped out the nose. The upper half of the vehicle comprises the electronic and high-voltage electrical housings to convert power to lower voltages for on-board use, and to monitor and control the CCHWD system. The rear of the current Phase I vehicle terminates in an off-the-shelf oceanographic high-voltage cable providing power and spooling from the surface. In Phase II, the vehicle will be extended from this point with the addition of the science payload and on-board tether spool sections, in which the external tether will be replaced with a much smaller-diameter custom cable.

The CCHWD system forms the core of the THOR technologies developed for Phase I. The goal of the system is to enable THOR to achieve high penetration rates by using high-pressure water jets to rapidly transfer heat into the ice from a novel direct high-voltage heater system called HOTSHOT (Figures 2.28, 2.29, and 2.30). The HOTSHOT technology is a critical enabler for terrestrial cryobot demonstrators which are not able to use nuclear power sources. In order to reduce the size of the tether bringing power from the surface to the cryobot, high voltage is required. However, the use of standard electro-resistive water heating elements capable of operating at high voltage while having sufficient surface area to allow rapid heat transfer leads to an impossible geometry that is not compatible with fitting into a cryobot. The HOTSHOT heater developed for THOR employs a novel solution by passing high-voltage, low-current AC power through a moving conducting fluid. This creates resistive heating in the fluid with 100% efficiency without inducing electrolysis. The

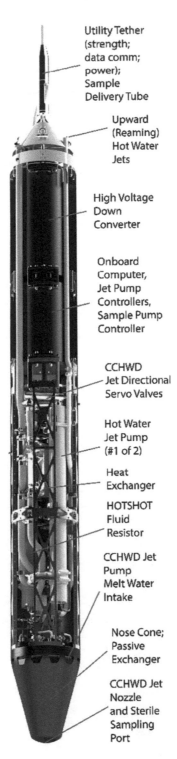

Utility Tether
(strength;
data comm;
power);
Sample
Delivery Tube

Upward
(Reaming)
Hot Water
Jets

High Voltage
Down
Converter

Onboard
Computer,
Jet Pump
Controllers,
Sample Pump
Controller

CCHWD
Jet Directional
Servo Valves

Hot Water
Jet Pump
(#1 of 2)

Heat
Exchanger

HOTSHOT
Fluid
Resistor

CCHWD Jet
Pump
Melt Water
Intake

Nose Cone;
Passive
Exchanger

CCHWD Jet
Nozzle
and Sterile
Sampling
Port

FIGURE 2.22 Computer solid model of THOR, a full-scale cryobot designed around the Kilopower 43-kW fission reactor. A surrogate power source, using 10-kV AC electrical power, is used to simulate the reactor. The vehicle can operate in both active CCHWD jetting and passive thermal probe modes.

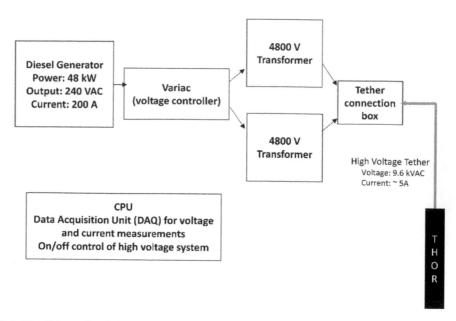

FIGURE 2.23 Schematic of the surface-based single phase AC high-voltage power system used for THOR deployments.

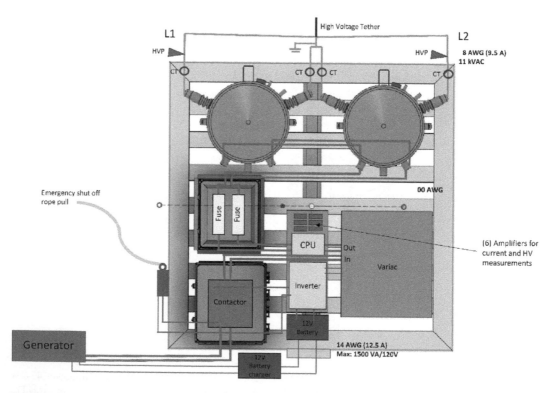

FIGURE 2.24 Annotated vertical view of a solid model of the THOR surface power generation and conversion system.

FIGURE 2.25 Assembled field transport pallet containing the THOR high-voltage power transmission system.

resistivity of the process fluid can be tuned over a wide range by controlling the concentration of polar molecules in the fluid. This novel high-voltage closed-cycle hot water drill architecture was developed in lab testing to TRL4 by Stone Aerospace in the NASA SPINDLE project and has been patented (Harman et al. 2018). Figure 2.26 presents the elements of the Phase I THOR CCHWD system, which transfers heat from HOTSHOT to the ice. Power densities as high as 600 kW/L have been achieved with the HOTSHOT design. Figure 2.29 shows the full scale HOTSHOT unit for THOR Phase I. All black-colored elements in Figure 2.29 are machined Delrin (a non-conducting acetal homopolymer). There are four annular electrodes, one at each end and two at the quarter points. The end contacts are neutral; the quarter point contacts are held at +5 kV and −5 kV, respectively. The entire stack is inserted into the non-conducting Delrin tube shown at left in Figure 2.29. This unit has been sized for 100-kW service. The overall system consists of two loops: a completely internal fully closed process loop to allow for tight control of the fluid resistivity in HOTSHOT, and the environmental loop which heats the surrounding ice via jetting.

The environmental or "jetting" loop transfers heat from the process loop into the ice to melt it. Meltwater is drawn in from the nose and passed through a bubble trap and sediment filter before entering the pump and heat exchanger, where it is heated by the HOTSHOT process loop. From here, the warmed water is jetted out of a nozzle at the front of the vehicle nose to melt the ice. Passive mode melting is implemented in a heat exchanger included in the vehicle nose (Figure 2.27). This brings warm water into contact with the aluminum vehicle skin to transfer heat into the ice via conduction without the forced convection of the meltwater induced by jetting. This mode of melting is less intense than the default jetting mode—in that less power can be transferred into the ice via conduction than via convection—so the power delivered to HOTSHOT will need to be scaled back accordingly when operating in this mode. The nose of the vehicle includes a forward camera and light to allow for the evaluation of the melt pocket and anything else in front of the vehicle—such as sediments, other obstacles, or the glacier bed. The nose is designed to be interchangeable, and the Phase II design is envisioned to add a port at the tip to transport pristine meltwater samples back to the science payload section.

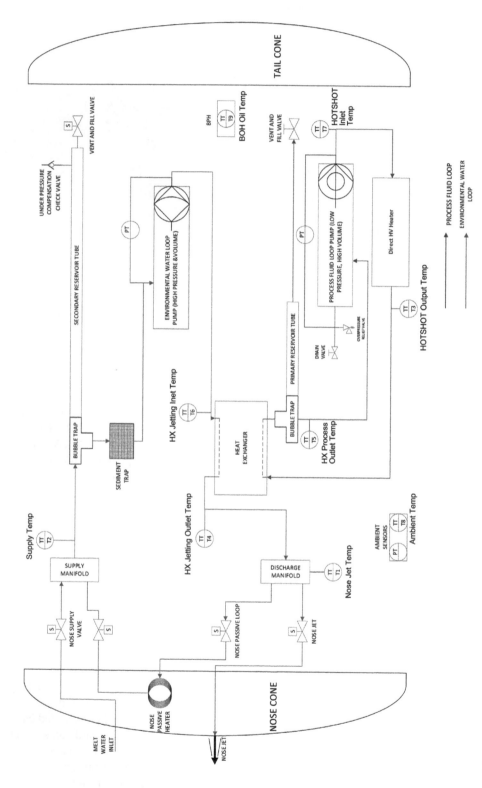

FIGURE 2.26 Internal schematic for the THOR cryobot showing fluid management details of the closed-cycle hot water drill.

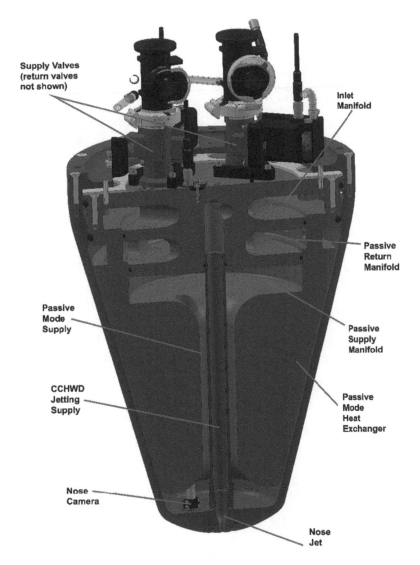

FIGURE 2.27 Solid model cross section of the THOR cryobot nose cone, showing CCHWD and passive thermal probe elements.

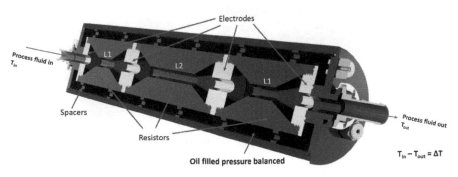

FIGURE 2.28 Solid model cross section of the HOTSHOT liquid resistor power dump for THOR.

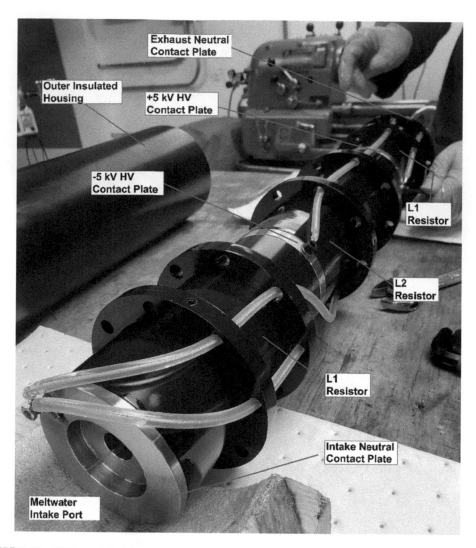

FIGURE 2.29 Solid model of the assembled HOTSHOT liquid resistor power dump for THOR.

FIGURE 2.30 The prototype 100-kW core assembly for the THOR HOTSHOT power dump.

The THOR vehicle structure (Figure 2.22) supports components during all phases of the THOR deployment: transport, assembly, and maintenance, lifting into and out of deployment position, penetration, and recovery. The THOR frame has been designed in a modular fashion to allow access to components in the field as well as to add on new sections in Phase II (such as the science payload and water sampling systems).

Pressure housings (rated to 1-km water depth) also protect THOR electronics while providing structural rigidity to the vehicle. The THOR pressure housings are custom-made for this unique application, providing the pressure and temperature environment for the electronics, and routing paths for all cabling and tubing along the vehicle surface. Care has been taken to ensure conducting contact at mating surfaces in the back power housing to provide shielding from electromagnetic noise emanating from the transformer and high-voltage electronics.

2.4.1 ONBOARD TETHER AND SPOOLERS

In an eventual planetary cryobot mission, water heating will almost certainly be accomplished using an onboard nuclear thermal power source. However, testing precursor technologies like THOR in analog environments on Earth means delivering power to the vehicle from the surface. As a cryobot descends, the water in the melt hole behind it will refreeze. Thus, any power and communication tethers cannot be fed down from the surface. Instead, it is necessary to have these lines spool out from the vehicle as it travels. THOR Phase II will use two servo-controlled tether spoolers: a dedicated strength spooler for descent and ascent and a dedicated power and communications spooler (Figures 2.21 and 2.31). This approach reduces overall vehicle diameter, reduces internal resistive heat buildup in the power spooler, and reduces differential thermal stress. While the use of two tethers in a single hole may appear to pose a challenge, the approach is not unprecedented. The IceCube project (Benson et al. 2014, Rack 2016) had multiple down-hole lines and did not have twisting problems in the course

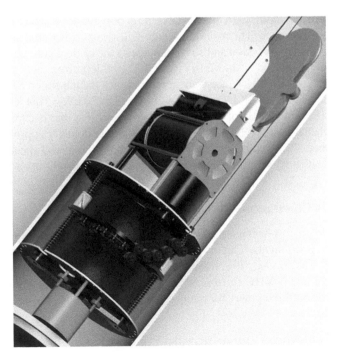

FIGURE 2.31 Computer solid model of THOR servo spoolers for (lower left) power and communications and (at center) the strength tether used to control vertical position.

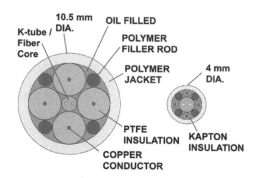

FIGURE 2.32 THOR high-voltage power tether (right side of image) compared with traditional ocean working ROV power tether rated for the same power (30 kW).

of drilling 86 2500-m-deep holes. To mitigate any twisting following breakthrough into open water, the IOP will contain tangential spin thrusters that will maintain vehicle yaw angle at all times.

Given a power tether stored in an onboard spooler, minimizing the tether volume becomes critical in order to maximize vehicle descent efficiency and achievable depth. Thus, high-voltage power delivery is desirable to reduce the required current and thus the wire diameter for a given amount of transmitted power. There are limits, however, because as voltage increases, the increasing thickness of insulation required to prevent arc-over between conductors in the spool begins to outweigh the gains from thinning conductors. In SPINDLE (Stone et al. 2018) tests were conducted on a wide variety of novel, low-volume wire insulators. This work produced a power tether design that THOR will implement. It consists of four Kapton-insulated conductors arranged in a square cross section. The center of the cross section is occupied by an armored K-tube fiber optic communications link. All interstitial spaces are filled with a low-viscosity, high-dielectric-strength oil to reduce friction during flexure and provide additional voltage standoff. A polyethylene jacket provides mechanical integrity and abrasion resistance. The resulting design will resist hydrostatic stresses and the stainless core will prevent ice compression and shear damage to the critical communications fibers. The dramatic size reduction achieved is shown in Figure 2.32: the left cross section shows a traditional ROV power cable designed for 30 kW, the right graphic shows the lab-tested SPINDLE design with custom-fabricated conductors and Kapton insulation (tested at 30 kW). This tether diameter reduction enables the design of a vehicle of reasonable deployment size that is capable of reaching any of Earth's deep subglacial lakes. THOR will validate this power delivery approach in the fully integrated Phase II cryobot. The strength spooler will use a compression- and shear-insensitive Spectra 4-mm recovery line with a tensile safety factor of 25 over vehicle weight in air.

2.4.2 Back Power Conversion and Low Voltage Power System

THOR requires 2 kW of low-voltage power onboard the IOP vehicle to run pumps, computers, winches, and science instruments. This power is derived from a series of ultra-high-permeability toroidal transformers. The 50 VAC, 40 A output of this transformer system is rectified and ultimately down-converted to 12 and 5 VDC to run onboard systems. An onboard lithium-ion battery and supercapacitor provide buffered power for pump starting and high-power events.

2.4.3 Vehicle Specifications and Anticipated Performance

Tables 2.1 and 2.2 provide THOR specifications for the Phase I and Phase II vehicles, respectively. As shown in Figure 2.22 the Phase I vehicle is focused on testing the power generation, transmission, and vehicle CCHWD systems. For Phase II (Table 2.2, and as envisioned in Figure 2.21) the science payload, water sampling system, and servo spoolers will be added as modular stages located

TABLE 2.1

Performance Specifications for the THOR Phase I Vehicle. This Vehicle is in Construction as of February 2020 and Will be Tested on the Matanuska Glacier in Alaska in Summer 2020

Specification	Value	Notes
Penetration depth	150 m	Determines component depth ratings, and tether length.
Round-trip transit time	3 days	Deployment window for full penetration depth, recovery.
Tether input power	48 kW	Electrical power supplied at surface.
Tether voltage	10 kV	AC line-to-line voltage supplied to tether.
Science Payload Size	0.0 m Ø x 0.0 m length	Payloads not integrated in Phase I.
Descent rate	4.5 m/hr @ 0 C 2.5 m/hr @ -50 C	Estimates based on current modeling.
Vehicle diameter	0.33 m	Results from penetration depth, speed, and available power.
Rear jetting	NO	A Phase II technology development.
On-board spooling	NO	A Phase II technology development.
Vehicle Length	3.3 m	CCHWD and backpower systems, only.

TABLE 2.2

Performance Specifications for the THOR Phase II Vehicle

Specification	Value	Notes
Penetration depth	1000 m (300 m ice, 700 m water)	Determines component depth ratings, and tether length.
Round-trip transit time	7 days	Deployment window for full penetration depth, recovery.
Tether input power	100 kW	Electrical power supplied at surface.
Tether voltage	10 kV	AC line-to-line voltage supplied to tether.
Science Payload Size	0.25 m Ø x 0.75 m length	Affects weight and length.
Descent and Ascent rates	6 m/hr @ 0 C 4 m/hr @ -50 C	Estimates based on current modeling.
Vehicle diameter	0.33 m	Results from penetration depth, speed, and available power.
Rear jetting	YES	Enables recovery from re-frozen borehole.
On-board spooling	YES	Allows sounding of sub-glacial waters, and recovery from re-frozen borehole.
Vehicle Length	6 m	All vehicle components, including science payload and on-board tether spooler.

behind the power down-conversion canister. These will increase the length of the vehicle to approximately 6 m, while maintaining a vehicle diameter of 35 cm.

When operated at the nominal Kilopower output of 43 kW (thermal), we have modeled the HOTSHOT-powered CCHWD flow temperatures as shown in Table 2.3. The term "primary" in Table 2.3 refers to temperatures in the closed-loop HOTSHOT fluid resistor system. "Secondary" refers to the high-pressure jetting loop, which is open loop and draws meltwater in from the nose of the vehicle. The power density in HOTSHOT can be increased or decreased by changing the pump flow rate. The operating points shown in Table 2.3 were selected to achieve the highest melting efficiency (kW-hours/meter of descent).

2.4.4 Discussion

THOR represents the highest power cryobot ever built. It will be field tested in the summer of 2021 on the Matanuska glacier in Alaska to validate performance metrics. Based on the VALKYRIE field tests and thermal model validations of those tests we are predicting flight vehicle performance as shown in Figures 2.33 and 2.34. We have bracketed a range of vehicle diameters to account for

TABLE 2.3

Flow parameters and output temperature simulations for the operation of the THOR CCHWD. Primary flows are input and output from the HOTSHOT liquid resistor power dump; secondary flows are meltwater input and jet output temperatures. Co-current and Counter-current refer to the directions of flow through the common heat exchanger that connects the HOTSHOT loop to the jetting loop

OPERATING MODE 1 Counter-Current flow

Power (kW)	43.5	primary out	22.8	C
Flow l/m (Process/jetting)	75/50	primary in	14.4	C
Oversurface %	0	second in	3	C
Jetting Inlet Temp (C)	3	second out	15.6	C
HX flow connection	counter-current			

OPERATING MODE 2 Co-Current flow

Power (kW)	43.5	primary out	26.1	C
Flow l/m (Process/jetting)	75/50	primary in	17.75	C
Oversurface %	0	second in	3	C
Jetting Inlet Temp (C)	3	second out	15.45	C
HX flow connection	co-current			

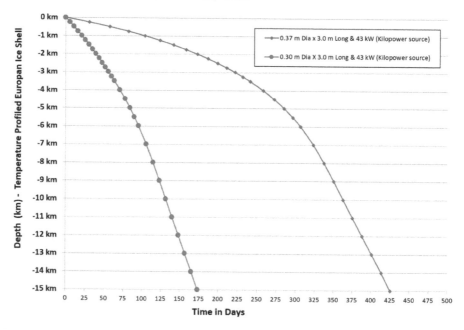

FIGURE 2.33 THOR ice penetration performance plot for 15-km-thick Europan ice shell using Kilopower fission reactor energy source. Temperature variance through the ice shell is accounted for in the calculations, starting at 77 K and reaching −3°C at ocean breakthrough.

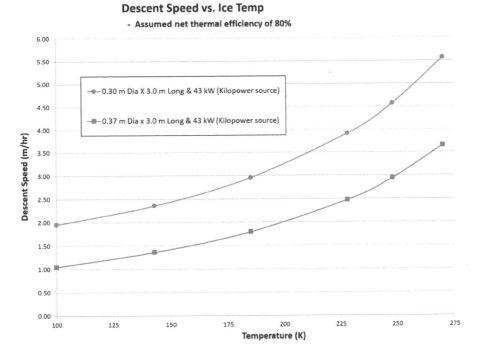

FIGURE 2.34 THOR performance metric: descent speed (m/hour) as a function of ice temperature for two different diameter vehicles.

variation in the ultimate design of the onboard reactor. The current diameter of THOR (35 cm) is controlled by the dimensions of the present version of the Kilopower reactor. Predicted descent rates vary from 1 to 2 m/hour at Europan surface conditions to as high as 3.6–5.6 m/hour at ocean breakthrough. The time-to-breakthrough from the start of melting, for a 15-km-thick Europan ice shell varies significantly as a function of the vehicle diameter—from 175 days for a 30-cm diameter vehicle to as much as 425 days for a 37-cm diameter vehicle. There is, therefore, a strong premium for a few extra centimeters. A planetary cryobot like THOR will still have to maintain a servo spooler or spoolers for communications and for a strength tether for vertical motion control—else the mission will terminate at the exact moment of ocean breakthrough. Environmental chamber tests of the THOR architecture are currently underway as part of the PROMETHEUS project under NASA SESAME program funding. This is a reduced-scale version of THOR that will operate in passive mode at startup (on the surface) and will switch to CCHWD once the hole closes from refreeze and sufficient vapor pressure has been developed to support the presence of liquid water. Assuming these tests are successful, the next phase in any mature cryobot development program would involve full-scale testing in deep ice to access a subglacial lake (most likely in Antarctica) and to validate thermal models and long-term performance of vehicle subsystems and components. We believe, however, on the basis of progress over the last 10 years in cryobot design and testing, that a subsurface mission to an Ocean World could be accomplished in approximately 7–10 years.

ACKNOWLEDGMENTS

ARCHIMEDES was supported by NASA via COLDTech Grant NNX17AF64G. The initial work on a DLP prototype was supported by the SPINDLE Project, NASA Grant NNX15AT32G. ARCHIMEDES and the DLP approach are one of the latest developments to come out of 10 years of ice-penetrator work at Stone Aerospace. THOR work is funded by NASA PSTAR Grant 80NSSC18K1738. PROMETHEUS is funded by NASA SESAME grant 80NSSC19K0612. We would like to thank Mary Voytek, Janice Buckner, and Ryan Stephan for their generous guidance in the conduct of these projects.

The authors would like to thank Rohit Bhartia, Photon Systems Inc., Covina, CA; Dara Sabahi, Honeybee Robotics, Altadena, CA; and Mircea Badescu, Jet Propulsion Laboratory/California Institute of Technology, Pasadena, CA, for reviewing this chapter and providing valuable technical comments and suggestions.

REFERENCES

Aamot, H.W.C., "Heat Transfer and Performance Analysis of a Thermal Probe for Glaciers," CRREL-TR-194, Cold Regions Research and Engineering Lab, Hanover, 1967.

Benson, T., Cherwinka, J. et al., "IceCube enhanced hot water drill functional description," *Annals of Glaciology* 55(68): 105–114, 2014. DOI: 10.3189/2014AoG68A032.

Biele, J., Ulamec, S., Hilchenbach, M., and Komle, N.I., "In situ analysis of Europa ices by short-range melting probes," *Advances in Space Research* 48: 755–763, 2011. DOI: 10.1016/J.ASR.2010.02.029.

Chu, P., Spring, J., and Zacny, K., "ROPEC–Rotary Percussive Coring drill for Mars sample return," in: *Proceedings of the 42nd Aerospace Mechanisms Symposium*, NASA Goddard Space Flight Center, Reno, NV, May 14–16, 2014, 195–210.

Dachwald, B., Mikucki, J., Tulaczk, S., Digel, I., Espe, C., Feldmann, M., Franke, G., Kowalski, J., and Xu, C., "IceMole: A maneuverable probe for clean in situ analysis and sampling of subsurface ice and subglacial aquatic ecosystems," *Annals of Glaciology* 55(65): 14–22, 2014.

Goodge, J.W. and Severinghaus, J.P., "Rapid access ice crill: A new tool for exploration of the deep Antarctic ice sheets and subglacial geology," *Journal of Glaciology* 62: 1049–1064, 2016.

Hand, K.P., Murray, A.E., Garvin, J.B. et al., "Europa Lander Study 2016 Report," JPL D-97667, Task Order NNN16DO11T, Europa Lander Mission Pre-Phase A, NASA, Washington, DC, 2017.

Harman, J., Smith, F., and Stone, W., Direct High Voltage Water Heater, United States Patent and Trademark Office, Patent-Pending: US-2018-0087804-A1, March 2018.

Hogan, B. and Stone, W., Direct Laser Ice Penetrator, United States Patent and Trademark Office, Patent: 9,963,939, Issued: May 2018.

Hsu, J., "Dual drill designed for Europa's ice," *Astrobiology Magazine*, April 15, 2010, http://www.astrobio.net/news-exclusive/dual-drill-designed-for-europas-ice/

Humphrey, N. and Echelmeyer, K., "Hot-water drilling and bore-hole closure in cold ice," *Journal of Glaciology* 36(124): 287–298, International Glaciological Society, January 1, 1990.

Kaufmann, E. et al., "Melting and sublimation of planetary ices under low pressure conditions: Laboratory experiments with a melting probe prototype," *Earth Moon Planet* 105: 11–29, 2009. DOI: 10.1007/s11038-009-9296-9.

Lorenz, R.D., "Thermal drilling in planetary ices: An Analytic solution with application to planetary protection problems of radioisotope power sources," *Astrobiology* 12(8): 799–802, 2012. DOI: 10.1089/ast.2012.0816.

NASA 2018-2, "Kilopower concept for Europa, NASA facts," 2018, https://www.nasa.gov/directorates/spacetech/kilopower

Philberth, K., "Une Methode Pour Mesurer Les Temperatures a L'interieur d'un Inlandis," *Comptes Rendus des Seances de l'Academie des Science* 254(22), May 28, 1962.

Rack, F.R., "Enabling clean access into Subglacial Lake Whillans: Development and use of the WISSARD hot water drill system," *The Royal Society, Philosophical Transactions* A374: 20140305, 2016. DOI: 10.1098/RSTA.2014.0305.

Sakurai, T., Chosrowjan, H., Somekawa, T., Fujita, M., Motoyama, H., Watanabe, O., and Izawa, Y., "Studies of melting ice using CO2 laser for ice drilling," *Cold Regions Science and Technology* 121: 11–15, 2015. DOI: 10.1016/j.coldregions.2015.09.014.

Schuler, C.G., Winebrenner, D.P., Elam, T.W., Burnett, J., Boles, B.W., Wogan, N., and Mikucki, J.A., "In situ contamination of melt probes: Implications for future subglacial microbiological sampling and Icy Worlds life detection missions," in: *AbSciCon*, Seattle, WA, 2019.

Stone, W.C., "Accessing oceans beneath ice," in: *Ocean Worlds 2 Conference*, Center for Marine Robotics, Woods Hole Oceanographic Institute, Woods Hole, MA, August 25–26, 2016.

Stone, W.C. and Hogan, B., Optical Energy Transfer and Conversion System, United States Patent 9,090,315, 2015.

Stone, W.C. and Hogan, B., Optical Power Transfer System for Powering a Remote Mobility System for Multiple Missions, United States Patent 9,383,520, 2016.

Stone, W.C., Hogan, B., Siegel, V., Lelivre, S., and Flesher, C., "Progress towards an optically powered cryo-bot," *Annals of Glaciology*, 55(66), 2014. DOI: 10.3189/2014AoG65A200.

Stone, W.C. et al., "Project VALKYRIE: Laser-powered Cryobots and other methods for penetrating deep ice on ocean worlds," in: Badescu, V. and Zacny, K. editors, *Outer Solar System: Prospective Energy and Material Resources*, Springer, Cham, Switzerland, 2018, pp. 47–165 (Chapter 4), ISBN: 978-3-319-73844-4, https://doi.org/10.1007/978-3-319-73845-1.

Talalay, P.G. et al., "Recoverable autonomous sonde for environmental exploration of Antarctic subglacial lakes: General concept," *Annals of Glaciology* 55(66): 23–30, 2014. DOI: 10.3189/2014AoG65A200.

Thorsteinsson, T. et al., "A hot water drill with built-in sterilization: Design, testing, and performance," *JÖKULL* 57: 71–82, 2008.

Ulamec, S. et al., "Access to glacial and subglacial environments in the Solar System by melting probe technology," *Reviews in Environmental Science and Biotechnology* 6: 71–94, 2007. DOI: 10.1007/s11157-006-9108-x.

Vago, J. et al., "ExoMars: Searching for life on the Red Planet," *ESA Bulletin* 126, 16–23, May 2006.

Vasiliev, N.I., Talalay, P.G. et al., "Twenty years of drilling the deepest hole in ice," *Scientific Drilling*, 11: 41–45, 2011. DOI: 10.2204/iodp.sd.11.05. 2011.

Weiss, P. et al., "Study of a thermal drill head for the exploration of subsurface planetary ice layers," *Planetary and Space Science* 56(9): 1280–1292, 2008.

Winebrenner, D.P. et al., "A thermal ice-melt probe for exploration of Earth-analogs to Mars, Europa and Enceladus," in: *Lunar and Planetary Science Conference*, Vol. 44, The Woodlands, TX, 2013.

Wirtz, M. and Hildebrandt, M., "IceShuttle Teredo: An ice-penetrating robotic system to transport an exploration AUV into the ocean of Jupiter's Moon Europa," in: *67th International Astronautical Congress (IAC)*, Guadalajara, Mexico, September 26–30, 2016, IAC-16-A3.5.2.

Zacny, K. et al., "Wireline deep drill for exploration of Mars, Europa, and Enceladus," in: *2013 IEEE Aerospace Conference*, Big Sky, MT, 2013.

Zimmerman, W., "Extreme electronics for in situ robotic/sensing systems," in: *2000 IEEE Aerospace Conference Proceedings*, Vol. 7, Big Sky, MT, 2000.

3 Scientific Rationale for Planetary Drilling

H. M. Sapers, L. W. Beegle, C. M. Caudill, E. Cloutis, J. Dickson, and P. Hill

Jet Propulsion Laboratory (JPL)/California Institute of Technology (Caltech), Pasadena, CA

CONTENTS

3.1 INTRODUCTION

One hundred and twenty-three kilometers or 76.26 miles. A distance that spans a little over half of the way between San Diego and Los Angeles, California. The sum total of human or robotic exploration on the surface worlds beyond Earth as of January 1, 2020 (Siddiqi, 2018; National Academies of Sciences, Engineering, and Medicine, 2018) doesn't even cover a fraction of a percent of the United States. And yet each of these missions represent monumental leaps forward in our understanding of planetary geology, propulsion technology, instrumentation, and our place in the solar system. On November 17, 1970 the Soviet Luna 17 spacecraft landed on the Moon deploying Lunokhod 1, the first successful rover to explore another world (Karachevtseva et al., 2013; Shevchenko et al., 2015; Vinogradov et al., 1971) and humankind officially began the landed exploration of our solar system: a dynamic population of rocky, icy, and gaseous bodies in gravitational orbit. Each target with its own technology and engineering challenges and a unique set of scientific questions driving exploration.

Our nearest neighbor, the Moon, is the only body beyond Earth explored directly by humankind. During the Apollo era, 15 missions saw six Moon landings (Apollo Program Summary Report, NASA, 1975) where astronauts explored the lunar surface. Samples were acquired during rover traverses and planetary drills were used to obtain samples from down to 3 m below the surface. By the end of the Apollo program in 1972, a total of 12 astronauts had collected 2196 lunar samples during a combined total of 80 hours of exploration returning approximately 382 kg of material to

Earth (Meyer, 2016, Heiken et al., 1991). By 2008 Europe, Japan, China, and India had all deployed successful orbiter and/impactors to the Moon. Continued robotic missions probe the lunar surface in a quest to further understand the formation of the Earth-Moon system, answer questions about early Earth, ground truth of inner solar system dynamics, and search for resources such as water to test the feasibility of establishing long-term lunar bases for continued exploration. Section 3.4 details the scientific questions driving discussions of exploratory drilling on the Moon.

Mars has long been a target of intense exploration, having had 37 orbital and landed missions from 1960 to 2019. Sharing an early planetary evolutionary history with Earth (e.g., Nisbet and Sleep, 2001; Zahnle et al., 2007), Mars was likely habitable in the past and may still harbor extant traces of past life or even extant life in putative deep subsurface aquifers (Stamenković et al., 2019, Cockell, 2014; Michalski et al., 2018; Onstott et al., 2019). In addition to an astrobiology focus, missions to Mars may help resolve mysteries of Earth's early history (Fairén et al., 2010; Westall, 2012). Unlike Earth, there are no active crustal recycling processes on Mars such that over 50% of the Martian crust over 3.5 Ga is still preserved (Tanaka et al., 2014; van Thienen et al., 2004). In contrast, only a small fraction of the Archean crust is preserved on Earth (e.g. Cottin et al., 2015; Westall et al., 2006; Wilde et al., 2001). The search of life beyond Earth is one of the main motivations for exploring the Martian subsurface through drilling campaigns and is discussed in more detail in Section 3.5.

Several surface probes and landers have obtained a wealth of information from additional planetary targets such as Mercury, Venus, Titan, and other small rocky bodies such as Itokawa, 4Vesta, Ceres, and Comet 67P/Churyumov. While these targets represent a wealth of fundamental knowledge about our solar system, a detailed review of all landed planetary missions and the scientific rationale for future exploratory drilling is beyond the scope of this review. Here we highlight the framework for integrating orbital and landed data to identify candidate sites for subsurface exploration and present a detailed summary of the exploration of the Moon, Mars, and small bodies and the driving scientific questions for subsurface exploration on these targets. Finally, we will summarize the current instrumentation that could be utilized on putative planetary drilling campaigns and the possible data that could be returned.

To date, humankind's exploration of the solar system has quite literally just scratched the surface (see Table 3.1). The rotary percussive drill on Curiosity, the Mars Science Laboratory rover, represents the most sophisticated sample acquisition system deployed to date allowing for drilling as deep as 5 cm to obtain samples minimally affected by oxidative damage and weathering, and obtained three full-length drill cores over the prime mission (Abbey et al., 2019). The Russian Lunnik 16, 20, and 24 lunar missions launched between 1970–1976 acquired the first subsurface samples beyond Earth. Automatic drills obtained lunar regolith from a depth of 27 cm (Lunnik 16), the 27–32 cm horizon level (Lunnik 20), and from 2 m (Lunnik 24) (Akhmanova et al., 1978; Crotts, 2011; Pillinger et al., 1977). The first autonomous drills to be used on another planetary body were carried by the Venera missions to Venus in the 1980s launched by the Soviet Union, capable of achieving a depth of down to 3 cm (Surkov et al., 1983, 1984). Both of NASA's Mars Exploration Rovers, Spirit and Opportunity, were equipped with Rock Abrasion Tools (Gorevan et al., 2003) designed to abrade the top 5 mm of the surface to expose material not affected by surficial weathering processes. ESA's ExoMars Rover (Vago et al., 2015), Rosalind Franklin, planned to launch in 2022, is equipped with a drill (Magnani et al., 2010) capable of acquiring soil samples to a maximum depth of 3 m. Lunar exploration has benefited from more extensive subsurface sampling with the Apollo Lunar Surface Drill used to obtain one deep drill core in each of the last three Apollo missions, Apollo 15, 16, and 17 for a total of 765 cm of material (Meyer, 2007a; NASA, 1968). But why drill? Drilling requires considerable technological advancement, heavy equipment taking up precious mass on launch, a high risk of failure, and the infrastructure associated with drilling will likely preclude long distance roving and surface exploration. Here we discuss the three main drivers for drilling: life and habitability, resources, and planetary evolution.

TABLE 3.1

Summary of Drilling Operations in Solar System Exploration

Target	Agency	Mission	Instrument	Depth	Technology	Information
Venus	Kosmicheskaya programma SSSR	Venera 13 and 14	Soil Drilling apparatus	30 mm	Rotary Drill	First successful lander on another planetary body
Venus	Kosmicheskaya programma SSSR	VEGA 1 and 2	Soil Drilling apparatus	30 mm	Rotary Drill	
Mars	NASA	MER	Rock Abrasion Tool[a]	5 mm	Diamond dust grinding wheel	Designed to get under weathering layer of rocks for APXS and Mossbauer spectroscopy.
Mars	NASA	MSL	Planetary Acquisition Drilling System[b]	50 mm	Rotary Percussive Drilling system[c]	Acquired data on rock hardness during acquisition. Acquired samples that were delivered to CheMin and SAM analytical instruments.
Mars	NASA/ESA	InSight	Heat Flow Probe	~35 cm	Self-propelling Mole	Probe was designed for 5 m, but only achieved 35 cm depth to date. Developed to measure heat transport on the planet
Moon	USSR	Luna 16, 20, 24		27 cm–2 m	Autonomous rotary percussive drill	First drill samples acquired beyond Earth.[d]
Moon	NASA	Apollo 15–17	Apollo Lunar Surface Drill (ALSD)	~3 m	Human held hollow rotary corer with 40-cm segments can reach 3 m	Acquired up to 3 m of lunar regolith.[e]

[a] Gorevan et al. (2003).
[b] Peters et al. (2018).
[c] Anderson et al. (2012).
[d] Crotts (2011).
[e] Apollo Lunar Surface Drill (ALSD) final report.

The rocky and icy bodies in our solar system are dynamic and the surface of every solar system body is altered by weathering processes obscuring nature of the surfaces when they first formed. Oxidative weathering either through exposure to oxygen such as on Earth or the generation of oxidative radicals through radiation processes such as on Mars has significant degradative effects to organic matter (Cockell et al., 2000; Dartnell et al., 2007, 2012; Poch et al., 2013). Additional weathering on airless bodies include solar wind, UV, and galactic cosmic ray radiation, micrometeorite bombardment. Surficial detection of definitive biosignatures may be precluded by destructive surface processes that hinder organic preservation and hostile environmental conditions that may have precluded surface life all together (Michalski et al., 2018; Onstott et al., 2019).

In situ resource utilization is a strategy to localize, assess, and acquire resources that can be used on future missions, usually assumed to be for human exploration. The nature of these resources depends on the mission(s) goal. They may be water and oxygen to support a human presence; propulsion components to allow for regular launches, or solid materials required for physical infrastructure. Regardless of the resources and the end use goal, all three stages of ISRU require subsurface access in the form of drilling from prospecting to acquisition and storage.

In each of the following sections we will expand on the current knowledge and missions to various targets in the solar system and how some of the current outstanding questions can be answered by drilling missions. We then turn our attention to how to plan a drilling mission and how to integrate information at different scales from planetary to sample to begin to answer some of most fundamental questions about our solar system and our place in it.

3.1.1 Planetary Evolution

Steno's four laws are one of the fundamental underpinnings of geology. A 17th-century Danish geologist, Nicolaus Steno, described four laws of stratigraphy to universally interpret patterns of deposition: law of superposition, law of original horizontality, law of cross-cutting relationships, and law of lateral continuity (Steno, 1669). Steno's laws were fundamental to the development of the principle of uniformitarianism, perhaps the most important and universal doctrine in geology. James Hutton first put forward ideas and observations that would become known as the principle of uniformitarianism: the idea that on a whole, geologic processes do not change over time; that the processes shaping the Earth today, i.e. erosion and deposition have been at work throughout Earth's history. This theory was described by Hutton as "from what has actually been, we have data for concluding with regard to that which is to happen thereafter" and published in two volumes as *Theory of the Earth* in 1788.

Hutton's hypothesis that gradual changes accumulating over time led to the formation of the Earth we see today was formalized as the principle of uniformitarianism and widely popularized by the revolutionary works of Charles Lyell. Between 1830 and 1833 Lyell wrote *Principles of Geology* published in three volumes, a work that explored Hutton's uniformitarianism in detail succinctly summarizing Hutton's hypothesis as "The present is the key to the past." By explaining that the Earth was formed by the same processes that operate today his work lead to the wide acceptance of "deep time" and a long, indefinite age to the Earth. Lyell's work greatly influenced Darwin's ideas on evolution, providing a geological timeframe for which evolution could exist and wrote in *On the Origin of Species*, 1859:

> He who can read Sir Charles Lyell's grand work on the Principles of Geology, which the future historian will recognize as having produced a revolution in natural science, yet does not admit how incomprehensibly vast have been the past periods of time, may at once close this volume.

The important result of this principle is that the vertical extent of a rock formation is therefore a temporal record of the processes that acted upon that formation: the deeper down the section, the further back in time we go. The principle of uniformitarianism can be extended over space and time to other planetary bodies. We can assume that the fundamental laws of geology are true on other bodies in the solar system, and that a temporal record of planetary history is preserved in the icy or rocky crusts of other worlds.

On Earth active crustal recycling: continued cycles of erosion, sedimentation, and uplift constitute the modern rock cycle such that, with a few notable exceptions, the exposed crust is very young. Drilling is required to access ancient strata informing paleo processes including climate. While crustal recycling no longer occurs on Mars and much more of the ancient crust is exposed on the surface (Tanaka et al., 2014), radiative damage is likely to have significantly degraded surficial organic matter. The harsh surface conditions may have precluded the evolution of surface life entirely, rendering the Martian subsurface the largest and longest lived potentially habitable environment on Mars (Michalski et al., 2018; Onstott et al., 2019). In addition to the search for biosignatures, drilling into Martian formations may help quantify erosion and sedimentation rates and provide data for interpreting past climatic history. On Earth geologic drilling has led to climate interpretations, quantitative data of geological strata, and the discovery of a significant subsurface biosphere. The terrestrial geologic time scale is broken into detailed epochs based on detailed geochronology. The time scale is used to interpret the timing and relationship between major events in Earth's history. The lack of sample material for dating and the lack of drilling on other planets precludes a detailed time scale extending beyond Earth. In contrast, the planetary time scale for Mars is largely based on extensions of Steno's laws and relative crater counting anchored in uniformitarianism. Essentially, when it comes to defining epochs on other planets, we're still very much in the 17th century. Accessing subsurface deposits on other planets will not only inform local planetary evolution, but will also provide clues to Earth's early history by extending the basic principles of geology to comparative planetary evolution.

Meteorite impact events are the only ubiquitous geological process in the solar system and have undoubtedly played a major role in the planetary evolution of every rocky and icy body that orbits the Sun. The rate or flux of impacts with respect to time is a subject of intense debate and with significant implications. The only samples that exist for dating early impact basins retaining geological context come from the Apollo samples. A relatively large concentration of ~3.9-Ga ages in these samples (Cohen et al., 2000; Kring and Cohen, 2002; Papanastassiou and Wasserburg, 1971; Stöffler and Ryder, 2001; Tera et al., 1974) lead to the concept of the Late Heavy Bombardment or lunar terminal cataclysm where a spike in cratering rates was proposed to explain the abundance of ~3.9-Ga ages (Chapman et al., 2007; Norman, 2009). However it has also been argued that that the LHB is an artifact of sample bias (Hartmann, 2003; Zellner, 2017), or age resetting (Boehnke and Harrison, 2016) favoring a monotonic decay of impact bombardment since ~4.5 Ga. Regardless of the distribution of late impact flux spikes, it is generally assumed that impact flux in the inner solar system over the first billion years was much higher than it is today. Obtaining additional samples through drilling from other planetary bodies would allow for additional quantitative chronologies of impact flux.

Impact chronology determined by crater counting, an analysis of the crater size-frequency distributions, is used to date virtually every rock body in the inner solar system (Baldwin, 1964; Hartmann, 1965; Neukum and Wise, 1976; Shoemaker, 1961; Soderblom et al., 1974). Assuming that all bodies in the inner solar system would have experienced similar impact cratering rates, terranes with fewer impact craters are assumed to be younger than terranes with many craters. In this way every planetary surface is placed in a relative chronology. The importance of precise dating of geologic features to unraveling planetary evolutionary histories cannot be underestimated. Following from the ideas stemming from the principle of uniformitarianism, different geological processes operate over different temporal scales and integrating these scales over physical distances allows us to estimate rates. With accurate estimates of rates of formation for different geological processes we can begin to build up a picture of planetary evolution. Was the surface ever habitable? Was climate stable enough for life to evolve? Was liquid water persistent long enough for life to gain a foothold? Is there evidence of climatic cycles?

But why drill? Dating of geological materials requires a precise contextual understanding. The minerals, or more accurately, the average of the molecules being dated, provide an age for that particular sample, not necessarily the geological formation that it was derived from. It does however constrain the formation age. We can interpret the geologic context of samples through stratigraphy, the science of the order and relative position of geological strata and their relationship to the absolute geological timescale. By drilling we retrieve a vertical record of past events that lead to the current geological expression at the surface. By interpreting this record, we are able to place dates of specific samples in context. The idea of provenance is essential to the interpretation of geological dates. A date of ~4.2 Ga for a zircon found in a mountain does not mean that the mountain is ~4.2 Ga. It means that the mountain contains minerals that are ~4.2 Ga old. If this mountain is a sedimentary sequence, we know that the material in the particular horizon that mineral came from was derived from ~4.2 Ga source. This is why dating primary material is important to pin an absolute date on a feature. If we date an igneous rock, we are acquiring the age of the rock cooling past its closure temperature, the temperature at which the "radiologic age" is set. Several things, including re-melting, thermal metamorphism, and shock metamorphism can "re-set" this age so that it appears younger and what we are actually dating is the age of the resetting event. The stratigraphic context provided by drill cores allows for the assessment of both relationships between different rock types as well as potential resetting events leading to accurate environmental reconstructions.

Dating of planetary surfaces has significant implications for exploration. Furthermore, the same geological process, for example sedimentation, can occur on vastly different timescales in different environments. One inch of sediment from a core in the middle of an abyssal plane in the ocean may represent several thousand years (Schott, 1955), because deposition rates are so slow whereas one inch of sediment from a core into the soil in a rainforest may only represent one or two seasons. We know from Earth that not all geological processes operate on the same temporal scale. In addition to putting specific dates on cataclysmic events providing an accurate relative chronology for the inner solar system, drilling into strategically selected formations will allow for refining the rates of geological processes on other planets, such as erosion and sedimentation. Understanding rate variation is of fundamental importance to accomplishing specific mission goals. For a life detection mission targeting sedimentary strata on Mars at least 3.5 Ga old, we would need to know the sedimentation and erosion rates within a particular setting in order to target a sufficient depth. Dating different strata within a sequence provides rate estimates improving targeting and allowing for more informed subsequent interpretation. Once specific strata and environments are dated and process rates determined, applying Steno's laws allows for the extrapolation of those dates along lateral gradients.

3.1.2 Habitability and the Search for Life

What is life and are we alone in the universe? Perhaps the most fundamental question humankind has pondered. While the definition of life remains elusive, there is, however, some consensus on the requirements for life to exist. The co-occurrence of these requirements forms the basis of the concept of habitability. The criteria for habitability must be satisfied to begin the search for life. Life requires both a constant source of energy and a constant supply of raw materials or nutrients in addition to a solvent. It has been argued that at its most fundamental level, an active geological cycle in some form is a prerequisite for life. A geologically "dead" planet has no intrinsic energy and chemical recycling system to sustain life. Based on these initial and ongoing requirements for life, a set of habitability conditions must be met for life to be plausible in any given environment. Life requires more than a solvent, and the simplistic definition of the habitable zone as the region in space around a host star with temperatures permitting the stability of liquid water in no way implies colonization or the existence of life. Yes, life fundamentally requires a liquid solvent such as water and the identification of regions in the solar system where liquid water is physically possible is a necessary first step in the identification of environments that could potentially harbor life. But life also requires a supply of organic carbon (several theories for non-carbon based life exists, but based on the stability, chemical promiscuity, and availability of carbon, carbon-based life remains the most probable (Rothschild, 2008; Rothschild and DesMarais, 1989; Zeki et al., 1999), a supply of bio-essential elements including N, O, S, P, and a sustained energy source. The latter is particularly important and often overlooked. If a particular life form acquires energy from the oxidation of a reduced metal, there must be another process either re-reducing the bio-oxidized substrate or re-supplying the reduced form. Without a recycling mechanism the system will inevitably be driven to an equilibrium state without a sustained source of energy. The existence of these prerequisites requires a rather specific set of physicochemical conditions to be satisfied for a sufficiently long period of time. The duration of time that these conditions must be met is a philosophical argument with practical implications: the conditions for life must have existed long enough for life to both evolve and either acquire enough biomass or to be sufficiently different from abiotic processes to leave a distinguishable signature in the rock record that is both identifiable and not subject to attrition. The amount of time a particular set of conditions existed for can be estimated by integrating the vertical extent of a particular rock sequence that is by drilling and acquiring samples at depth.

The subsurface is an under-appreciated environment for not just life on Earth, but for putative life on other planets. Our view of life in the universe has an undoubtedly terrestrial bias shaped by the Earth's surficial biosphere. However, recent exploration of the subsurface has uncovered a surprising terrestrial subsurface biosphere. Subsurface life on Earth is present to depths of 4–5 km in the continental crust (Moser et al., 2005) and 2.5 km in sub-seafloor sediments (Inagaki et al., 2015). The habitable volume of the Earth's subsurface is estimated to be 2–2.3 billion cubic kilometers, over twice the volume of the world's oceans. The deep subsurface (at least 8 m below the surface) is host to over 95% of all bacterial and archaeal biomass on Earth representing ~70 Gt of carbon (Bar-On et al., 2018; Parnell and McMahon, 2016). In other words, less than 5% of extant microbial life on Earth is on the surface. This bears consideration: microbial life on Earth is distributed in subsurface, today, with a perfectly habitable surface. There is no reason to suggest that the microbial life would have a significantly different distribution on other rocky bodies that share similar early geological histories with Earth such as Mars. Furthermore, the surface of Mars today is uninhabitable. Endolithic, or rock-loving microorganisms dominate hostile Earth environments and in regions where the Earth's surface is uninhabitable such as the Antarctic Dry Valleys (Archer et al., 2017; Torre et al., 2003) or the Atacama Desert (Meslier et al., 2018; Wierzchos et al., 2006). Microbial life is found in the subsurface where niches in rocky crevasses provide protective microclimates that meet the minimal physicochemical parameters required for life (Nienow et al., 1988; Wierzchos et al., 2015).

If life existed on Mars, or potentially still exists, it is likely to be found in the subsurface (Michalski et al., 2018; Onstott et al., 2019; Stamenković et al., 2019). Regardless of either an early warm wet or cold dry Mars, by the time the earliest evidence for life on Earth was recorded in the rock record, due to loss of the magnetic field by ~3.9–4.1 Ga (Acuña et al., 1999) and the subsequent significant loss of atmosphere by 3.7 Ga (Bristow et al., 2017; Edwards and Ehlmann, 2015; Hu et al., 2015; Wordsworth et al., 2015, 2017), the surface was largely inhospitable due to the ionizing radiation and instability of surface water by 3.5 Ga.

The Martian subsurface remains the largest and longest lived habitable environment on Mars. There is evidence of intermittent surface water throughout the first 1.5 Ga of Mars' history (Fassett and Head, 2011) with a more spatially and temporally extensive subsurface presence (Clifford et al., 2010; Clifford and Parker, 2001; Cockell, 2014; Des Marais 2010; Ehlmann et al., 2010). The earliest evidence for life on Earth appears between 4 and 3.5 Ga (Bell et al., 2015; Mojzsis et al., 1996; Nutman et al., 2016). During this time there is evidence of widespread surficial terrestrial oceans (Valley et al., 2002). The dominant surficial biomass on Earth is arguably due to the evolution of oxygenic photosynthesis at ~2.5 Ga (Dismukes et al., 2001; Soo et al., 2017). This singular evolutionary event was a consequence of over 2 billion years of continual evolution on a stable and habitable surface, conditions that were not present on Mars. It can be argued that as a result of the lack of continued habitable conditions on the surface of Mars, a surface biosphere never evolved.

Rather early chemosynthetic metabolisms independent of organic photosynthate likely dominated in the relative refugia of the subsurface. On Earth, active crustal recycling has led to a paucity of the early geologic record whereas over 50% of the ancient (>3.5 Ga) geological record on Mars remains preserved (Tanaka et al., 2014).

A significant portion of this subsurface biomass is independent of surface organic photosynthate, fixing CO_2 and obtaining energy through chemosynthetic metabolic reactions in terrestrial environments, such as hydrogen-driven metabolism, analogous to those predicted to occur in the Martian subsurface (e.g. Onstott et al., 2019). The distribution of terrestrial subsurface life is largely dependent on available energy and areas representing energy gradients such as active hydrothermalism and mineral alteration that are suggestive of out-of-equilibrium environments (Hoehler and Jørgensen, 2013; LaRowe and Amend, 2015; Onstott et al., 2019; Osburn et al., 2014; Parnell and McMahon, 2016). On Earth, these chemoautotrophic microbial communities may fix atmospheric gasses in vadose zones gaining energy from reduced metals in the substrates (Onstott et al., 2019 and refs therein). Hydrogen produced through mineral weathering (Mayhew et al., 2013; Stevens and McKinley, 1995), radiolysis of water (Lefticariu et al., 2006; Li et al., 2016; Lin et al., 2006), and cataclasis (Kita et al., 1982) is likely a key electron donor in deep subsurface communities (Nealson et al., 2005). The higher porosity at a given depth on Mars resulting from lower gravity (Dzaugis et al., 2018; Onstott et al., 2006; Tarnas et al., 2018) suggests that the H_2 flux in the Martian subsurface via radiolytic H_2 production may be greater than on Earth. A global cryosphere on Mars would limit surface/subsurface communication suggesting that even ephemeral surface habitability later in Martian history may not have had both the spatial and temporal connectivity to a putatively inhabited environment to establish colonization.

3.1.3 Instrument Design Considerations

Throughout this chapter, we will discuss the science behind the need for planetary drilling. In the development of the mission concepts and the instruments designed to be part of the payload, the scientific rational for why the mission needs to be flown is always the driving force. The need to answer specific science questions leads to requirements on specific measurements that need to be made. In general, these high-level requirements define the instruments that are used as part of the in situ package that would analyze samples either obtained by a subsurface drill or surfaces exposed by a drill. However, there are engineering requirements that need to be considered in the development of these payload packages and drill design. In planetary science a Decadal Survey is produced every 10 years that outlines future mission priorities, including open questions that need to be addressed (National Research Council, 2011).

The interplay between science requirements and available mission realities is a main consideration as to what missions are selected for future flight. Mission design usually consists of programmatic, engineering, and science-based requirements that all need to be fully addressed during every stage of a mission from initial concept through operations. Programmatic requirements usually can be defined after initial design of a mission is developed. There is a complex interplay between engineering and science requirements that sometimes is not fully understood until right before launch. The reality is that both scientists and engineers usually work together to overcome these issues, but hard decisions are usually required to make a successful mission. Future planetary drill missions are no exception.

Programmatic requirements are perhaps the most basic. These consist of cost limitations and scheduling issues that can determine which mission is flown and when. The world's space agencies are on a fixed budget, and although considerable, do not allow for more than a few flagship level planetary missions to be developed each decade. Additionally, smaller more focused cost-capped missions are chosen through peer review in the New Frontiers (~$1 billion) and Discovery programs (~$600M). Scheduling is also a consideration when planetary missions are selected. Historically the cadence has been initial fly-by-mission, orbital reconnaissance mission, followed by simple lander and then by a more complicated mission concept. This is done in order to constrain both scientific and engineering requirements in order to fit within available costs and engineering feasibility.

Engineering requirements are complex and sometimes not necessarily understood in the initial mission design. Obvious mission constraints are mass and power. As of this writing, landed mass on Mars is roughly constrained to 1000 kg that is achievable by current launch capabilities and entry-decent-landing (EDL) hardware. For example, this means that for a deep Martian drill, the use of a segmented drill to achieve multiple 100s of meters is most likely not feasible without significant development in lift capability and EDL technologies. Power is also a major concern, with on the order of 100s of watts per hour available with proven landed power systems including solar panels for Martian exploration and radio thermal isotopic generators (RTGs). The power required for drilling constrains a critical resource for operations of science instruments. Many drilling concepts require large power draws that leave very little for instruments. Some other drilling concepts are impossible with current technology. This includes melt probes on Europa that require megawatts of heat generation to overcome high thermal conductivity of the ice on that body in order to reach the subsurface ocean.

Additional engineering constraints include accommodations issues, including volume and thermal constraints. For thermal, these constraints can be both from local conditions such as diurnal temperature fluctuations on the Moon and Mars, temperature extremes on Titan, Europa, and Enceladus, as well as thermal constraints on instrument measurements. For example, some CCD detectors need to be cooled in order to reduce dark noise so that extremely sensitive measurements can be made. This is an issue with borehole instruments that operate in an environment where dumping heat can be problematic. Volume constraints on instruments are also complex and can limit the success of missions. Many instruments, such as UV/Vis/IR spectrometers, can potentially overcome volume issues by bending light, but such alterations make them complex and thus increase costs due to their packaging.

Perhaps the most complex issues involve scientific requirements. Scientific goals are carefully mapped against mission objectives and specific measurements are identified that allow the science goal(s) to be realized. Science questions are developed into a science traceability matrix that defines the measurements that need to be made. In most cases this is usually done before an instrument payload is developed. The science questions are developed into goals then objectives. From these objectives, instrument concepts are chosen that can meet these objectives and feed into the mission concepts such as depth the mission needs to reach to be scientifically valuable.

On a deep drill mission, two different types of instruments are generally possible: ones that are integrated into the drill stem or those require a sample be brought to the surface. Cameras and spectrometers such as IR spectrometers and laser-based Raman spectrometers have been integrated into drill stems and successfully deployed into the field (Eshelman et al., 2019). Neutron spectrometers are also readily able to be integrated into drill stems and acquire valuable data on subsurface elemental abundances (Andrews et al.). Other analytical instruments, such as mass spectrometers and chromatographic investigations, tend to require samples to be brought to them on the surface. Additionally, analytical instruments tend to place a requirement on the sample acquisition hardware to return either an intact core or fines in order to analyze samples. This complicates design and usually drives mission requirements including drilling depth and potential instrument sensitives.

The following subsections take a detailed look at the scalar integration of data required to lead a drilling campaign on another planetary body and the history and future of drilling campaigns on Mars, the Moon, and solar system small bodies.

3.2 UNDERSTANDING GEOLOGIC CONTEXT

3.2.1 SCALE OF ASSESSING INFORMATION

Geologic context exists in a nebulous region in the broader study of the evolution of planetary crusts: rarely can specific, detailed scientific questions be answered with low-resolution morphological information alone, yet at the same time, high-resolution remote sensing data, drill cores, and other in situ analyses with landers, rovers, and astronauts depend on geologic context to differentiate among hypotheses and to understand the narrative within which detailed measurements are made. This symbiotic relationship is fundamental to the study of planetary evolution: information obtained at an outcrop does not *replace* the lower-resolution data that pointed you to that outcrop in the first place, rather it *fills in a gap in knowledge* that was specifically defined by analysis of the context data. Ideally, a feedback loop is achieved: geologic context informs in situ analyses, which then provides high-resolution data that further clarifies the contextual data and leads to a greater understanding of a planetary landscape or a planetary body as a whole.

Geology is principally conducted in a *forensic* manner: the physical and chemical processes that shape and alter a planetary body have already occurred, and it is the job of the scientist to follow the breadcrumbs in the rock record to understand what those processes are, their relative magnitude, and the order in which they operated. Drilling into the surface of a planetary body allows scientists to access this archival information directly and, ideally, collate the data acquired into a narrative of how that specific site has changed over time. These data can then be used to generate models of how the surface, subsurface, and atmosphere are likely to change in the future.

Following differentiation, planetary crusts evolve as a function of time, and geologic context provides the storyboard for that evolution: a framework within which more detailed observations are able to connect the comparatively blurry panels of the storyboard that describes the full history of a planet's crust. A snippet of dialog extracted from a film loses much of its meaning if the characters, setting, time period, and narrative arc of the story are all unknown. Of course, that line of dialog can still have value on its own, in the same way that an isolated rock with diverse mineralogy and weathering rinds certainly has value, but its value in accomplishing the goal of developing a full narrative of a landscape's evolution is diminished if its broader context is unknown.

The scale of geologic context can change as a function of the specific question being investigated. For an outcrop geologist studying precipitation of minerals and the shape of vugs where concretions used to be, the outcrop itself is the context. For an orbital spectroscope, that same outcrop viewed in a hyper-spectral image is at the limit of that sensor's resolution, such that it is the primary target and the context is the crater wall within which the outcrop is found. Understanding the scale of context being investigated is essential for proper framing of what question is being addressed, and geologic context provides the scale at which proper geologic questions are asked. Drilling is expensive and labor-intensive, such that scattershot sampling without planning with the hope of sampling something of interest is inefficient and likely to fail; this is why the first vertical cores extracted from Mars or any other planet will require extensive geological reconnaissance before execution. While a planetary enthusiast would say that any core extracted from any site on any planetary body will provide valuable information, and they would be right, the context of that core promotes the data returned from the realm of trivia to geologic information pertaining to the origin and evolution of the unit under investigation. Was the core extracted from the central peak of an impact crater, representing uplifted material that existed *before* the impact event, or from a resurfaced crater floor, such that accessed material are units formed *after* the impact event? Geologic context is what gives the data extracted from the core *meaning*.

The concept of scale of geologic context plays a fundamental role in our understanding of planetary surface ages: the scale at which an observer views a planetary surface can alter the age of that surface by, in some cases, billions of years. This is especially true on planets and moons with an atmosphere (Chamberlain and Hunten, 1990), where aeolian activity can reset surface ages within topographic lows while the host landform within which they occur is still, in essence, preserved. A vivid example of this phenomenon is the valley networks on Mars (Figure 3.1A) (Baker et al., 1992; Carr, 1995; Fassett and Head, 2008; Hynek et al., 2010). Valley networks, dendritic troughs extending hundreds of kilometers that resemble desiccated river channels on Earth, have long been considered the strongest morphologic evidence that Mars was once a planet capable of supporting rivers of liquid water at the surface (Carr, 1996). Stratigraphic relationships (e.g. Pieri, 1976)

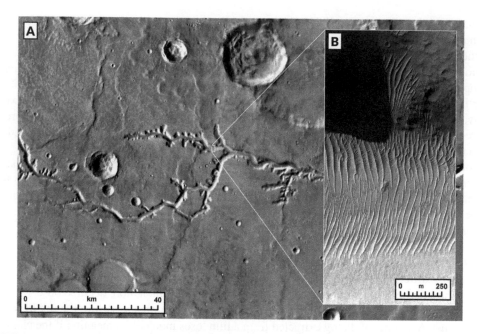

FIGURE 3.1 (A) THEMIS IR daytime mosaic of Nirgal Valles, Mars. Nirgal is a classic valley network with superposed crater populations reflective of a late-Noachian, early-Hesperian age. (B) False color HiRISE image (ESP_035304_1525) of the floor of Nirgal Valles. Aeolian resurfacing has reset the age of the valley floor to late Amazonian.

and quantitative analysis of their superposed crater populations (Pieri, 1980; Fassett and Head, 2008) have converged to the broadly accepted conclusion that this fluvial activity was prevalent in early Mars' history, but ceased at the independently calculated transition from the late-Noachian to early-Hesperian epochs of Mars' history, ~3.7 Gyr ago (Tanaka, 1986). This determination was made with what would now be considered low-resolution visible (>200 m/px Viking imagery) and thermal (100 m/px THEMIS imagery) data (Figure 3.1A). A natural follow-up study would be to use the terabytes of meter-scale resolution imagery now available of these features to produce crater size-frequency distributions of the floors of valley networks: if they really are as old as the low-resolution data suggest, higher-resolution data should support this finding and tighten the error bars, as well. A study that set out to do this would reach the opposite conclusion: the floors of valley networks retain a superposed crater population indicative of *Amazonian* (young) surface processes. Except, the geologic context shows that these two studies, using identical techniques but at different resolutions, do not measure the same *process*. Valley floors that were once the site of significant fluvial erosion that carved the valleys 3.8 billion years ago have ever since been the site for aeolian deposition of sand that has collected and is now represented by ripples and dunes that are likely still active today (Figure 3.1B). This is an exaggerated example for effect: a focused, objective study that simply and earnestly counted the superposed impact crater population on valley network floors and ignored geologic context would reach a provocative conclusion worthy of 25 pages in *Science*: Martian valley network floors retain minimal large craters, this is statistically robust across the planet, therefore while Martian river channels were formed billions of years ago, they are still active enough today to erase craters and the southern highlands of Mars are as temperate as the mid-latitudes of Earth.

Of course, this is not the case: geologic context tells us that valley network floors have been *resurfaced* by aeolian processes (Figure 3.1B), such that the *age of the surface* is a function of the resolution at which the observer is sampling, as images of the same surface at different resolutions reveal different *units*, formed by different processes. Thus, dispassionate and detailed data collection that ignores context, while feeling more objective, will generate results that do not truly reflect the history of the landscape, as improved resolution introduces distinct units that were not observable before. In almost all circumstances, on all planets and moons, *increasing resolution from global, to regional, to local scales results in the increased recency of the formation of the units being resolved.*

This phenomenon, common generally across planetary crusts, introduces a counter-intuitive relationship: there are times where an *increase* in data resolution results in a *decrease* in the amount of pertinent information for the question being asked. This is particularly true for processes that manifest at the regional and even global scale. For example, suppose a student is charged with determining whether a simple crater on the Moon is a primary crater (formed directly by impact of a bolide from space), or a secondary crater (formed via the ejection and re-impact of material from a primary crater). This is important because secondaries contaminate crater size-frequency distribution plots and thus will introduce error into calculations of the age of the host unit (Bierhaus et al., 2005). She is provided a moderate-resolution (by today's standards of lunar data) image of the crater at 100 m/px (Figure 3.2A). Should she zoom out or zoom in? The answer is the same as it is when that same student reaches a new field site in their introductory geology class field trip: climb to the highest point. This strategy, to survey an entire landscape before taking your hand lens out of its case, has been a staple of field training for over a century. While a reasonable impulse may be to process spectacular high-resolution imagery of the crater to provide as much data as possible (Figure 3.2B), these images reveal small slope processes that do not differentiate between primary and secondary craters. Rather, zooming out and evaluating the geologic context of the simple crater shows that it is found within distinct chains of other impact craters (Figure 3.2C), all of comparable size, radiating away from the Orientale multi-ring basin on the Moon's western limb (Head, 1974). The crater is clearly a secondary from the Orientale impact event and should not be included in crater counts (Wilhelms, 1976). In this instance, lower resolution *over a wider area* results in more valuable information than more literal information *over a smaller area*. The question being asked defines the scale of information that is of value to the scientist.

In modern times, orbital imagery and topography can be instantly accessed anywhere on Earth and over much of the inner solar system, figuratively allowing a field team to climb to the highest point before arriving in the field. *This is where geologic questions about large-scale planetary evolution are asked.* These questions cannot be answered if they are not posed in the first place, and geologic context provides the storyboard within which they can be posed.

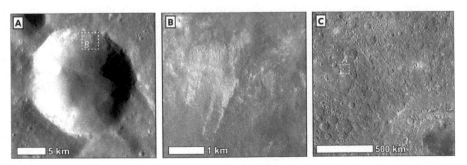

FIGURE 3.2 (A) Lunar Reconnaissance Orbiter Camera (wide-angle camera, WAC) view of an unnamed simple crater on the Moon. (B) LROC (narrow-angle camera) view of the south-facing wall of the same crater. (C) LROC WAC mosaic of the northwestern ejecta deposit of the Orientale multi-ring impact basin. Context imagery reveals that the crater in (A) is part of a secondary crater cluster, formed by re-impact of material ejected from Orientale.

3.2.2 PLANETARY SCALE

3.2.2.1 Geology/Geophysics

Geologic context, while being most intuitive on the x- and y-axes, is equally valuable as we try to understand planetary interiors, which are far more challenging to study in detail and, naturally, a target for drilling projects that require deep access to the crust. Context is pivotal for the three-dimensional understanding of a differentiated planet's interior, and the engines that drives its evolution, recorded in the crust. Mantles and cores are well beyond the limit of what contemporary drilling technology can access. Thus, we look for clues that have been recorded at varying depths through the crust and integrate them until they are sufficient for developing hypotheses that can be tested with samples acquired from depth through drilling.

Geophysical sensing techniques, primarily seismology, provide our highest quality data of the Earth's mantle and core *in their current state*. The *history and evolution* of the Earth's interior, and the interior of any differentiated body, is recorded in the crust and the composition of a body's magmatic interior can be analyzed when that magma ascends, erupts, and crystallizes at the surface. Thus, studies of stratigraphic sequences of lava layers in the crust provide our only time series of the composition of that body's magmatic interior, or at least the component capable of achieving full ascent to the surface. This makes drilling through thick sequences of basaltic rock a compelling target on all differentiated bodies to provide direct comparisons of how the interior drivers of planetary evolution have evolved on all rocky planets and satellites.

The evolution of a planet's core is most directly studied through the magnetization state of rocks in the crust. During crystallization, if a magnetic field is present, the rock is magnetized with the polarity of the host body preserved, provided that the magnetic source is that planetary body's interior, and not a local field (Hood et al., 1979). Information can certainly be gleaned from one sample, including the petrologic history of that rock and its chemical weathering history, but while movies are not produced based on one line of dialog, the geophysical evolution of a planet is not derived from one igneous rock. One sample could reveal the behavior of the core *at the time that sample crystallized*, not how it evolved over time. Imagine picking up a rock at a field site. Determine its polarity. Does a rock 10 m away show the same polarity? 100 m away? 100 km away? Why does the polarity rhythmically alternate? What is the topography/bathymetry of this site? How do the ages of the samples change as a function of distance? If these data are mapped out, what patterns do they reveal? Of course, this was the process that lead to the discovery of seafloor spreading and contributed to the broader understanding of plate tectonics on Earth (Vine and Matthews, 1964). This is only made possible through the careful integration of detailed studies of material on or near the surface integrated with regional and global context.

Plate tectonics facilitates our understanding of the geologically recent evolution of Earth's interior, while at the same time impeding our ability to study the Earth's interior early in its history. Continuous generation of new crust within moving plates allows for analyses of mantle composition and the magnetic state of the core at the time of crystallization using samples *at the surface*, alleviating the need for drilling. The evolution of the mantle plume below Hawaii can be studied as a function of time by tracing the chain of Hawaiian Islands to the northwest towards Asia across the seafloor. Yet these same samples eventually get subducted beneath plate boundaries, destroying this information. Single-plate planets and moons (Solomon, 1978) behave differently: warming in the interior generates crustal extension and permits volcanism at the surface, followed by cooling, contraction, and cessation of volcanism. This model for planetary evolution (Solomon, 1978) appears typical throughout the solar system for silicate bodies, and its record is preserved sometimes in basins like the lunar mare (Figure 3.3) or on the flanks of shield volcanoes like the Tharsis Montes of Mars. But unlike mid-ocean ridges or sites of mantle plumes on Earth like Hawaii, volcanic units on single-plate planets accumulate vertically instead of being transported laterally.

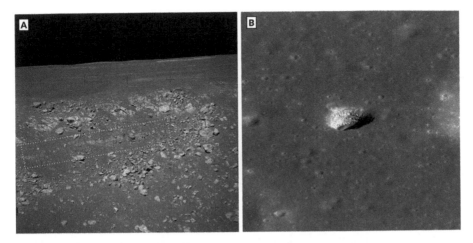

FIGURE 3.3 (A) Wall of Hadley Rille on the Moon, which cuts through the mare units that fill the Imbrium basin. Distinct layers of ancient lavas are observable exposed in the rille wall (white box). Apollo 15 image AS15-89-12104. (B) LROC NAC oblique view of a rimless pit within Mare Ingenii, showing separate layers that record the volcanic history of the lunar mare at this location. Pit is ~130 m in diameter. *Source:* Courtesy of LROC NAC image M1152670325R.

Thus, drilling through sequences of lava flows will be considerably more valuable on other planets/moons than on Earth for fully characterizing the evolution of the host body's magmatic and magnetic systems. A stack of lava flows with the Imbrium Basin on the Moon (Figure 3.3) records magmatic history *since the formation of the basin itself*, with each successive generation of lava protecting the preceding layers below. While exceptionally challenging to access, the lack of plate tectonics elsewhere in the solar system helps to preserve unique records of planetary evolution, accessible through drilling, no longer recorded on the Earth.

3.2.2.2   Habitability

With the proliferation of exoplanets now positively documented (Lissauer et al., 2014), the concept of the "habitable zone" has entered the scientific lexicon to provide a first-order assessment of whether a planetary body could sustain liquid water (Kasting et al., 2014 and references therein). This classification is broad but is a convenient binning system to guide future study. *This is geologic context extrapolated to the scale of a solar system* and provides a useful model by which we can determine habitable zones on a planetary surface or within its atmosphere.

The surface of Venus experiences a climate capable of melting spacecraft, but should that same spacecraft launch itself up, traveling at 60 mph, it would escape the Venusian atmosphere in about an hour. Along the way, it would experience a steep gradient in temperatures and encounter ambient temperatures that we would broadly consider hospitable. An astronaut on Mars dropped in the caldera of Arsia Mons could take an ambitious hike to Amazonis Planitia in the northern lowlands, and somewhere around the Medussae Fossae Formation (Figure 3.4) will encounter the Martian datum, or "sea level," defined by the elevation at which the Martian atmospheric pressure at the surface equals the triple point of H_2O (Smith et al., 1999), and from then on would experience a climate under which transient liquid water could theoretically exist during specific times of year before rapid boiling (Hecht, 2002). Does this type of broad zoning provide conclusive evidence of life? Of course not, but it does allow us to remove the blindfold when playing pin-the-tail-on-Mars: decisions about where to send a mission to drill into a planetary body depend largely on informed analysis of the regional and global context of the sites in question.

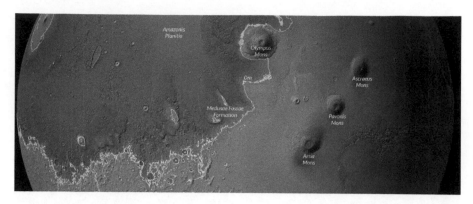

FIGURE 3.4 Global topography of Mars centered on the Medusae Fossae Formation, which falls just below the Martian datum (0 m). Terrain below the datum (blue and purple) experiences average surface pressures that surpass the triple point of liquid water (6.11 mb), such that transient liquid water could exist under contemporary conditions.

3.2.3 LANDING SITES

Drilling operations that will eventually be conducted on other bodies in the solar system will require extensive reconnaissance of sites: drilling cannot be performed in a random "unbiased" manor due to the practical challenges involved. By analogy, similar reconnaissance has been conducted in preparation for the return of surface and near-surface samples from Mars, and this provides another example of the value of geological context, this time at the outcrop scale.

The value of geologic context can be crudely quantified by assessing the lengths to which we will go to return samples of known context from other planetary bodies back to the Earth. NASA is investing $2.4 billion in the Mars 2020 Rover mission (Williford et al., 2018), which will collect and cache ~40 small cores, each the size of a stick of chalk, from outcrops within Jezero Crater (Fassett and Head, 2005), an ancient 50-km-wide open-basin lake with two sedimentary fans that recorded conditions, and possibly preserved organic material, when it was deposited in the late-Noachian (~3.7 Gyr). The $2.4 billion figure is just to *core* the samples and cache them on the surface and does not include the cost to *return* those samples. This is a significant price for a total mass of samples considerably short of what could be acquired by an introductory field course over one weekend on Earth. Add to that the opportunity cost of not using these resources to do something of scientific import with far less risk, considering that landing a spacecraft softly on Mars is hard and Mars 2020 uses an ambitious coring technology never implemented on the surface of another planet before.

Scientifically, this investment is curious to non-specialists when considering the fact that *we already have samples of Mars* that were transported for free. Martian meteorites have informed us of the igneous processes that operate on Mars (Bridges and Warren, 2006), allowed lab-based analyses of the mineralogy of the crust (McSween, 1985; Papike et al., 2009), helped temporally constrain the death of Mars' magnetic field (Lapen et al., 2010), and catalyzed the field of astrobiology by hypothesizing that sample ALH84001 could host evidence of primitive life on early Mars (McKay et al., 1996).

Yet the top priority of the Mars science community is Mars Sample Return (MSR) (Beaty et al., 2019), and this is because the Mars meteorites do not come with a label of where they are from with any more precision than they are from Mars itself, or potentially a volcanic terrain on Mars. Recently found Northwest Africa 7034 ("Black Beauty") (Agee et al., 2013; Cassatta et al., 2018) contains more water than any other Martian meteorite (Agee et al., 2013), but did it derive from the floor of the northern lowlands, where oceans may have been in Mars' past, or from the flank of one of the Tharsis volcanoes, or from the ice-rich polar regions, or from weathering on Earth? The Martian meteorites are lines of dialog in a screenplay that can be put in broad chronological order, but we have no clue which character is speaking which line.

While a sample's context is critical for understanding its evolution, selecting drill sites on a planetary body using context data alone without high-resolution mineralogical and radar data will lead missions astray, and cautionary tales highlight the importance of integrating broad contextual information with high-resolution morphologic and, particularly for Mars, mineralogic data. Long before the mineralogic diversity of the Martian crust had been mapped (Bibring et al., 2005), the Viking 1 Lander was sent to the mouth of Aram Chaos, a major outflow channel that debouches into the broad Chryse Planitia (Figure 3.5A). From morphologic information alone, it is challenging to argue with the hypothesis that this location was likely to have once been the site of catastrophic flooding from the west and south, most likely from enormous amounts of liquid water (Binder et al., 1977). Yet the surface revealed a lunar landscape: dusty, volcanic plains littered with basaltic float rocks (Figure 3.5A). The Spirit Rover (MER-A) landed on the floor of Gusev Crater, a large impact basin that had been breached along its southern rim by a valley network, almost assuredly due to the action of liquid water early in Mars' history (Figure 3.5B) (Cabrol et al., 1998). Gusev was once the site of a massive lake, but the surface, once again, resembled the Moon more than it did the desiccated portions of the Great Salt Lake (Figure 3.5B) (Squyres et al., 2004). With more data, we now understand that each of these sites has likely been resurfaced by Hesperian lavas, burying ancient sediments and potential evidence for life on Mars.

FIGURE 3.5 (A) Viking image mosaic of Chryse Planitia, Mars, site of the Viking 1 Lander. Massive outflow channels, likely carved by liquid water, drain from the west to the smooth plains. The surface, however, is characterized by Hesperian ridged basaltic plains (inset). (B) MOLA rendition of Gusev Crater, Mars, an ancient paleolake in the southern highlands. Similar to Viking 1, sedimentary lacustrine deposits are buried beneath Hesperian lavas (inset).

Each of these two sites, Chryse Planitia (Figure 3.5A) and Gusev Crater (Figure 3.5B) amplify the importance of drilling and of characterizing the three-dimensional structure of planetary crusts, instead of inferring geologic history at the surface. The surface typically records conditions when a process *stopped*, not when it began or went through significant transformations in its character. Drilling through planetary crusts at carefully chosen sites allows us to more fully evaluate how planets evolve in their entirety, not simply their most recent expression at the surface. This history of planetary evolution and the record of habitability in the solar system is best preserved *at depth* within the drillable crusts of the terrestrial one-plate planets and moons.

3.3 SMALL BODIES

3.3.1 HISTORY OF SMALL BODY EXPLORATION

The interiors of small bodies (comets, asteroids) are largely unexplored territory. While surfaces can be easily interrogated by Earth-based and Earth-orbiting telescopes, spacecraft flybys, and rendezvous missions, how their surface relates to subsurface properties is still largely a matter of speculation, inference, and modeling. For example, bulk densities of a number of small bodies are known or constrained using techniques such as mutual gravitational perturbations (for larger asteroids; e.g., Britt et al., 2002), or their effects on proximate spacecraft (e.g., Ceres—Konopliva et al., 2018) when combined with determinations of size and shape. However, such determinations allow for various internal structures to be present (i.e., the nature of macroporosity, variations in density, composition, or porosity with depth).

In addition, ample observational data indicates that the optically accessible surfaces of small bodies may not be representative of their bulk, largely due to the spectrum-altering effects of space weathering (e.g., Clark et al., 2002). For example, touchdowns by the Hayabusa-2 spacecraft on asteroid Ryugu showed that the near subsurface is optically darker than the surface (Figure 3.6). There is also direct evidence of space weathering operating on the surface of asteroid Itokawa from analysis of the Hayabusa1 returned samples (Noguchi et al., 2011).

For these and other reasons, there are compelling scientific rationales for more direct investigations of small bodies, including surface operations (such as landers) and sample return. This is evidenced by the fact that small body (asteroid) sample return was the primary goal of the now completed JAXA Hayabusa1 mission, and the ongoing JAXA Hayabusa-2 and NASA OSIRIS-REx asteroid sample return missions. It was also a primary goal of the proposed Triple-F comet nucleus sample return mission (Küppers et al., 2009), the shortlisted (but ultimately not selected) NASA New Frontiers CAESAR comet sample return mission (Lauretta et al., 2018), a recently announced CNSA near-Earth asteroid sample return mission (Jin et al., 2019), and the proposed JAXA MMX Phobos sample return mission (Ogawa et al., 2018).

The complexity and sophistication of such missions has also evolved over the years, and more sophisticated sample acquisition methods (such as drilling) have been proposed, based on advances in the necessary technology (e.g., Okada et al., 2018; Zhang et al., 2017). These have been driven largely by advances in small body surface operations and the compelling scientific nature of contextual sample return.

3.3.2 SCIENTIFIC RATIONALE FOR DRILLING SMALL BODIES

Arguably, the most compelling scientific rationales for drilling small bodies are that drilling provides: (1) links to a specific asteroid, (2) spatial context, and (3) depth information (the first two points are also addressable by just sample return). Examining past or current small body sample return missions, we know that these missions obviously provide direct information on a specific asteroid (i.e., Hayabusa1 (asteroid 25143 Itokawa), Hayabusa-2 (162173 Ryugu), OSIRIS-REx (101955 Bennu)). This is a major advance over previous studies, where approaches such as relating

FIGURE 3.6 Image captured near the Hayabusa-2 touchdown site immediately after touchdown. The photograph was taken with the Optical Navigation Camera—wide angle (ONC-W1) on February 22, 2019 at an onboard time of around 07:30 JST. *Source:* Courtesy of JAXA, University of Tokyo, Kochi University, Rikkyo University, Nagoya University, Chiba Institute of Technology, Meiji University, University of Aizu, AIST.

spectral reflectance properties to specific asteroids, plus dynamical arguments, may be compelling but not proven (as in the case of Vesta—McSween et al., 2013). Assigning a specific meteorite to a single parent body is complicated by the fact that spectroscopic studies of asteroids have shown that almost all asteroids belong to larger taxonomic spectroscopic groups; i.e., no asteroids are spectrally unique (DeMeo et al., 2009), although such classification schemes may necessarily group together spectra which have petrologic and compositional diversity (e.g., Rivkin et al. 2000).

Sample return missions also provide scientifically important spatial context. All asteroids visited by sample return missions (Itokawa, Ryugu, Bennu) show albedo variations across their surface down to small (cm) scales (e.g., Jaumann et al., 2019; Figure 3.7), suggesting that they are not homogeneous objects. Therefore, knowing the location of sample acquisition is important for providing geological context: is the sample location possibly anomalous and represent exogenic material, or is it likely representative of the bulk of the asteroid? Perhaps the most compelling argument for small body drilling is that it provides crucial depth information, also in a spatial context.

3.3.2.1 Space Weathering

Earth-based and spacecraft observations of small bodies (Binzel et al., 2010) and the Moon (Pieters and Noble, 2016), analysis of returned lunar and asteroid samples (Adams and McCord, 1970; Hapke, 2001; Noguchi et al., 2011), and asteroid orbital data (Nittler et al., 2001) have all demonstrated that what is observable using spectroscopic techniques (i.e., the uppermost few millimeters at most of the surface of a small body) differs from what is present in the interior. The main cause of this is what is commonly referred to as "space weathering" (Hapke, 2001). This term encompasses a number of processes that alter the optical (observable) properties of a surface. Space weathering can

FIGURE 3.7 Near-surface image of asteroid Ryugu obtained by the Mobile Asteroid Surface Scout (MASCOT) after separating from the Hayabusa-2 mother craft at an altitude of 41 m on October 3, 2018. *Source:* Courtesy of MASCOT/DLR/JAXA: https://www.dlr.de/content/en/images/stills-videos/still-video-mascot-free-falling-towards-ryugu.html

be severe enough that the mineralogical make-up of many small bodies can become unknowable by spectroscopic techniques or can introduce large ambiguities into possible compositions (Abell et al., 2010; Binzel et al., 2010).

A manifestation of this is that many asteroids for which we have spectroscopic data are feature-less in terms of lacking mineralogically diagnostic absorption features and differ only in terms of albedo and spectral slope (DeMeo et al., 2009). The mineralogical significance of these differences is largely unknown. This is in contrast to spectroscopic data for meteorites derived from asteroids: only a few classes of meteorites are spectroscopically featureless (e.g. iron meteorites—Britt and Pieters, 1998). It seems likely that the "overabundance" of spectrally featureless asteroids is a result of as-yet not fully understood space weathering processes.

The one successful, to date, small body sample return mission—the JAXA Hayabusa1 mission to near-Earth asteroid Itokawa—showed that space weathering causes darkening (lowering of albedo), spectral reddening (an increase in reflectance with increasing wavelength), and a reduction in the depths of mineral absorption bands (Binzel et al., 2010). These effects have also been documented for the Moon (Adams and McCord, 1970; Hapke, 2001; Pieters and Noble, 2016). More indirect evidence suggests that the relative importance of these effects may vary among different asteroids (Gaffey, 2010; Pieters et al., 2012), and is a function of factors such as heliocentric distance, pres-ence or absence of a magnetic field, etc. (Vernazza et al., 2006). Laboratory simulations of space weathering also support the fact that space weathering can cause darkening, reddening, and reduc-tion in absorption band depths, and that the spectral changes are different for different meteorites (e.g., Gillis-Davis et al., 2017; Lantz et al., 2017; Strazzulla et al., 2005).

However, contrary to most expectations, before and after images from the touchdown of the Hayabusa-2 spacecraft on Ryugu showed that the subsurface is darker than the surface. Whether this unexpected result is due to some as-yet unknown form of space weathering or heating/desicca-tion of the asteroid's surface is not yet known. Drilling of its surface would provide valuable depth-dependent contextual information on the cause of this subsurface darkening.

An additional unexpected result from the other dark asteroid sample return mission (OSIRIS-REx) is an apparent disconnect between imagery and thermal properties. The imagery suggests a rock-rich, fine-grained regolith-poor surface, but low thermal inertia values are inconsistent with the imagery (DellaGiustina et al., 2019). Drilling of the asteroid would help to resolve the causes of this apparent discrepancy.

The not yet completely understood effects of space weathering have been a compelling driver of sample return missions. As mentioned, one of the goals of the Hayabusa-2 and OSIRIS-REx asteroid sample return missions is to bring samples to Earth for the purpose of understanding space weathering and determining the mineralogy of these bodies. To this end, the Hayabusa-2 mission is intended to collect samples of the uppermost surface of its target asteroid Ryugu, as well as a sample of the subsurface, from a crater created by the spacecraft (Watanabe et al., 2017). OSIRIS-REx intends to collect a bulk sample that will include the uppermost space-weathered surface and interior material fluidized by the sampling mechanism (Bierhaus et al., 2018; Lauretta et al., 2017).

While these missions will provide valuable insights into the nature of space weathering, their sample collection approaches will not preserve subsurface stratigraphy. While this can be inferred from the returned samples, information such as depth of space weathering will be poorly constrained.

3.3.2.2 Thermal Evolution

Another compelling argument for contextual sample collection (i.e., drilling) are the effects of asteroid dynamic evolution. For example, dynamical modeling of asteroids Bennu (Delbo and Michel, 2011) and Ryugu (Michel and Delbo, 2010), suggest that in the course of their evolution from the main asteroid belt, they may have experienced periods of surface heating beyond 500 K. Such temperatures are known to significantly alter the spectral reflectance properties of carbonaceous chondrites, changing spectral slopes, albedo, and potentially mineralogically diagnostic absorption bands (e.g., Cloutis et al., 2012; Hiroi et al., 1993). Drilling into a small body can provide important depth information to constrain models of asteroidal thermal-dynamical evolution. For example, to what temperature and to what depth was an asteroid affected by close passages to the Sun and how well does this contextual information agree with dynamical models?

3.3.2.3 Desiccation

The vacuum of space will affect the spectral properties of any hydrated meteorites (Garenne et al., 2016; Takir et al., 2019), and by extension, asteroids. Current spectral interpretations of both Ryugu (Kitazato et al., 2019) and Bennu (Hamilton et al., 2019) suggest that the shallow depth of a 2.7-μm region hydroxyl-associated absorption feature in their spectra are indicative of possible thermal metamorphism. However, as mentioned, laboratory experiments also indicate that absorption bands in this region are also affected by exposure to vacuum (Garenne et al., 2016; Takir et al., 2019). Such laboratory experiments are necessarily of short duration, and it is not known whether prolonged exposure (in isolation or in combination with space weathering) would lead to gradual reduction in such hydroxyl/H_2O-associated absorption bands.

3.3.2.4 Particle Ejection Events and Outbursts

A number of asteroids exhibit evidence of surface activity, such as unexpected brightening (e.g., Phaethon: Li and Jewitt, 2013) and, in the case of Bennu, particle ejection events (Scheeres et al., 2019). A number of mechanisms could explain these phenomena, such as sublimation of subsurface volatile ices. Drilling would provide insights into the causes of these ejection events, as well as determinations of whether seemingly desiccated asteroids do in fact possess subsurface volatiles that can sublimate due to processes such as exposure by impacts.

For comets, drilling would provide valuable knowledge on any near-surface layering, such as formation of a desiccated crust, and mechanisms driving outbursts, formation of comas, gas and dust jets, and other surface modification processes (Hughes, 1975).

3.3.2.5 Layering

A number of lines of evidence for small bodies suggest that they may be primordial (e.g., comet 67P/Churyumov-Gerasimenko: Davidsson et al., 2016) or reaccreted rubble piles (Sugita et al.,

2019; Walsh et al., 2019). Images of their surfaces show surface laminations that may be indicative of stratigraphic layering (Figure 3.8), although the cause of this layering is not yet known. Drilling of such targets would enable determination of whether such layering is due to compositional or physical effects. In the former case, compositional variations with depth could provide supporting or refuting evidence for so-called "onion-skin" models of the internal structure of carbonaceous chondrite parent bodies (Consolmagno et al., 2010).

3.3.2.6 Ponds

The few near-Earth asteroids imaged at high resolution by spacecraft show a diversity of shapes and surface features. One of the most intriguing of these are so-called smooth areas and ponds. These seem to be composed of finer-grained material than rougher areas, and have been seen on multiple asteroids, such as Itokawa (Miyamoto et al., 2007) and Eros (Cheng et al., 2002) (Figure 3.9). Various mechanisms have been proposed to explain their presence (Roberts et al., 2014). If the identification of a possible lithified ponded deposit in a carbonaceous chondrite meteorite is correct (Zolensky et al., 2002), drilling one of these fine-grained deposits could help constrain models of smooth area/pond formation, and possibly provide a sample of a wider range of materials than would be derived from bedrock or a boulder (Zolensky et al., 2002).

3.3.2.7 Summary

These factors indicate that the next generation of landed or sample return missions will or should include techniques designed to preserve stratigraphic information. The most effective approach to this is drilling. The appropriate drilling techniques would allow for information to be obtained that relates to the processes that affect small body surfaces as a function of depth. These could include coring or augering (as discussed in more detail to follow).

3.3.3 SCIENTIFIC VALUE OF DRILLING

As mentioned previously, the surfaces of asteroids and comets are likely not fully representative of their bulk, due to factors already described. These are strong drivers of sample return. The nature of the science goals and objectives that a mission is designed to address would drive the drilling/sampling requirements. The science objectives would need to be balanced against technical feasibility, and also informed by what science objectives could be addressed by in situ downhole investigations versus extraction of a subsurface sample for inspection by a landed asset, versus sample return. Sample fidelity and the level of precision of stratigraphic or depth information will also inform drilling choices. Some types of science questions may be sufficiently addressed by samples whose subsurface depth is only approximately known, such as via augering. Drilling provides valuable contextual depth information, and this is a primary motivator for drilling versus bulk sampling. Among the questions that drilling could most robustly address:

- Depth of space weathering.
- Depth of thermal excursion effects.
- Cause and mechanisms of particle ejection events and cometary sublimation.
- Mechanism of formation of smooth areas and ponds.

While in a different context, the MA_MISS drill on the ESA Mars 2020 Rosalind Franklin Rover (De Sanctis et al., 2017) combines the precision of drilling with the extraction of samples for onboard analysis. MA_MISS is designed to drill up to 2 m below the Martian surface. Downhole stratigraphy will be investigated directly by an onboard reflectance spectrometer, and unsorted drill cuttings will be interrogated by multiple instruments aboard the rover (Vago et al., 2017). This drilling strategy is being implemented because of the desire to investigate the subsurface to sufficient depth that processes (such as galactic cosmic rays) that can alter organic molecules and biomolecules are minimized.

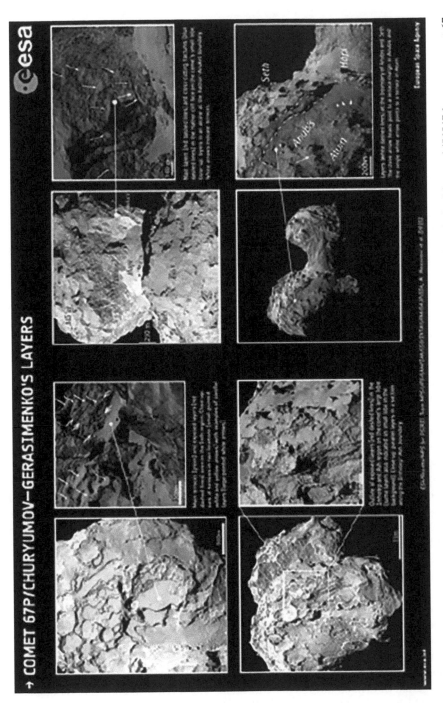

FIGURE 3.8 Views of layering on comet 67P/Churyumov Gerasimenko. *Source:* Courtesy of https://sci.esa.int/web/rosetta/-/56547-layers-on-comet-67p-surface

FIGURE 3.9   View of ponded deposit on asteroid Eros. The image was taken of a point 13.5 km (8.4 mi) away; the scene is about 550 m across. *Source:* Courtesy of NASA, https://www.jpl.nasa.gov/spaceimages/details.php?id=PIA03138

In a similar vein, the NASA Mars 2020 rover will acquire short (~10 cm long) drill cores for eventual return to Earth for analysis. In this case, coring is intended to secure more pristine and dust-free samples than would be possible with a scoop or other sampling mechanism. By focusing on recently exhumed terrains, it is hoped that these samples are minimally affected by exogenic alteration (Williford et al., 2018).

There are both benefits and disadvantages to return of a drilled sample. If the presence and nature of volatiles is of interest, it may be beneficial to perform in situ analysis, so that sample alteration is minimized, and the needs to preserve sample integrity for return to Earth are obviated.

Balanced against this are the technical limitations of what sorts of analyses can be performed by a lander versus on-Earth analysis. Lander-hosted payloads are becoming increasingly sophisticated, so that the science to be done in situ versus on Earth is narrowing. As mentioned, there are a number of options for small body drilling, each with some advantages and disadvantages. One of the simpler options is direct downhole investigation (as implemented on the MA_MISS instrument on the ESA 2020 Rosalind Franklin Rover). It will demonstrate the ability of downhole spectroscopy to investigate subsurface stratigraphy. The advantage is that no core needs to be extracted, and subsurface samples will not be exposed directly at the surface prior to analysis. The technology readiness level of downhole investigations is advancing and is driven to some extent by the desirability of such downhole investigations by the terrestrial resource extraction industry (e.g., LoCoco, 2018).

As demonstrated by MA_MISS, a number of investigative techniques can be adapted for downhole use. In addition to reflectance spectroscopy, these could include Raman spectroscopy and laser-induced breakdown spectroscopy (e.g., Moreschini et al., 2010; Stoker et al., 2006; Zacny et al., 2018). These techniques are robust and their optical requirements could be accommodated into a drill for in situ downhole sample interrogation.

Extraction of a drill core for examination by a lander is a technically more complex undertaking, while interrogation of drill cuttings is more tractable. However, drill cuttings lose most contextual information, which is one of the more compelling science drivers for drilling.

Extraction of cores can allow for a wider array of analytical techniques to be applied to a sample. However, it is likely that analytical techniques that would be applied to a drill core, and which could not be adapted for downhole deployment, may require mechanisms for delivering core samples to onboard analytical tools, adding an additional level of technical complexity to an already more complicated procedure (core extraction) than downhole interrogation. However, as mentioned, there are analytical techniques that cannot be easily adapted for downhole use.

The science questions and target type will also drive whether in situ analysis is sufficient versus sample return. For instance, in situ analyses such as optical spectroscopic techniques (reflectance, Raman) and laser-induced breakdown spectroscopy are arguably sufficiently robust that they can be applied in situ with high fidelity. Elemental techniques, such as mass spectrometry (elemental composition, isotopic analysis) and wet chemistry, are currently best suited to extraction of a subsurface sample, and a significant precision gap exists between onboard and Earth-based instruments.

Also as mentioned, sample type must be balanced against science requirements. A volatile-rich sample may require sophisticated sample handling to preserve its integrity on the small body for analysis by lander-mounted instruments or for return to Earth.

Of the analytical tools mentioned, they fall into a few categories: mineralogical and elemental, and address different aspects of mineralogy and composition, with varying levels of specificity. A review of these techniques is beyond the scope of this article, but to cite one example, ultraviolet-visible near-infrared spectroscopy is generally most sensitive to the presence and oxidation state of transition series elements, while mid- and far-infrared spectroscopy are more sensitive to determining the types of molecules that form a mineral's framework. Thus, they are highly complementary, and together provide a fuller picture of mineralogy and associated composition than either one alone) (Burns, 1993; Farmer, 1974).

3.3.4 SCIENTIFIC FACTORS CONTROLLING DRILLING

Locations for drilling on a small body will be driven by mission science drivers and technological capabilities and limitations (such as payload mass, instrument technology readiness level, capability of onboard instruments). Next we outline some basic requirements.

3.3.4.1 Bulk Composition

A common goal of asteroid sample return missions is to understand the origin and evolution of a targeted asteroid or comet. In the case of asteroids Bennu and Ryugu, they may consist of rubble piles: reaggregations of fragments from a catastrophic impact on their progenitor parent body (DellaGiustina et al., 2019). In the case of asteroid Itokawa, Kuiper Belt object MU69, and comet 67P, they appear to consist, to first order of at least two larger fragments that gently reaccreted to form their current irregular forms (Davidsson et al., 2016; Mazrouei et al., 2014; Stern et al., 2019). This can complicate determinations of where to drill for bulk compositional determinations. Likely targets would include the largest boulders or suspected bedrock. Pre-drilling data sources, such as imagery and thermal spectroscopy can help to guide such determinations.

Imagery from Ryugu and Itokawa in particular, indicate that their surfaces are composed of different major components (crumbly versus angular lithologies on Ryugu—Jaumann et al., 2019) and may contain exogenic surface materials (e.g., "Black Rock" on Itokawa—Hirata and Ishiguro, 2012).

3.3.4.2 Age

The dynamical evolution and age of an asteroid are also often key science drivers for sample return missions, and the sampling location requirements are similar—acquiring a representative sample. Once again, imagery and other investigative techniques can identify likely areas that are, for example, unaffected by possible subsequent impact-induced resetting or disruption of radiometric systems.

3.3.4.3 Space Weathering

Space weathering effects on asteroids occur over a variety of time scales, some of them quite rapid (Shestopalov et al., 2013). This alteration is one of the main drivers for acquiring a surface-to-subsurface sample, so that the intensity and kinetics of space weathering can be better understood. It also relates to the goals of acquiring a suitable sample for bulk composition and age dating. It seems likely that space weathering is a global phenomenon, so to address this goal, sample site location requirements may be quite relaxed.

3.3.4.4 Thermal Effects

Thermal effects can include Sun-induced and impact-associated localized heating. The former will be global in scale but may vary across the surface depending on pole orientation, while the latter will be more localized, but may extend to a deeper depth. For solar-induced heating, orbital information and surface age dating could help to identify areas least affected by it. For impact-induced heating, imagery and other remote sensing data could help to identify areas least affected by it. In any case, remote sensing data can help to constrain the degree to which possible sampling sites have been affected by one or both of these processes.

3.3.4.5 Desiccation

Vacuum desiccation of an asteroid or cometary surface will be global in scale. The depth and degree of desiccation will likely vary as a function of surface age. Recently exposed surfaces will obviously host less desiccated materials. This information may be available largely from imagery. For comets, areas of recent outbursts would provide the least desiccated samples, while for asteroids, age dating of surfaces and possible particle ejection events can serve as a guide for sample target selection. It can be argued that understanding how exposure to the space environment affects small body surfaces is not a primary goal for landed investigations, but such information will derive as a result of surface investigations for other science objectives. Laboratory experiments suggest that desiccation is a rapid (on the order of days) process (Garenne et al., 2016).

3.3.4.6 Particle Ejection Events and Outbursts

Cometary surfaces appear to be able to undergo rapid changes in surface morphology and albedo (e.g., scarp retreat on comet 67P—El-Maarry et al., 2017). Some asteroids also show evidence of rapid and large-scale changes in their observable properties, perhaps as a result of cometary-like activity (e.g., Chiron: Hartmann et al., 1990), while asteroid Bennu exhibits particle ejection events of as-yet unknown origin (Scheeres et al., 2019). Imagery, primarily, can be used to identify such areas and could contribute to the goal of understanding their underlying mechanisms as well as identifying areas of more-pristine materials for sampling.

3.3.4.7 Layering

As mentioned, areas on comets and asteroids have been identified as showing layering which may be due to a number of causes. Imagery and other remote sensing techniques may help to constrain the causes and drilling them would provide contextual information to help ascertain the cause.

3.3.4.8 Ponds and Smooth Areas

The origin of ponds and smooth areas on asteroids such as Itokawa and Eros are not well understood. A possible lithified ponded deposit has been identified in one carbonaceous chondrite meteorite (Zolensky et al., 2002) and it appears to be more compositionally diverse than the bulk meteorite. Drilling of a pond deposit or smooth area would help to understand mechanisms of formation (e.g., seismic shaking), and time scales of resurfacing events. Preservation of stratigraphy would be important for this type of investigation.

3.3.5 Summary

Drilling is the logical evolution of small body sampling, but technical challenges remain (e.g., Garrick-Bethell and Carr, 2007). As discussed above, there are a number of science questions that are best addressed, or exclusively addressable, by drilling, and three broad categories of drilling investigations that can be undertaken on small bodies: downhole, extracted core/cuttings, and core/cuttings sample return. The advantages and disadvantages of each approach are related to the science objectives and goals that are being addressed and each involves tradeoffs between attainable science and technical feasibility. As the technical capabilities advance, drilling on small bodies will become increasingly feasible, allowing new science questions (not addressable by bulk sample return) to be addressed.

3.4 LUNAR DRILLING

3.4.1 INTRODUCTION

Unlike any other planetary bodies, the Moon has a rich history of both robotic and human exploration. It is one of the most explored planetary bodies in the solar system, with numerous remote sensing missions and 382 kg of sample returned to Earth. It has also experienced one of the most advanced drilling programs ever implemented beyond the Earth, with over 7 m of core returned to the Earth (Lucey et al., 2006; McKay et al., 1991; Meyer, 2007a). This section will provide a brief overview of that history before looking at the scientific rationale that will drive future drilling programs. Several excellent summaries on lunar geology and geochemistry have already been written (e.g., Lucey et al., 2006; McKay et al., 1991) so a focus has been made to summarize those efforts in the context of planetary drilling. Additionally, as we look to return to the Moon, three potential planetary landing sites will be explored to demonstrate how drilling could be implemented in future lunar exploration.

3.4.2 LUNAR DRILLING TO DATE AND PLANNED

Understanding the structure and nature of the Moon's subsurface has always been a key scientific question in investigating the geology and geophysics of the Moon. This has been driven by two main types of scientific exploration programs: drilling programs and geophysical investigations. To date, lunar drilling programs have been limited to the upper portion of the crust through the Apollo and Luna missions (McKay et al., 1991). Though push cores were utilized as early as the Apollo 11 mission, reaching a depth of 30 cm (Meyer, 2007b), the Apollo Lunar Surface Drill (ALSD) allowed for a much greater depth to be reached by the Apollo 15, Apollo 16, and Apollo 17 missions (Allton and Waltz, 1980; McKay et al., 1991; Meyer, 2007a). The overall objective of the system was to extract a soil column and emplace heat flow probes in the lunar surface. With each core segment measuring about 40 cm in length, the drilling program was able to reach up to 3 m depths (Kribs, 1969). Comparatively, the Soviet space program—Luna 16, Luna 20, and Luna 24—reached depths of 0.36 m (Vinogradov, 1971), 0.25 cm (Vinogradov, 1972), and 1.60 m depth (Barsukov, 1977; Florensky et al., 1977). The drilling proved more difficult than expected during the Apollo program, with numerous complaints of removing the core segments and overheating of the instrument (Zacny et al., 2008); however, the three missions returned vertical profiles that were integral to understanding the evolution of lunar soil and the structure of the regolith.

The multi-layered nature of the lunar regolith was not fully understood until the return of the Apollo 12 push core as the Apollo 11 site left an incomplete record of the 30 cm depth it sampled (McKay et al., 1991; Meyer, 2007b). A more complete view of the Moon's near-surface structure came from the return of the ALSD samples where grain size was shown to vary greatly between layers (Heiken and McKay, 1974; Heiken et al., 1973, 1976; McKay et al., 1974, 1980). Within the Apollo 15 drill core, Heiken (1975) identified 42 distinct textural units over the 2.36 m of core with all the layers observed to be poorly sorted. The observed graded bedding within the core was proposed to have resulted from base surge deposition from impact events or volcanic activity (Heiken, 1975; Heiken et al., 1976). Heiken (1975) proposed that each of the 42 layers was deposited layer by layer and that each layer could have been reworked by micrometeorite bombardment. Additionally, Basu and Bower (1977) analyzed the >250 µm size fraction of the Apollo 15 drill core to understand the provenance of the particles within the regolith. The analysis of mineral fragments indicated a 40:60 ratio of mare to highland material, with quartz-normative basalts dominating the mare basalt component (Basu and Bower, 1977). The non-mare component was predominately KREEP (material enriched in K—potassium; REE—rare earth elements; P—phosphorus) material, which was proposed to have a different source than the other highland components (Basu and Bower, 1977).

Similarly, several major subdivisions in the Apollo 16 and 17 cores were made. In the case of the Apollo 16 site on the Cayley Plains, the 2.21-m section (Allton and Waltz, 1980) was subdivided into three major subdivision (Meyer and McCallister, 1977) and 46 textural units (Duke and Nagle, 1976). McKay et al. (1977) noted that to a depth of ~65 cm there was a general increase in mean grain size, except for the bottom layer, with the top of the core (first 20 cm) showing the most in situ reworking and maturation. In the case of Apollo 17, the 2.84 m was divided into five stratigraphic units (Vaniman et al., 1979) and eight stratigraphic units (Taylor et al., 1979) reflecting different degrees of reworking (Duke and Nagle, 1976; McKay et al., 1974). Additionally, a push core was taken from the rim of Shorty Crater; however, that core does not represent a soil profile but rather a pyroclastic deposit (Heiken et al., 1974). Variations in grain size in that core are reflective of eruption processes rather than maturation processes (Heiken et al., 1974; McKay et al., 1974).

As previously mentioned, the Apollo program was not the only drilling program that has returned lunar samples from depth. One of the major achievements of the Soviet Union's Luna program was the returning of 0.326 kg of lunar samples through the advance robotics of the Luna 16 (September 1970), Luna 20 (February 1972), and Luna 24 (August 1976) (Barsukov, 1977; Florensky et al., 1977; Vinogradov, 1971, 1972). As with the Apollo drill core, multiple textural units were identified within the cores (Basu et al., 1978; Vinogradov, 1971, 1972). Interestingly, Luna 24, which landed in the Mare Crisium, shows a decrease in the average grain size with depth; the opposite of what is observed in Apollo 16 (Basu et al., 1978; McKay et al., 1977). This is believed to be the result of reworking with depth (Basu et al., 1978; McKay et al., 1977). Though Luna 16 and Luna 24 both sampled mare basalts, Luna 20 distinctly sampled the lunar highlands. This makes Apollo 16 and Luna 20 the most representative sampling of the lunar highlands acquired to date.

As we enter into the 21st century, a renaissance of lunar exploration has flourished as more nations began to utilize remote sensing to investigate and map the lunar surface. Missions including the European Space Agency's SMART-1 (2003); China National Space Administration's Chang'e 1 (2007), Chang'e 2 (2010), Chang'e 3 (2013), and Chang'e 5-T1 (2014); the Indian Space Research Organization's Chandrayaan-1 (2008); Japan Aerospace Exploration Agency's Kaguya (2007); and NASA'S LCROSS (2009), LRO (2009), and GRAIL (2011). Before Chang'e 3 and the Yutu Rover, no mission had returned to the surface since Luna 24 in 1976. China sees their lunar exploration as a means to "improve knowledge of the Moon and improve abilities to utilize the lunar resource. It has strategic importance for China to safeguard legal rights on the Moon" (Zheng et al., 2008). Likewise, ESA's lunar exploration is looking toward the Moon as the next target for human exploration after the International Space Station (ISS) with PILOT, SPECTRUM, and PROSPECT as three preliminary missions for hazard detection, resource investigation, and ground communication (2019). Even the Trump administration is interested "in returning American astronauts to the moon for the first time since 1972 for long-term exploration and use" (The United States, 2018). The Moon is now universally seen as the next target for human exploration, with drilling destined to play a predominate role as we further explore the lunar surface and subsurface. An example of this can be seen in the upcoming Chang'e 5 mission planned for 2020. Though exact details for this planned mission are still to be determined, one of Chang'e 5's mission goals is to return a drill core sample from the Rümker region at the South Pole of the Moon (Qian et al., 2018).

3.4.3 Science Rationale and Future Exploration

Given the strategic and political motivation driving the return to the Moon, numerous reports and studies have detailed the scientific rational and priorities of future lunar missions. A lunar drilling program will complement these priorities by providing scientifically important spatial context. The constant reworking of the lunar surface through impact events means that the Moon's surface is exceptionally heterogeneous with depth. The uppermost portion of the lunar crust (~10 km) is the product of impact events and known as "megaregolith" (Hartmann, 1973). Whether interlayering is from episodic volcanism or the Moon's constant bombardment of impact events, understanding variation through the lunar crust will be essential to understanding the Moon's geologic history. This has not only been demonstrated by the Apollo and Luna programs, but also through the Chang'e 3 and Chang'e 4 missions, which recently utilized lunar penetrating radar (LPR) to investigate variation in regolith and mare stratigraphy (Fa et al., 2015; Xiao et al., 2015). For greater context, examples of the scientific rationale that can be advanced by drilling the lunar crust will be explored. The objectives and priorities listed next are taken from the Lunar Exploration Analysis Group's (LEAG 2016) *The Lunar Exploration Roadmap: Exploring the Moon in the 21st Century* and the National Academies of Science, Engineering and Medicine's (Council, 2018) *The Scientific Context for Exploration of the Moon.*

3.4.3.1 Regolith Processes and the Lunar Environment

Taken from Lucey et al. (2006) and McKay et al. (1991), regolith is the term for the layer or mantle of fragmental and unconsolidated rock material, whether residual or transported and of highly varied character, that nearly everywhere forms the surface. The lunar regolith completely covers the underlying bedrock, with bedrock only exposed in very steep crater walls or lava channels, and as a result, none of the Apollo or Luna missions sampled in situ bedrock (Lucey et al., 2006). The lunar regolith is ultimately the product of more than 4 billion years of impacts from meteoroids shaping the Moon's surface. The lunar regolith not only contains important information regarding the Moon's evolution but also acts as an important recorder for extra-lunar processes. Due to the absence of any active magnetic field or atmosphere, the lunar surface is an exceptionally hostile environment due to ionizing radiation and constant micrometeoroid bombardment (Lucey et al., 2006). Within the lunar regolith, however, this primarily manifests itself through the addition of micrometeorite components and implantation of particles that originate from cosmic rays and solar winds. This process is called *space weathering* and refers to the modification of the lunar surface that occurs because of this environment (Lucey et al., 2006; McKay et al., 1991). The scientific significance of understanding the lunar environment and the effect it can have on the lunar surface cannot be understated. The energetic-particle environment chemically and physically alters the lunar crust through the implantation of ions from solar winds and cosmic rays and the sputtering of surface atoms by interactions with said cosmic rays. The constant bombardment causes the breakup, lithification, melting, and vaporization of crustal material, as well as contributes macro- and micro-meteorite components to lunar regolith. It was through the three-dimension profiles in the Apollo drill cores that insight into the maturation of airless bodies was gained.

Over time, two main physical processes were recognized as controlling the maturation of the lunar regolith: comminution—the breakdown of material into smaller fragments—and agglutination—the formation of agglutinates (Lucey et al., 2006; McKay et al., 1991). Agglutinates are glass-bonded aggregates produced through melting of the lunar soil (Adams et al., 1975; Papike et al., 1981; Rhodes et al., 1975). Petrographically then the two main components found within the lunar regolith are material derived from the breakdown of underlying bedrock and material derived from the meteorite bombardment (Lucey et al., 2006; McKay et al., 1991). With a freshly exposed surface, such as a fresh impact ejecta deposit or mare basalt eruption, McKay et al. (1974) proposed that maturation brings with it a decrease in the grain size as finer material is produced from comminution. Once the soil matures, the fine material is converted into coarser material through agglutination. This process was called *soil evolution path 1*, where reworking of the material is dominating any mixing. Eventually an equilibrium would have been reached and there would no longer be any large particles because of comminution and all small particles would have undergone agglutination. McKay et al. (1974) noted that material generally coarsened with depth (e.g., Apollo 16) and proposed that an additional mechanism is required. During *soil evolution path 2*, large impact events, which penetrate deep layers of the regolith or the entire regolith, excavate fresh material across the surface (McKay et al., 1974). Here mixing dominates reworking of the material.

Generally, mature lunar soils have a high percentage of agglutinates, have a high density of energetic particle tracks, are finer grained than immature soils, and have an abundance of solar-wind-implanted ions. The constant solar wind implantation combined with everlasting bombardment of micrometeorites is known to reduce Fe^{2+} to pure Fe, allowing for the creation of a maturity index (Morris, 1980). Additional nanophase Fe is produced from the condensation of vapor created through micrometeorite impact of the regolith (Keller and McKay, 1997). In this process, solar winds irradiate soil grains generating amorphous rims with nanophase Fe. Micrometeorites melt and vaporize the soils grains reaching temperatures capable of breaking the element-O bound, resulting in the condensations of Fe reduced in its metal state. The amorphous rims are also broken off by micrometeorite impacts and melted, allowing for the coagulate and coarsening of nanophase Fe (Keller and McKay, 1997). The covariance between agglutinates and pure iron allows for the investigation of the lunar regolith history (Morris, 1978). Through deeper drill cores (>10 m), a greater understanding of the various processes that the lunar regolith has undergone can be deciphered. In order to access material that has not been contaminated by solar wind and galactic cosmic ray particles after burial drilling to depths of greater than 20 m will be required (Crawford and Joy, 2014; Crawford et al., 2010). This would not only provide geologic context of the regolith but is essential for certain isotopic studies of the Moon (Crawford and Joy, 2014).

The return of drill core from the Moon will also contain important information regarding the temporal variation in the Sun, variation in cosmic radiation through time, spatial variation in volatile concentrations, and provide insight in the other planetary bodies through meteoritic components found in the regolith. Understanding how the lunar surface interacts with space is an essential question that planetary drilling can address in the very near future. Borehole temperature measurement would be extremely useful in determining the solar constant over time, and past records of solar winds would be preserved within the lunar regolith. By implementing a drill system that could access older layers of regolith, the abundance of material implanted through solar winds and cosmic rays can be correlated with absolute age measurements at different depths to provide a stratigraphic history of the solar winds. Even the constant meteoritic bombardment of the lunar surface provides the opportunity to explore other planetary bodies. Achondritic and chondritic components within the lunar crust mean that any drill core sample would contain meteoritic component with the possibility to extract material that could have originated from the asteroid belt, the other inner solar system planets, or even the Earth.

3.4.3.2 Impact Processes and Chronology

The impact cratering process is the predominate geologic process shaping the lunar surface, affecting its regolith, crust, and mantle (Stöffler et al., 2006). Unlike the Earth, where erosional forces have reshaped and removed material, impact structures on the Moon are better preserved (Lucey et al., 2006; Stöffler et al., 2006). From remote sensing missions, such as the Lunar Reconnaissance Orbiter, we have already been able to investigate the impact cratering process. Baker et al. (2011) for example was able to categorize the progression from simple craters to complex craters with peak-rings and multiple rings. It cannot be overstated how important the return of lunar material has been in understanding the chronology of the early solar system. The return of more lunar material will be essential in furthering our understanding of planetary differentiation, the evolution of the lunar crust and mantle, and better age constraints of the early impact history of the solar system. Understanding the early bombardment history of the Moon has been highlighted as an important goal by not only by *The Scientific Context for Exploration of the Moon* and *The Lunar Exploration Roadmap*, but also numerous independent scientists (e.g., Crawford, 2006; Crawford and Joy, 2014; Crawford et al., 2012; Flahaut et al., 2012; Jaumann et al., 2012; Kring and Durda, 2012; Neal, 2009).

The spatial context that comes with planetary drilling will be essential in furthering our understanding of the impact process at all scales. The fundamental issue of how impact ejecta is distributed and how ejecta varies from distance sources, both laterally and vertically, can only be achieved through a well-thought-out drilling program. Through this kind of investigation, determining the degree of vertical mixing and determining the connection between the megaregolith and underlying bedrock will be essential in advancing our understanding of the evolution of the lunar crust. Additionally, the combination of drilling programs and geophysical profiles will enable the investigation of how the original, underlying igneous rocks relate to the present state of the megaregolith. Answering these kinds of questions can only truly be understood through a vertical investigation of the lunar surface as constant bombardment and the creation of the megaregolith has left little bedrock intact. By expanding drilling on the lunar surface from the 1–3 m that occurred in the Apollo era to the 10s of meters we will truly understand how the impact process has radically shaped the lunar surface.

Additionally, the return of cored material and absolute age dating of that material will continue to advance our understanding of the flux of material to the Moon with time. One of the most important results of the Apollo era was revealing the bombardment history of the inner solar system (Stöffler et al., 2006). Radiometric dating of the lunar material in combination with crater counting led to relative age dating that is now routinely used throughout the inner solar system. The return of more material from the Moon would provide means to understand the cataclysm hypothesis of the Moon, while providing a precise absolute chorology of the lunar surface. Through a vertical examination of the lunar surface, our best estimates of impact flux through geologic time can be made. Through absolute age dating and geochemical analyses, the provenance of overprinting impact events within a single geologic region of interest could be assessed. This would be essential in advancing understanding how the flux of impacts has changed over the past 3.0 Ga. In essence, through a sample return drill program, a much deeper understanding of the both the early and recent impact flux within the inner solar system could be increased.

3.4.3.3 Volcanic Processes

Volcanic lithologies from the Moon can be broadly broken into two main categories: mare basalts and pyroclastic deposits. Both of these lithologies have been crucial in understanding the Moon's mantle composition, evolution, and thermal history. The mare basalts makes up over 17% of the lunar surface often infilling impact basins (Head, 1976; Head and Wilson, 1992). Dominantly present on the near side of the Moon, only younger craters on the lunar far side are observed to contain mare basalts, due to the thicker crust on the far side (Kaula et al., 1972, 1974; Toksöz et al., 1974). It has been proposed, however, that this could actually be the result of a non-uniform distribution of heat-producing regions within the lunar crust post-differentiation (Haskin et al., 2000). To further complicate our understanding of volcanism on the Moon, it became apparent following the Lunar Prospector mission (1998) that the Apollo landing sites all had contributions from a geochemical distinct terrane known as the KREEP terrane (Jolliff et al., 2000). To date no sample from the far side has been geologically sampled and returned to Earth, with lunar meteorites constituting the only materials we have from that side. The pyroclastic deposits on the Moon are observed as glass beads and fragments in the lunar soil (McKay et al., 1991) and have become of particular interest due to their volatile content (Chou et al., 1975; Meyer Jr. et al., 1975; Saal et al., 2008). Recent studies have drawn into question the role of volatiles on lunar volcanism with more evidence suggesting that the Moon is not as dry as we previously thought.

It is apparent that numerous questions remain with regards to lunar volcanism; however, to see how planetary drilling can be utilized to advance these questions one only has to look at the planned Chang'e 5 mission to the far side of the Moon. Chang'e 5 is now scheduled to launch in 2020

(Qian et al., 2018), with the intent to collect 2 kg of lunar surface regolith and drill core within the northern Oceanus Procellarum region (Qian et al., 2018). The objective of this mission is to investigate the origin of the Mon Rümker volcanic complex, a large plateau structure (~4000 km^2) that has numerous lava flows, rilles, and domes. This volcanic complex is of particular interest because of the relatively young age of volcanism this structure represents. Qian et al. (2018) propose that by returning young mare basalt (1.21–1.51 Ga; Hiesinger et al., 2003, 2011; Qian et al., 2018) a better understanding of the size-frequency distribution ages in the latter half of lunar impact chronology can be gained (Crawford, 2006; Stöffler and Ryder, 2001b; Stöffler et al., 2006). In addition, understanding the origin of low-Ti and very-low Ti basalts, the episodic nature of volcanism, the formation of lunar volcanic domes, and understanding the late stage magmatic history of the Moon are scientific goals that Chang'e 5 hopes to address with its returned drill core sample (Qian et al., 2018).

3.4.3.4 In Situ Resource Utilization and Preparing for Exploration on Other Bodies

For many national space agencies, a sustained human presence on the lunar surface is the next desired international effort. Establishing a human presence on the Moon is by no means an easy task, but the Moon's accessibility from Earth means that it provides a useful staging ground to testing equipment and drilling strategies that can be implemented on other bodies. To overcome the drilling difficulties encountered during the Apollo program and Luna program, lunar simulants have been used to test the drilling system that will be used by Chang'e 5 (Qinn et al., 2020); however, the true test of modern drilling systems will only come with implementation on the Moon. The lessons learned from Chang'e 5, and future missions that implement drilling, will be directly applicable to drilling on asteroids or Mars. Though each environment is distinct, the Moon provides a unique proving ground for planetary drilling and the development of advanced equipment and techniques. This is especially true as we look to go beyond 3 m of depth. Though the Apollo program's drill core did provide excellent insight into processes in the lunar regolith and surface, to put it bluntly, we have only scratched the surface. As we look to implement technologies already being utilized in the geosciences on Earth, our capabilities will only expand (Sawaryn et al., 2018).

As we look to explore the lunar surface, the concept of in situ resource utilization has arisen to reduce cost of operating in space for further exploration of the Moon or beyond. This would ultimately mean utilizing lunar regolith to produce products such as oxygen, hydrogen, water, and building materials (Duke et al., 2006). As with all resource prospecting, understanding the three-dimensional nature of a given resource will be essential for implementing any production strategy. Two examples of how drilling can assist in our understanding and development of in situ resource utilization can be seen in oxygen and water. Though there are numerous proposed methodologies for the production of oxygen on the lunar surface, by far the most popular is through the reduction of ilmenite (Gibson and Knudsen, 1985; Schwandt et al., 2012). Understanding variation in ilmenite with depth by investigating the nature and emplacement of ilmenite-bearing lithologies will be essential in developing efficient production. Similarly, the suggestion of water at the lunar poles (Arnold, 1979; Feldman et al., 2001; Li et al., 2018; Mitrofanov et al., 2010; Starukhina and Shkuratov, 2000; Watson et al., 1961) presents a significant source of hydrogen and oxygen that could be utilized in the production of rocket fuel (Schwandt et al., 2012). By implement a drilling program with a neutron spectrometer, understanding the three-dimensional nature of these deposit can be explored. With the possibility of human exploration of the Moon within the next decade, a planetary drilling program will be an essential asset for both the development of the Moon and eventual exploration of other planetary bodies.

3.4.4 Examples of Potential Drilling Sites

The goal of this section is not to suggest that the following targets of interest are the best or most scientifically important destinations. Rather three sites on the Moon have been chosen to demonstrate how a drilling component could advance the scientific objectives in a given mission. They have been chosen to demonstrate specifically how drilling can advance the scientific objectives of a mission that looks to explore the research themes of regolith processes, impact processes, and lunar volatiles. The locations were taken from Kring and Durda (2012), who list case studies of landing sites that satisfy strategic scientific priorities in lunar exploration. Regardless of the site of interest, the implementation of drilling and sample return missions provides the opportunity to address the many questions in lunar science that remain after Apollo and Luna missions.

3.4.4.1 Regolith Processes

Kring and Durda (2012) propose that the Moscoviense Basin would provide the opportunity to examine regolith processes on the Moon. A multi-ring impact basin on the lunar far side (26°S 147°E), it is located within the Feldspathic Highland Terrane (Jolliff et al., 2000; Thaisen et al., 2011). Within the basin, Kring and Durda (2012) pay particular attention to the boundary between two mare units and highland units as an ideal landing site. Given that drilling capabilities on the Moon are still limited, accessing ancient regolith layers could be made easier by exploring the crater walls of simple impact structures. Kring and Durda (2012) point to a rim of a crater that exposed regolith layers believed to be of Eratosthenian age. The argument for this location with regards to regolith processes is made easily by the presence of numerous different terranes including mare basalts, pyroclastic deposits, impact melt bearing units, and highland lithologies (Kring and Durda, 2012; Thaisen et al., 2011).

Additionally, remote sensing indicates the presence of magnetic anomalies within the Moscoviense Basin and the presence of lunar swirls (high-albedo sinuous features on the lunar surface; Thaisen et al., 2011). As pointed out by Kring and Durda (2012), the formation mechanism for lunar swirls is still poorly understood; however, it is thought that an interaction between a magnetic anomaly and fine fraction of regolith is involved (Garrick-Bethell et al., 2011; Kramer et al., 2011). To address this unknown process a combination of a magnetic geophysical survey and determination of fluxes in cosmic rays and solar winds concentration with depth is required. Investigating both the present interaction between space and lunar surface and how that interaction changed through time will enable insight into how this region has and continues to be affected by space weathering and micrometeorite bombardment.

3.4.4.2 Impact Processes

As previously stated, the Moon provides a pristine laboratory for investigating the impact cratering process. Schrödinger Basin at the Moon's South Pole (75°S 132°E), provides the opportunity to explore a pristine peak-ring basin and geologic process that formed it (Kring and Durda, 2012). At 312 km in diameter, Schrödinger Basin is of particular interest because it is situated within the much larger South Pole-Aitken (SPA) basin. As a result, it presents the opportunity to study the impact cratering process over a range of scales, i.e., formation of the SPA basin itself, peak-ring formation and the cratering process that led to Schrödinger Basin, simple craters that occurred post-Schrödinger, and the modern-day flux affecting the uppermost portion of the lunar surface.

To conduct this investigation, however, a clear stratigraphic understanding of the various different impactites must be obtained. As with the recent proposal of SPA ejecta found by Chang'e 4 (Lin et al., 2019), the central peak-ring at Schrödinger presents the opportunity to sample uplifted SPA material (Steenstra et al., 2016), but confirmation will only be possible with age dating of returned samples. In addition, given that subsequent impacts have overprinted the basin floor, a well-placed drill hole could be utilized to create a clear geologic history of the region. If a drill was place on the ejecta of a small simple crater within the crater floor, the returned material could contain impactites from SPA, Schrödinger, and the simple crater.

If more than one drill hole is placed, then the understanding of how ejecta is distributed spatially could be gained and compared to drill samples collected at terrestrial craters, such as the Ries impact structure (Arp et al., 2019; Stöffler, 1977), Chicxulub (Gulick et al., 2017; Morgan et al., 2016), or

Chesapeake Bay (Gohn et al., 2006, 2008). The comparison between a well-documented impact crater on the Moon with one on Earth would provide great insight into the impact cratering process. A more expansive drilling and geophysical survey, with multiple drill sites across the basin floor, would provide a cross section of the impact structure and better constrain its geologic structure. If peak-ring material was drilled, there is the possibility to obtain not only SPA-sourced material but also potential lunar mantle material that was brought to the near surface during the uplift that formed the peak-ring (Kring et al., 2016; Steenstra et al., 2016).

Returning datable samples from the SPA basin, Schrödinger, and any earlier simple craters, would also greatly aid in fine-tuning the size-frequency distribution ages and our ability to date planetary surfaces through crater counting. It would return samples that would be crucial in understanding the early bombardment history of the solar system and the evolution of the early Earth-Moon system. With regards to depth of drilling, any expansion beyond the ~3 m that was achieved during the Apollo program would advance our understanding of the megaregolith and its structure within a crater basin; however, numerous technical difficulties would have to be addressed before a drilling campaign on the order of 10s to 100s of meters could be achieved.

3.4.4.3 Lunar Volatiles and In Situ Resource Utilization

With the discovery of deposits of volatile-rich areas at lunar poles, a desire to understand the nature, both physically and chemically, of the lunar regolith in these areas has arisen. Permanently shadowed regions (PSRs) within craters are thought to allow temperatures to reach sufficiently low enough that cold traps are present that could host volatile phases like water and carbon dioxide ices (Sefton-Nash et al., 2019). Kring and Durda (2012) proposed that Amundsen Crater (84°S 83°E) at the lunar South Pole, provides the opportunity to investigate a permanently shadowed area of regolith that is adjacent to warmer diurnal regions.

Concerning drilling, accessing these PSRs provides the opportunity to investigate many uncertainties surrounding these deposits. Firstly ground-truthing Lunar Prospector Neutron Spectrometer and Lunar Reconnaissance Orbiter Diviner data of these regions, both at the surface and how those measurement change with depth, will be essential for future exploration of other PSRs. The nature of the volatile materials and how they change with depth is only possible through a drill core sample of regolith in PSRs. Even the basic physical properties of lunar regolith at such cold temperature is not well understood so investigations into regolith processes in PSRs would further advance our understanding of processes shaping the lunar surface. Amundsen Crater facilitates all of these investigations while also providing non-PSRs to compare conditions and mitigate technical difficulties of such cold environments (Kring and Durda, 2012).

3.5 MARS

3.5.1 Martian Exploration to Date and Planned

Mars has been a target of exploration and public interest since the 1800s, when Schiaparelli's maps of the surface spurred the imagination. This was further fueled by Percival Lowell's observations in 1894, misinterpreting natural landforms as artificial canals that littered the surface. Mars is our closest neighbor, and from the perspective of planetary exploration, it seems the most attainable goal for robotic and human advancement into the solar system—and the interest and imagination that Mars has captured has not been unwarranted. As we continue to explore the planet's surface, the story of a past that was once wet, warm, and indeed habitable—if not by canal-building Martians, but potentially microbes—has come full-circle. The discovery that Mars may have been once much like Earth in terms of climate, atmosphere, and surficial water has focused efforts on the search for life on Mars. This fundamental question—*Are we alone?*—is a deeply held, existential question, and answering this is considered by Mars scientists, and the public alike, to be a goal worthy of continued exploration. In recent decades, the increase in data and the resultant extraordinary discoveries of the surface of Mars has fueled the debate regarding past climates that could sustain Earth-like, habitable conditions in the past. While orbiting and surface robotic mission have transformed our understanding of the past and present processes, little is known about the subsurface of the planet. It has been suggested that the most tantalizing details of the planet's past—indeed, biologically significant terrain—still lay beneath the surface and are highly sought after for exploration.

Mars was an early target for interplanetary travel when the technological foundation of rocketry made robotic expeditions plausible after World War II. In 1964, NASA's Mariner 4 was the first spacecraft to successfully reach Mars and provided a few dozen images of the cratered surface. The Soviet Union answered in 1971 with the Mars 3 orbiter, and the first successful, if short-lived, landed mission. However, it was images provided from NASA's 1971 Mariner 9 mission that revealed a surface with a complex history—not simply cratered as Earth's Moon—but with impressive dormant volcanoes and great rift valleys. These were the largest geologic features in the solar system. The first extended missions to explore the surface were a pair of landers from the 1975 Viking program; these landed missions revealed the isotopic composition of Mars' atmosphere, confirming the origin of Martian meteorites found on Earth. Another landed mission wouldn't come until 1997 (Pathfinder and Sojourner), following the highly successful Mars Global Surveyor (MGS) orbiter. Two decades of orbitally derived data from MGS produced rich Mars science regarding the surface processes and provided a deeper understanding of its potential past. Spacecraft accelerations due to gravity and magnetometer readings from MGS showed that, although a magnetic field does not exist in the present day, Mars once had a dynamic interior that sustained an early magnetic field. As for Earth, the convective dynamo on early Mars would have driven plate tectonics, volcanism, and crustal recycling and evolution. Furthermore, scientists discovered the dynamic surface processes, including both present-day seasonal changes and long-term climactic changes, were recorded in the layered carbon dioxide (CO_2) ice (known as "dry ice") in the polar regions and other geologic terrains. MGS imaged geologic landscapes that were formed in the past through water-involved processes, including deep gullies and large-scale debris flows—mineralogical evidence for a history of water included hematite. Hematite is an iron-oxide mineral often found as precipitated out of water but is also formed under other geologic conditions. The finding of hematite by MGS was significant as the first mineralogic evidence of the potential for water-involved processes in Mars' past. To investigate the hematite and its formation processes, NASA sent one of the two Mars Exploration Rovers (MER), Opportunity, to Meridiani Planum. What was found there was incredible. The region surrounding the landing zone was littered with iron-oxide spheroids, dubbed "blueberries." (The mineralogy was confirmed by the rover's on-board Mössbauer spectrometer which irradiates surface materials and measures gamma ray emittance to determine iron-bearing minerals.) The "blueberries" indicated that a dynamic subsurface aqueous environment precipitated solid, iron-rich concretions over long spans of time: as water moved through the subsurface, iron was preferentially mobilized and concentrated to form the iron-rich pebbles. The chemistry of the subsurface water was determined to likely have been acidic, though, and difficult for life to have thrived. The blueberries were later eroded out of the unit, as they were iron-rich and more resistant to erosion than the surrounding materials. Meanwhile, the second MER rover, Spirit, was busy exploring Gusev Crater, providing evidence that hot springs or steam vents once percolated water through to the surface, leaving a fossilized mineral vent. Spirit also identified magnesium and iron carbonate minerals at one key location, known as Comanche, that were formed in a near-neutral pH environment and thus was potentially habitable. Opportunity also found that habitable conditions were once present at Endeavor Crater with the discovery of calcium-rich gypsum mineral veins. Although the MER rovers were incredibly successful, the evidence of long-lived, global habitable conditions remained elusive, yet the search of extraterrestrial life was still tantalizingly close.

The Mars Reconnaissance Orbiter (MRO; McEwen et al., 2007a) arrived at the planet in 2006 and began operations of its instruments, obtaining the highest-resolution orbital images and other data to date. Still in operation as of this writing, the orbiter has returned more data than all other previous Mars missions combined. Thus, this powerful asset for studying the planet has fundamentally changed, and deepened by leaps and bounds, our understanding of the history of the planet and its present-day processes (McEwen et al., 2010). Dynamic, present-day processes on the surface include recent impacts events—some of which expose water-ice just below the surface—avalanches, dust storms, and the ever-changing surfaces of polar regions concomitant with CO_2 accumulation and sublimation (McEwen et al., 2007b). One of the many major discoveries from the MRO's High Resolution Imaging Science Experiment (HiRISE) camera came on its 10-year anniversary from launch: the possibility of seasonally present, liquid, briny water (recurring slope lineae, or RSL) on the present-day surface of Mars (McEwen et al., 2014). The discovery of RSL features were incredibly exciting, as they indicated present-day habitable environments (if ephemeral by nature) may be at the surface and in reach for *in situ* exploration. However, due to the pressure and temperature of the surface of Mars, water is not stable and quickly sublimates upon interaction with the atmosphere. The RSLs are likely sourced from deeper (~750 m) within the Martian cryosphere—the ~2.5- to 6-km deep ice-rich, frozen ground and polar deposits—as a discharge of briny aquifers along geologic structures like fractures or faults (Abotalib and Heggy, 2019).

The search for present-day, stable liquid water—and indeed, the search for markers of past life—has been driven underground. The cache of liquid groundwater present in the cryosphere may provide the best chance of having preserved evidence of extinct life, but it has also been suggested that microbes could still be alive there (Stamenković et al., 2019). Ground-ice of the cryosphere was detected at more than a kilometer thick near the equator, as observed in 2005 during the commissioning phase of Mars Advanced Radar for Subsurface and Ionosphere Sounding (MARSIS, aboard Mars Express) at several kilometers of depth (Picardi, 2005). A 2018 MARSIS study found evidence that strongly suggests that, below the thick carbon-dioxide ice deposits of the South Pole of Mars, subterranean water systems are active, with a lake ~20 km in diameter (Orosei et al., 2018). The authors of the study further indicate that the MARSIS data gives constraints on the subsurface Martian cryosphere, showing that stable liquid water persists below the South Pole at depths of 1.5 km (Orosei et al., 2018). On Earth, lakes are commonly observed at depth below glaciers and other thick deposits of ice; the sixth largest lake on Earth—Lake Vostok—is stable at over 3 km below the surface. On Mars, the conditions for stable liquid water are met subsurface partly due to its depths; the increased pressure of the depth and ice overburden lower the freezing point of the water. Importantly, the water is likely also a brine; substantial amounts of dissolved solids like salts (magnesium, calcium, and sodium perchlorates)—previously detected by NASA's Phoenix Lander wet Chemistry Lab (Hecht et al., 2009) and present in the minerals left behind from RSLs—would also serve to strongly suppress the freezing point of water.

Subsurface water on Mars has important implications for an exploration program, beyond the search for life in the solar system outside of Earth. Water could prove vital to future human colonies on Mars, useful for basic life-support, shielding against solar and cosmic radiation, and as a source of hydrogen fuel. The amount of water on the surface of Mars has varied under many circumstances, including the potential loss of the planet's internal dynamo and is not stable at present conditions. It is clear, however, that groundwater and vast reservoirs of CO_2 and water-ice persist to the present, but their extent, composition and chemistry, active circulation, and habitability potential remain elusive. The subsurface has been interrogated through remote-sensing but it is completely unexplored territory. Exploration of the Martian subsurface has the potential to fundamentally change our understanding of subsurface processes on cold, dry planets and its habitable conditions. The compelling scientific rationale for exploration of the Martian subsurface is evidenced by the primary goal of the future ESA Mars Rosalind Franklin Rover (Vago et al., 2017) and the NASA Mars 2020 Rover (Williford et al., 2018) missions—to drill into the subsurface to acquire samples for return to Earth—as well as the ongoing international efforts in developing architecture, protocols, and mission and technological concepts to explore the subsurface as the new frontier in Mars exploration (Stamenković et al., 2019).

3.5.2   Science Rationale and Future Exploration

It has been suggested that the return of drilled core samples—Mars Sample Return (MSR)—is the key to answering the question of life on Mars, and to broadly deepen our understanding of the fundamental evolution of the solar system. Robotic orbiters and landed missions have proven powerful tools to explore the surface, however, more complex analyses necessitate a laboratory environment on Earth. NASA's Mars 2020 and ESA's ExoMars Rover missions are both designed as the first in a series of missions, with the explicit goal of characterizing, acquiring, and caching samples for return to Earth. The ExoMars Rover will be ESA's first mission to carry an exobiology payload—a set of instruments specifically designed to search for life—with the explicit goal of drilling into the subsurface to retrieve samples, studying the physical environment, and looking for evidence of biomarkers. Biomarkers will be clear signs that life has existed on Mars in the past, or even survives to the present day; though it has been recognized that biomarkers or geomorphologic indicators are likely to remain suspect until the samples have been returned to Earth for laboratory study.

The priorities and objectives for MSR were defined in the 2011 report *Planning for Mars Returned Sample Science* (McLennan et al., 2011), illustrating the rationale of the Mars science community for drilling and acquiring samples. In 2018, NASA and ESA formed the international MSR Objectives and Samples Team (iMOST) to re-evaluate science objectives and engineering targets based on science in the intervening years and the engineering realities of their planned respective rover missions (Beaty et al., 2019). These objectives are provided in Table 3.2. Chiefly among them are to assess the rocks for past environments that would have been conducive to habitability of ancient life and the reconstruction of past climactic events, the near-surface water-involved processes, and the early magmatic and magnetic history of Mars.

The sampling objectives prioritized by the Mars science community provide the architecture for guiding the science, and thus, the ground operations of the upcoming Mars rover missions (i.e., Mars 2020 and ExoMars). However, these missions have limited depths to explore (up to 5 cm and 2 m, respectively). Although it has been suggested that ~2 m depth may represent a stratigraphic zone that has been protected from the effects of surface oxidation and galactic cosmic radiation (Kminek and Bada, 2006; Vago et al., 2017), the depths of such damaging effects is likely highly locally variable and remains an outstanding question. Direct measurements of the flux of ionizing radiation effecting the surface of Mars were obtained by the Radiation Assessment Detector (RAD) instrument aboard MSL; it was found that ionizing radiation may exert only a negligible impact on surface microbes over a 500-year timeframe (Altan, 1973; Dartnell et al., 2007; Norman et al., 2014). However, Vago et al. (2017) justify the necessity for drilling depths of up to 2 m for the ExoMars Rover mission given that ionizing radiation penetrates the uppermost meters, with a slow but non-negligible effect over many millions or several billions of years that is likely to alter organic molecules beyond the detection sensitivity of rover-bound analytical instruments. It is generally accepted that a few meters of overburden depth is likely to offer preservation protection for organic molecules (Kminek and Bada, 2006; Pavlov et al., 2002, 2012). Even with the ability to drill into the subsurface and beyond potential oxidative and radiative effects, the challenge to find and recover a sample with well-preserved organics will remain. The ExoMars and Mars 2020 mission scientists will be searching for sampling locations that: (1) were formed in an aqueous environment; (2) have been relatively recently uncovered by wind erosion or other geologic process—to have been preserved from damaging surface exposure; and (3) are present as a receding scarp or other landform that is of an orientation to achieve a greater depth than is possible in a direct, downward drill hole (Farley et al., 2014; Vago et al., 2017; Williford et al., 2018).

NASA's MSL rover is equipped with the first robotic drill deployed on Mars; the Sample Acquisition/Sample Processing and Handling (SA/SPaH) subsystem can acquire powdered rock samples as deep as 5 cm (Abbey et al., 2019). The Mars 2020 Rover will be similarly capable of drilling up to 6 cm to retrieve and cache rock cores that are designed to be more secure, pristine, and dust-free as compared to samples acquired directly at the surface (Zacny et al., 2014). (Drilling

TABLE 3.2

MSR Science Objectives as Defined in Beaty et al. (2019)

Objective	Objective Description
1	**Geological environment(s):** Interpret the primary geologic processes and history that formed the martian geologic record, with an emphasis on the role of water.
1.1	**Sedimentary System:** Characterize the essential stratigraphic, sedimentologic, and facies variation of a sequence of martian sedimentary rocks.
1.2	**Hydrothermal:** Understand an ancient martian hydrothermal system through study of its mineralization products and morphological expression.
1.3	**Deep subsurface water:** Understand the rocks and minerals representative of a deep subsurface groundwater environment.
1.4	**Subaerial:** Understand water/rock/atmosphere interactions at the martian surface and how they have changed with time.
1.5	**Igneous terrane:** Determine the petrogenesis of martian igneous rocks in time and space.
2	**Life:** Assess and interpret the potential biological history of Mars, including assaying returned samples for the evidence of life.
2.1	**Carbon chemistry:** Assess and characterize carbon, including possible organic and pre-biotic chemistry.
2.2	**Biosignatures—ancient:** Assay for the presence of biosignatures of past life at sites that hosted habitable environments and could have preserved any biosignatures.
2.3	**Biosignatures—modern:** Assess the possibility that any life forms detected are still alive, or were recently alive.
3	**Geochronology:** Determine the evolutionary timeline of Mars.
4	**Volatiles:** Constrain the inventory of martian volatiles as a function of geologic time and determine the ways in which these volatiles have interacted with Mars as a geologic system.
5	**Planetary-scale geology:** Reconstruct the history of Mars as a planet, elucidating those processes that have affected the origin and modification of the crust, mantle and core.
6	**Environmental hazards:** Understand and quantify the potential martian environmental hazards to future human exploration and the terrestrial biosphere.
7	**ISRU:** Evaluate the type and distribution of in situ resources to support potential future Mars Exploration.

depths and potential science return are provided in Table 3.3.) The Mars Multispectral Imager for Subsurface Studies (MA_MISS) drill aboard the ExoMars Rover combines drilling, sample extraction, and onboard analysis. Downhole stratigraphy will be investigated directly by an onboard reflectance spectrometer, with a spectral range capable of characterizing clay, carbonate, and sulfate minerals (De Sanctis et al., 2017; Vago et al., 2017); these minerals are important indicators of aqueous activity during their formation.

Looking forward to future missions, it has been recommended that the most productive and important sample acquisition goals for Mars should be focused on drilling at depths of kilometers. NASA's Mars science advisory committee (MEPAG: Mars Exploration Payload Advisory Group) formally recommended that deep drilling be undertaken as a priority investigation to meet astrobiology and geology goals. The 2007 Feasibility Study *Science Rationale and Priorities for Subsurface Drilling* (Clifford et al., 2001) communicated guidelines from the Mars science community for drilling justifications and in situ analyses, and described the potential science return in such an endeavor (Table 3.3). The scope of the feasibility study was limited to depths less than 200 m; depths no less than 20 m were recommended, with diminishing science return suggested as the drill

TABLE 3.3

Drilling Depths, Potential Stratigraphic Units Reached by Drilling, and The Science Return that May be Expected at Given Depths

Depths (m)	Stratigraphic Unit	Potential Science Return
<1 1–2	Depth of diurnal and annual thermal wave; depth of eolian dunes and dust	Penetration below soil/regolith; penetration potentially below thin oxidative and other alteration rinds to interrogate "fresh" rock surfaces Soil/regolith petrology, regolith physical properties, atmosphere/regolith interactions on diurnal to seasonal timescales, and natural seismicity
5		Heat low, soil/regolith stratigraphy, and long-term (multi-year) volatile transport
10	Zone of surface/atmospheric interaction; variable volcanic and sedimentary interbedded units	Likely penetration below the surface-oxidized layer (enabling a reasonably definitive test for organics and other biomarkers), and improved access to near-surface stratigraphy
20	Below regolith to bedrock; gas stability zone	Potential access to ice-saturated frozen ground (depending on the latitude and local properties of the landing site), reasonable chance to sample bedrock (basic petrology and geology studies), improved seismic coupling and heat flow measurement; measurement of stable volatiles
>20	Bedrock and massive ice lenses; cryosphere reached at 2.5–5 km depths at the equator, and 8–13 km depths at the poles regions where crust is at temperatures <273 K); liquid water stability beneath the cryosphere	Probability of access to bedrock increases with greater depth. Accessing the source stratigraphy of the gullies, and possible near-surface liquid water, will require a hole with a depth of at least several hundred meters. Under current climatic conditions, and outside of local geothermal anomalies, access to liquid water appears unlikely for depths shallower than several kilometers. However, massive lenses of segregated ground-ice (representing the frozen discharge of the outflow channels or a relic of an early ocean), are another high-priority volatile target that may be present at depths as shallow as several tens of meters

Source: Modified from Clifford et al. (2001).

depth shallows. It was concluded that reaching at least 20 m depth would access bedrock, below the regolith—the soil and other disaggregated material present as a result of meteoritic impact, weathering, and erosion.

If the depth of drilling is not capped by potential technological and logistical constraints, depths of several kilometers is a preferred target—depths that might finally provide evidence of life in the solar system beyond Earth. It has been suggested that these depths into the subsurface may provide liquid, freshwater (not highly saline) environments that provide high potential for past and even present-day habitability (Clifford and Parker, 2001). Not only are environments at these depths likely to be protected from the harmful effects of ionizing radiation and oxidation but may be part of a long-lived, self-contained, subsurface cryosphere cycle—subject to geochemical, geophysical, and potentially astrobiological processes that were separate from those at the surface (Michalski et al., 2018). Drilling at these depths will also provide valuable information with regard to geologic context, and this is a primary motivator for drilling versus bulk sampling. Among the outstanding questions that deep drilling could most robustly address include:

- Depth of space weathering and alteration.
- Depth of thermal excursion effects.
- Depth of water table/cryosphere and depth of stable water-ice.
- Presence of currently habitable environments.
- Geologic and atmospheric evolution over time, and thus, habitability potential.
- Provenance of the abundant clays in the ancient Noachian Southern Highlands regions.
- Broad geologic context to support Mars science to date.

Arguably, the most compelling scientific rationales for drilling are that it may provide: (1) evidence of past conditions conducive to habitability and liquid water stability on the surface, (2) evidence of extinct or extant life, (3) spatial and geologic context for Martian meteorites and returned samples, (4) comprehensive geologic context for remote-sensing data and in situ rover-acquired data, and (5) depth information to further understand the internal dynamics of the planet. The science rationale for deep drilling are discussed in the following sections.

3.5.2.1 Evidence of Past Habitable Conditions, Surface Liquid Water Stability, and Extinct or Extant Life

Due to potentially highly inhospitable and variable conditions and damaging alteration at the surface, if life did develop on early Mars, the evidence may only remain intact within the subsurface. It is particularly important to note that open questions remain regarding whether or not the ancient Mars' climate would have been stable long enough to support and sustain habitable conditions on the surface. Evidence of long-term basin-filling lakes are common in Mars' ancient terrains and are thought to have provided a prime location where life might have developed. NASA's MSL rover arrived in Gale Crater in 2012 to investigate one such ancient lake deposit. The impact crater provided a catch-basin for sedimentary materials, some deposited by deltas, streams, and lakes (though most were deposited by wind erosion). MSL data and remote, orbitally derived observations suggest that although associated fluvial events were likely transient, a lacustrine system could have persisted in the crater basin for a period of up to 10,000 years (Grotzinger et al., 2015). MSL data show that the formation of the minerals in the lake deposits were driven by oxidation and elemental concentration and mobilization due to evaporation during periods of atmospheric perturbations (Bristow et al., 2018). Climactic perturbations and cyclical climate shifts on Mars are known to have occurred due to an unstable obliquity—the degree to which the planet tilts on its axis over very long timescales—that has led to episodic ice ages. The instability of surface water—due to temperature and pressure constraints—and the likely absence of a past Martian atmosphere capable of sustaining the necessary conditions is a looming problem, one which climate modeling has failed to reconcile. Our understanding of the early Mars climate is incomplete, but it is known that during Mars' past, the sun was only 75% as luminous as today; given that Mars is 50% farther from the sun than Earth, this presents a damaging blow to the capacity for early Mars to attain temperatures necessary for surface-water stability. Thus, the capacity for sustained surface water in valleys, channels, and lake beds, for sufficient periods of climactic stability so that life could develop and prosper at the surface, remains unknown.

It is clear, however, that groundwater and vast reservoirs of CO_2 and water-ice persist to the present, but their extent, composition and chemistry, active circulation, and habitability potential remain elusive. The subsurface has been interrogated through remote-sensing but it is completely unexplored territory. Thus, investigations into groundwater systems (Fisk and Giovannoni, 1999; Michalski et al., 2013) and the volatile molecules (e.g., water, hydrogen, or methane-based) that may support microbial metabolism (Boston et al., 1992) have been important drivers of Mars future exploration frameworks (Rummel et al., 2014). Models for generating chemoautotrophic microbial communities—organisms capable of growth through inorganic energy sources without photosynthesis—suggests the potential for these processes through water-rock interaction in the Martian subsurface (Lollar et al., 2007; Lyons et al., 2005; Oze and Sharma, 2005). Indeed, these chemolithoautotrophs are the key microbe type considered for potential habitability on Mars, given pockets of increased heat and water in subsurface (Rummel et al., 2014). The potential for such subsurface pockets to provide self-contained systems that cycle volatiles and provide oxidation-reduction (redox) reactions are an important key for habitability potential. The evidence of past life may be present as morphologic indicators—chemical biomarkers exist in specific configurations solidified in rock—but these may mimic inorganic mineral structures and are unlikely to be definitive. The best chance to identify biosignatures on Mars is suggested to be through molecular biomarkers—a wide range of substances including specific classes of organic compounds, isotopic ratios of C, N, and S, and fossilized remains of once living organisms (Clifford et al., 2001). However, these molecular structures and compounds are sensitive to oxidation and radiation which can destroy structural complexity that is indicative of biogenicity (Stamenković et al., 2019; Vago et al., 2017). This is crucial for preservation of biological material on the surface of Mars, which has been exposed to oxidative conditions and bombardment by cosmic and solar radiation for billions of years, since early Mars lost its protective magnetic field. It is thought that a number of meters into the subsurface, materials including potential biomarkers would be shielded from reactive chemical oxidants, desiccation, and ionizing radiation (Stamenković et al., 2019). As discussed earlier, the effects of space weathering—and their potentially detrimental effects to biomarkers, if life was ever present—remain unknown. A primary rationale for drilling at depth may be to understand the surface effects on microbial communities, but this is an important hazard to understanding for future human missions to Mars: does the regolith provide a measure of shielding from the radiative surface environment?

3.5.2.2   Spatial and Geologic Context for Martian Samples and Data

It is clear that the top concerns of the Mars science community are surrounding the conditions and environments that may have supported or currently support life; fundamental to understanding these conditions is a basic knowledge of the nature and chronology of the major planetary processes. Broad geologic context is critical to reconstruct the history of liquid water and its interactions with surface materials on Mars and is a primary objective for Mars exploration. Primary exploration objectives, including understanding the climate history and the evolution of the surface and interior of the planet, have been described in the 2013–2022 NASA Decadal Survey (NASA, 2013). The Decadal Survey, as the name suggests, is a National Research Council (NRC) survey of the broader science community, funded by NASA's Science Mission Directorate to identify and prioritize the leading-edge scientific questions. The Survey also details the concomitant research, observations, and notional missions that represent the best opportunities to gain insight into those questions. A common goal and motivation of sample return missions as outlined in the Survey is to contextualize current—largely, orbitally derived—observations and investigations on a global scale. Such samples would therefore offer understanding of the origin of the surface materials through understanding the subsurface composition, and thus, the geological and dynamical evolution of the planetary body.

The geologic implications of Mars 2020 and ExoMars Rover-acquired samples may not be limited to an understanding of the local surficial processes but may extend to deeper crustal processes from billions of years of crustal evolution. Crucial for understanding the geologic history of Mars, the Mars 2020 and ExoMars missions have focused their investigations on the most ancient terrains of Mars that are accessible on the surface: the ancient Noachian Southern Highlands. For the Mars 2020 mission, the terrain that will be explored is incredibly geologically diverse, and may afford the opportunity to study the evolution of the Martian crust over billions-year timespans (Mustard et al., 2007). The nominal Mars 2020 mission traverse through deltaic deposits in Jezero Crater will also allow the mission to explore Late Hesperian–Early Amazonian fluvially transported materials and lake deposits (Goudge et al., 2017) from a period of Martian history thought to be warmer and wetter, more like that of Earth (Bishop et al. 2018). In the potential extended mission

(Farley et al., 2018), Pre-Noachian materials are likely to be encountered as megabreccia-embedded bedrock in the Isidis basin (Mustard et al., 2007). The breccias were presumably emplaced as ejecta from the ~3.9 Ga Isidis impact event (1352 km diameter, (Caprarelli and Orosei, 2015)), thus representing more ancient rock exhumed from depth. The potential for investigation of such extended timescales will allow an unprecedented opportunity to study the evolution of the Martian crust. This geologic history is fundamental in understanding the larger implications of surface investigations that are limited temporally, spatially, or by depth. However, the surface materials are not likely fully representative of the bulk of the crust, and certainly only expose a very limited depth of stratigraphy—even given the extent of stratigraphy exposed in the Isidis basin region. Drawing global-scale conclusions regarding geology and thus climate history from limited rover investigations necessarily leads to limited implications of the findings.

Surface missions are necessarily limited in scope and require a greater geologic context to extrapolate to larger systems like planetary dynamics and evolution. Furthermore, the implications from even the most detailed laboratory investigations of the returned rover-acquired samples mean little without broader understanding of the geological context, age, and climactic environment. Such is the current state of Mars science: the data, investigations, and conclusions are largely based on orbitally derived data, from which geologic context is often difficult to discern. The challenge of rover-based findings is that they are spatially limited; the challenge of Martian meteorite studies is that provenance is not possible to obtain; the challenges of orbitally derived data are of resolution and scale.

The National Research Council's 1978 report *Strategy for Exploration of the Inner Planets 1978–1987* (Space Science Board, 1978) indicated that larger geologic context would necessarily be gained through assessments of a body's stratigraphy; access to the stratigraphy at depth may represent the only definitive avenue to acquire this data. Chronological assessments—and indeed, contextualization with relevance to past habitable climates—include (1) the determination of surficial materials in terms of cosmic-ray exposure and (2) determination of crystallization ages of igneous and metamorphic rocks, and depositional ages of sedimentary rocks.

3.5.2.3   Planetary Dynamics and Geophysical Evolution

A compelling argument for contextual sample collection (i.e., drilling) are the unknown internal processes and dynamic evolution of Mars. The 2013–2022 NASA Decal Survey indicates that deep subsurface observations will shed light on the evolution of the interior of the planet: no direct evidence for its interior composition and structure, and thus, planetary history is available. Planetary-scale geochronology is the reconstruction of the history of the planet—this can be characterized through deep subsurface radar sounding techniques but can only be verified through direct observation of in situ drill core sampling of the planet's stratigraphy. Radar Imager for Mars' Subsurface Experiment (RIMFAX) is a ground-penetrating radar instrument aboard Mars 2020 that will be able to investigate depths greater than 10 m, depending on the materials. The team explains that it is largely unknown what lies directly below the surface of Mars, and RIMFAX will provide an opportunity to "see the unexplored world" of the Martian shallow subsurface. What is known from orbitally derived radar observations (e.g., MARSIS, aboard Mars Express) is that evidence exists for subterranean water systems active at several kilometers of depth (Picardi, 2005)—a Martian cryosphere with interior pathways of stable liquid water (Orosei et al., 2018). How did these systems evolve? What is their composition, as determined largely by the lithologies within the subsurface?

Understanding the dynamical and crustal evolution of Mars is deemed fundamental to understanding planetary formation and the evolution of the solar system. Some outstanding questions about the fundamental nature of planetary accretion and the available supply of water and other volatiles during accretion: how did this effect chemistry (water, atmospheric, and rock chemistry), internal planetary differentiation, and the evolution of the atmosphere and potentially biological systems? Might the subsurface of Mars retain evidence of the role of meteoric and astroidal bombardment in the presence of water, atmosphere, and even life? As an extension from the investigations of these questions, can we better understand how geodynamics influenced the changing climate of Earth over time, and punctuated climactic perturbations like climate change?

Other critical subsurface environmental aspects may be acquired through exploration of the deep interior (km–scale) within the larger context of understanding planetary evolution. These aspects include internal heat flow of the planet and pathways of water-rock interaction as well as the cycling of volatiles (e.g., water, carbon dioxide) within the crust (NASA, 2013). These conditions greatly influence locally habitability potential for both the past and present. As mentioned earlier, the surface is not fully representative of bulk composition of the planet, and rover-based surface operations lack the scope for planetary-scale implications. Some types of science questions are sufficiently addressed by samples from the surface; the Mars rover surface missions thus far have provided unprecedented resolution and quality of data from another planetary body regarding local environments and the potential climactic conditions that supported those environments. These surface missions are incredible stepping stones that pave the way to understanding processes on another planet. However, data and samples from depth will be the only avenue to sufficiently address Mars' planetary internal evolution, and thus, surface-atmosphere interaction in the early stages of planet formation and over its billion-years history. The lack of fundamental geologic context provides a strong driver of sample return and rationale for deep drilling.

3.5.3 Examples of Potential Drilling Sites

3.5.3.1 Massive Bodies of Water-Ice

In the search for life, the northern latitudes are suggested as a particularly interesting target for drilling. The best locations for downhole drilling are likely to follow from radar observations regarding the putatively active subterranean waterways and large bodies of water and ice. Ground-ice in the northern latitudes may be as shallow as the regolith; regolith-ice mixed overburden may provide protection from surface oxidants and radiation, protecting organic signatures (Zacny et al., 2013). Results from wide-ranging missions launched between 1996 and 2003 confirmed that large quantities of water-ice remain near the surface. In fact, in 2016, radar data from the SHARAD (Shallow RADar) instrument aboard the Mars Reconnaissance Orbiter showed that a subsurface ice sheet covering 375,000 km^2 and up to 14,000 km^3 volume is present in southwestern Utopia Planitia (Stuurman et al., 2016). This volume of ice is roughly the equivalent to that of the water in Lake Superior. In neighboring Arcadia Planitia, the SHARAD team also found that a 40-m-thick sheet of water-ice is present just below the surface covering an area the size of Texas and California, US, combined (Bramson et al., 2015). Vast amounts of water-ice are also present in the high and low latitudes as particular features known as polar-layered deposits. The north and south polar regions of Mars host water-ice deposits that are 2 km thick, and as expansive as the area of Ontario, Canada. These polar deposits are comprised of many thousands of layers of ice and dust, each a recording of the climactic conditions of which it was deposited (Smith and Holt, 2010). Those layers thinner than one meter may yield a climate record at a resolution on the order of hundreds of years. This well-preserved and highly detailed climate record makes these deposits very compelling for exploration, with the potential to a provide answers about past habitability of Mars. On Earth, such accumulations of ice in glaciers and ice sheets are drilled and their cores studied to discover a number of clues to the past, including: thickness measurements for deposition/precipitation rates; examination of bound particles such as dust, salts, and volcanic ash; and discovering exact atmospheric composition from the time of deposition from tiny bubbles containing the ancient atmosphere. The 2007 Mars Science Working Group Feasibility Study *Science Rationale and Priorities for Subsurface Drilling* suggested that the uppermost kilometer of polar-layered deposits likely represent a highly accessible exploration target for evidence of past habitability with potential preservation of past life (Clifford et al., 2001). What atmospheric and geochemical processes are recorded, and what are the biological implications? Might ancient microbes also be preserved, if they ever thrived in the water-ice during the time of its deposition? Furthermore, it has been suggested that the examination of such deposits (e.g., for purity or salinity) will allow a deeper understanding of the potential of their use as water resources for future human explorers.

3.5.3.2 Building on the Success of Past Surface Missions

The most successful exploration strategy may be to broaden the types of targets (e.g., many different lithology types, many different environments and materials) and use the locations and geologic context gained from past surface operations as intellectual momentum for following missions. Due to the highly successfully series of past surface missions, Mars is uniquely situated to exploit this momentum. Downhole drilling of an area that has already been investigated by landed assets would provide an incredible background of data; investigations on the surface and the subsurface in a location would afford an incredible depth of context, powerful enough for the findings to extend temporally and spatially. The many highly successful landed missions on the surface of Mars pinpoint potential locations for drilling projects, as the surface and subsurface missions would prove mutually contextual.

Coupling a surface mission data with downhole drill data may represent an incredible leap forward in understanding geologic context on a planetary scale, and vastly broaden the types of lithologies and materials investigated in a given region. In the search for life, the 2013–2022 NASA Decadal Survey recommends broadening the potential targets of exploration, not relying on a particular lithology or geologic setting to provide a one-shot potential for a habitable environment (NASA, 2013). Another critical take-away from the Decadal Survey is "Go Deep": exploration of subsurface environments provide a broader perspective than missions focused solely on surficial activity and materials, and far-reaching implications. Relying on areas previously explored by surface missions, the context can be broadened without broadening a mission scope. Planned drilling efforts of kilometers into the Martian subsurface—drilling much deeper than the sampling by current and planned missions—may prove to position future missions to even better investigate the prioritized sampling objectives.

The compelling case for the Mars 2020 Rover exploration site—ancient and geologically diverse Noachian terrain of the Isidis impact basin—could be extended to the case for downhole drilling in these areas. Drilling in such a region, known to preserve varied stratigraphy through an extensive timescale, may provide a deep drilling target to maximize the science return and provide wideranging, potentially global-scale geologic context. The Isidis impact basin and the Northeast Sytris Major region, in particular, preserve incredible mineralogic and geomorphologic diversity that may be representative of major alteration (i.e., water-involved) epochs on Mars (Quinn and Ehlmann, 2019). It is thought that this region retains the greatest concentration of mineralogic diversity, and neatly organized into stratigraphy that can be read as pages in a book of geologic history. The stratigraphy of the Isidis impact basin has stratigraphic exposures that are unique on Mars, both for their diversity and their extent, capturing impact cratering, volcanism, and sedimentation, with complex depositional and erosional settings (Bramble et al., 2017). Enigmatic clay-rich stratigraphy may point to a period of climactic warming and water-involved formation; layered sulfates may indicate a period of climactic drying. Active groundwater—and potentially surface water—processes during the Noachian-Hesperian transition are supported by the presence of these hydrated minerals, and their investigation may lead to an understanding of a potential epoch of globally changing environmental, geological, and climactic conditions (Quinn and Ehlmann, 2019). A characteristic and fundamental change in the evolution of the crust is also recorded in the layers exposed in this region: volcanic materials from the Noachian, Hesperian, and Amazonian are broadly distinct from each other. The younger volcanic material is enriched in Ca and relatively lower in K; these chemical changes suggest that the older Noachian crust was formed through more geodynamically complex processes, and over time, the interior cooled and the mantle thickened (Schmidt et al., 2019). Surface rover operations may serve to pin-point locations for following deep drilling mission, which would then be positioned to fill in fundamental gaps in our understanding of (1) the effects of dynamical planetary evolution through igneous stratigraphy, and (2) the effects of impact cratering on dynamical evolution and crustal formation.

3.5.4 SUMMARY

The next step in Mars exploration will build on the successes of past missions and should expand geologic and planetary context with stratigraphic and deep cryospheric investigations—investigations that are powerful enough for the findings to extend temporally and spatially. The most effective approach to these investigations is with deep drilling. The nature of the science goals and objectives of a given mission will drive the location to drill, downhole instruments and investigation techniques, and sample extraction considerations.

The deep subsurface of Mars represents the most viable, most likely, perhaps the most exciting location to focus explorations efforts in the search for life outside of Earth. Deep subsurface environments on Mars are believed to be the most promising place for the next step of exploration where liquid water is stable, has been protected from harmful exposures for millennia or longer, and thus potentially sustains a self-contained water-rock nutrient cycle and redox gradients that fuel life (Stamenković et al., 2019). Missions into the Martian subsurface are the new frontier of exploration, and we cannot know what discoveries are waiting there to transform our understanding about planetary processes, and potentially, life.

3.6 INSTRUMENTS AND MISSION DESIGN CONSIDERATIONS

Instrument selection for drilling missions is something that requires a different set of thought from previous landed missions. There are two types of instruments that can be used for inclusions into drilling missions. The first are instruments that require the sample be returned to the surface where analysis can take place while the second require the instruments to be lowered into the hole either as a standalone string that is incorporated into a larger drill stem or as part of a drill stem when the drill is not assembled on the surface.

For the case where instruments are present on a surface platform, the drill obviously needs to be able to collect samples and return them to the surface. This can be done either through retracting the entire drill string to the surface or through a bailer where only a small fraction of hardware is brought to the surface. In this case the mission will have to contribute mission resources to hardware that can remove the sample and present it to instruments on the surface. Instruments may be surface operated because of a variety of unique challenges including volume considerations, need for stability or other accommodations issues that make it impossible to be included onto the drill string. Sample return missions by their very nature would need samples brought to the surface so they can be cached for eventual analysis in terrestrial laboratories. Terrestrial laboratories are generally much better equipped to analyze material, given that the physical parameters of the instruments are not constrained for things like mass, power, volume, and data volume. Instrument accommodations on surface also have to take into account harsh environmental conditions of places such as Europa (radiation) or Mars (dust and temperature variations). Discussion of some of the issues of in situ instruments can be found elsewhere (Beegle et al., 2009).

For the case where instruments are part of the drill string, it requires that the instruments be developed into the drill string, where accommodations can be difficult. Downhole examinations have the advantage of minimizing exposure of samples to the space environment, thus preserving phases that may be affected by such exposure, such as condensed volatiles. Balanced against this is the fact that not all investigative techniques can be adapted for downhole investigations. The most advanced of these techniques is downhole reflectance spectroscopy which will be deployed on Mars on the 2020 ESA Rosalind Franklin Rover (De Sanctis et al., 2017). Other optical-based techniques, such as reflectance spectroscopy in other wavelength regions, induced fluorescence, or Raman and laser-induced breakdown spectroscopy, could all conceivably be deployed for downhole investigations as they rely on similar technologies as those associated with MA_MISS (i.e., fiber optic fed). It is worth noting that these techniques are highly complementary. For instance, reflectance spectroscopy at different wavelength regions can probe minerals at the single element level, determine oxidation states, elemental site occupancies, and framework type (Burns, 1993; Farmer, 1974; Karr, 1975). Raman spectroscopy is considered complementary to reflectance spectroscopy because it can provide information about phases that are infrared-inactive, as well as about compounds that may be undetectable by reflectance spectroscopy, or provide more specificity about, and sensitivity to, organic compounds (Cloutis et al., 2016; Karr, 1975).

Downhole instruments will have to handle the ambient conditions including vibration, dust, and the unique thermal environment present in the borehole. Since the instrument segment of the drill almost has to be deployed behind the drilling part of the drill stem, instrument design will have to incorporate the power, data, and telemetry cabling for the drill to function. They will also be volume constrained in order to fit within the diameter of the drill string. While the length of the segment that contains the instrument can be variable, the length adds to the complexity of the drill string. Most landed payloads have a diameter that depends on the faring that the upper stage can fit and have to fit behind the heat shield of entry decent and landing systems. When the entire drill string is longer than the launch and landing system can handle, it requires the drill stem to be automatically assembled on the surface, increasing cost, complexity, and mission risk.

As with all missions to planetary bodies, drill mission payloads would be selected in order to answer specific scientific questions. These scientific questions are organized into a science traceability matrix that define the parameters of the payload instruments. The highest-level requirements constrain mission parameters such as cost, mass, power, and data rate box that the instrument has to fit within.

3.6.1 Instrument Categories

For landed missions there are usually five different high-level science themes that are the reason they are conducted, including drilling missions. These are studies of:

- **Atmosphere**: Studying the atmosphere of a planet including understanding local weather and long-term atmospheric parameters.
- **Geology**: Studying the stratigraphy of material being drilled through.
- **Mineralogy**: Identifying mineral assemblages within the drill hole.
- **Chemistry**: Quantifying chemistry and elemental makeup of the subsurface.
- **Biology**: Studying relevant biological properties, looking for biosignatures and biomarkers, and determining habitability potential of a planet.

While these are general categories, some instruments may make measurements that fulfill multiple different measurement classes. For example, the chemical composition of a sample can be used help elucidate biology, as would be done by advanced gas chromatography/mass spectrometers. Additionally, instruments such as Raman spectrometers can perform mineral, chemical, and biological measurements depending on how that instrument is designed. Even given this reality, it is useful to assign instruments to a primary category for science planning purposes and the development of requirements for the mission.

The need for development of mission requirements cannot be understated. The trade space on how to develop an instrument really depends on the mission requirements and how they flow down to the instruments so that hardware choices can be made. For life detection missions, this includes the development of exact search parameters and target species. There currently no infallible true/false instruments when it comes to life detection. However, there are instrument concepts that are developed to detect specific targets species such as DNA, RNA, or proteins. For smaller organics, we must move beyond simple detection and into quantification of species like amino acids. Such a search strategy is usually referred to as the LEGO principal where we assume of the hundreds of possible abiotically created amino acids, life only uses a subset (McKay, 1997, 2004). However, an instrument to quantitate all amino acids in a sample might not be able to identify DNA or proteins. This could lead to the non-detection of life-as-we-don't-know-it.

Many individual instruments are being developed for NASA under programs such as PICASSO, MATISSE, DAHLI, and others. Many of these programs fund orbiter and surface instrument development. Only recently have some of these developments moved into the realm of including instruments into drill strings. Here we discuss general classes of instruments that can be part of drill strings.

3.6.1.1 Geology and Mineralogy

Imagers are perhaps the easiest instrument to be included into a drill. These instruments can identify subsurface sedimentary structure as well as grain size which elucidates how the formation was created. Current technology makes images resistant to the temperature and vibration that is expected to be found on most drill strings. The advance of the borehole camera, complete with LEDs, has led the way, and accommodations are possible. The key to accommodations is to make sure that the borehole is able to be illuminated and the lens of the camera does not become caked with fines. This can be mitigated by including multiple lens and detector systems throughout the instrument string. One important issue is making sure that cutting fines don't obfuscate the borehole, so a design needs to incorporate a way of being able to see the walls, including using illumination.

Seismometers can be included as either part of the drill string or on the surface. During operations, the seismometer as part of the drill string would have to be secured. A seismometer is the only tool capable of furnishing detailed global and regional information on the structure of a planet's interior. The primary disadvantages of micro-seismometers for use on drill missions is that the science return from a single station is limited and each seismometer generates potentially very high data rates. In addition, extremely good mechanical coupling to the planetary body is required so that the seismometer signal is due to natural seismic events rather than lander events.

Minerals are solid materials that form through geologic processes and have a characteristic composition, crystalline structure, and physical properties. Minerals are formed through volcanic, sedimentary metamorphic recrystallization, hydrothermal activity and weathering, among others. For analysis of mineralogy, one could either analyze the borehole with IR, UV/VIS, or Raman spectroscopy as the drill progresses. Spectrometers can identify minerals through the effects that occur when light interacts with solid materials through the absorption, emission, or reflectance of photons. Different wavelength ranges between UV and the mid-IR provide different types of information about the solid material under analysis. For more quantitative analysis of all mineral phases, ingesting some fines for XRD analysis is possible.

- UV/VIS spectroscopy can identify elements based on electron transitions between different states.
- Near-IR spectroscopy identifies molecular overtones and combinations of molecular vibrations, some of which are forbidden.
- Mid-IR spectroscopy can identify compounds through molecular movement. This includes symmetrical stretching, asymmetrical stretching, wagging, and other vibrational modes. Typically, mid-IR spectroscopy is performed in the 2.5–25-micron region.
- Far-IR spectroscopy involves the detection of molecular rotations through low-energy photon analysis. This technique is significantly more challenging to implement on an in situ platform than near-IR and mid-IR methods, due to the need for an efficient far-IR source and low-temperature system components (to minimize the thermal IR signal). Far-IR spectroscopy is

not generally considered to be a promising technique for mineralogical analysis on planetary missions.

- Raman spectroscopy provides information about chemical bonding and crystal structure. The method is relatively rapid and selective. Raman spectroscopy can be used to identify rock forming minerals, accessory minerals, and secondary minerals. Raman spectroscopy can also be used to identify organic molecules. Raman spectroscopy relies on the inelastic scattering of light from a monochromatic laser source in which a small fraction of laser light is scattered at frequencies above and below the laser frequency (Abbey et al., 2017; Eshelman et al., 2019; Razzell Hollis et al., 2020; Wang et al., 2004).

If a sample can be ingested into an instrument in a drill string, powder X-ray diffraction (XRD) is possible. XRD provides definitive mineralogical analysis of powdered rock samples and can distinguish between different crystalline phases at ~1% levels. XRD is the preferred method for mineralogical analysis of unknown samples in terrestrial laboratories. A main issue with this technique is developing sample handling hardware that both ingests and then cleans the sample from the instrument so that another sample can be analyzed.

For many of these instruments two major concerns are thermal and volume environments. The thermal environment which they will be expect to operate in will affect both the spectrometer system and the detectors. Many detectors require cooling, to reach optimum detection limits. Inside a borehole in a wall consisting of insulating fines created in the drill process it will most likely be difficult to control the thermal environment so instruments can operate. Volume will be an issue as well, as clever optical paths will have to be developed to ensure spatial and spectral resolution is achieved.

3.6.1.2 Chemistry

The present state of planetary surfaces and conditions that existed in the past can be understood by understanding the chemical nature of material in the surface and subsurface. Understanding the chemistry is vital to understanding MEPAG goals I–IV for Martian exploration (MEPAG, 2020).

Several different types of instruments have flown that specifically target chemistry of surface material. This includes the MECA instrument on the Phoenix Lander which identified perchlorate in material scooped from the surface (Hecht et al., 2009). MECA utilized ion selective electrodes (ISEs) to identify ions that were leached into solution from surface material. This was a major discovery and helped explain some of the results of the Viking gas chromatography/mass spectroscopy (GC-MS) on scooped material (Biemann and Lavoie, 1979; Biemann et al., 1976; Klein et al., 1992). Preforming elemental analysis can be done through X-ray fluorescence though a side wall, in much the same way APXS on MER and MSL occurred (Gellert et al., 2009). The chemistry of isotopes is probably too complicated for including in the borehole.

Most instruments focusing on chemistry investigations require ingestion of samples. Instruments will have to be designed for either a fixed number of samples, such as TEGA and MECA on Phoenix that had one-time-use cells to do their analysis or develop a system to reuse cells. Additionally, thermal environments with ovens and instrument power will have to be well understood in order to operate at optimum proficiency to answer scientific questions.

3.6.1.3 Biology

Terrestrial-based biology is the study of life, where it studies the structure, function, origin, evolution, and distribution of life. On Mars (as well as for future missions to Mars Enceladus or Europa), identifying life as we-don't-know-it as well as understanding the potential habitability of Mars in the past would be key to mission science objectives. On Earth, virtually anywhere there is liquid water, an energy source and access to carbon, life exists from the dry valleys of Antarctica surviving on photosynthesis to deep sea hydrothermal vents existing on volcanic effluent kilometers underneath the Earth's surface existing on chemical gradients. At one point in the Martian past, Mars had all the ingredients that life needs to thrive. If life still exists on Mars, it could exist in the subsurface where a drill mission would target. If life exists on Europa or Enceladus, it could be present in the ocean under the surface.

There are a number of important considerations common to many biological investigations on other planetary bodies. What does non-earth-centric biology look like and would we know it if we found it? What is the range of possible concentrations of any organisms that may be found? What detection limits are required to observe organic compounds in a given planetary materials matrix? What is the long-term stability of any reagents involved in assays and analyses? How many different analyses are required on the same sample and neighboring samples to provide confidence in the measurement results? These (and other) questions need at least a reasonable answer in order to properly design a biology instrument and experiment.

There is a large range of biology experiments that have been proposed as well as a wide range of measurements. For drill missions, instrument considerations for biology would be the same as the instruments described earlier. Volume, thermal and data rates would have to be considered as well as the same sample handling issues that other instruments have (McKay, 2004).

Finally, planetary protection and contamination control will be vital for any mission targeting biology. Cleaning and maintaining a sterile environment prior to launch would have to be a main mission constraint.

3.7   CONCLUDING PROSPECTS

We have traveled a little over 100 km on worlds beyond Earth, with the vast majority of this exploration being robotic. In this chapter, we have highlighted the scientific rationale and significance of exploring deeper. To date, the deepest humankind has drilled is the Kola Superdeep Borehole, a 9-inch diameter hole reaching an astonishing 12,262 m (7.5 mi) taking almost 20 years to drill. This endeavor highlights the scale of planetary exploration: this borehole and nearly 20 years of drilling reaches not even halfway to the Earth's mantle: it penetrates approximately only a third of the crust, the thin skin comprising less than 1% of the Earth yet provides an unprecedented wealth of information about the Earth's crust. The deepest hole drilled on another solid body is the 305-cm hole drilled during Apollo 17—only 0.02% of the Kola borehole. The lunar samples collected from the Deep Lunar Drill Strings have been called "among the most important sample collected" (Meyer, 2007a). Samples acquired from depth provide a record of the geological processes that gave rise to the observed surface features. Depth yields a time dimension in addition to a record of events, such as potential biosignatures left behind by long extinct life, that surficial processes such as erosion and oxidative damage have erased. Careful selection of drilling targets will yield unprecedented samples that have the potential to ground truth planetary formation mechanisms, date major solar system events linking planetary chronologies, and reach ancient environments that may have recorded the origin of life. Interplanetary drilling represents the next frontier in planetary exploration.

> Wherever he saw a hole he always wanted to know the depth of it. To him this was important.
> —Jules Verne, *Journey to the Center of the Earth*

ACKNOWLEDGMENTS

Some of the research reported in this chapter was conducted at the Jet Propulsion Laboratory (JPL), California Institute of Technology, under a contract with the National Aeronautics and Space Administration (NASA). The authors would like to thank Alfred William (Bill) Eustes III, Petroleum Engineering Department at the Colorado School of Mines, Golden, CO; and Yang Liu, Jet Propulsion Laboratory/California Institute of Technology, Pasadena, CA, for reviewing this chapter and providing valuable technical comments and suggestions.

REFERENCES

Abbey, W., Anderson, R., Beegle, L., Hurowitz, J., Williford, K., Peters, G., Morookian, J. M., Collins, C., Feldman, J., Kinnett, R., Jandura, L., Limonadi, D., Logan, C., McCloskey, S., Melko, J., Okon, A., Robinson, M., Roumeliotis, C., Seybold, C., . . . Warner, N. (2019). A look back: The drilling campaign of the Curiosity Rover during the Mars Science Laboratory's Prime Mission. *Icarus*, 319, 1–13. https://doi.org/10.1016/j.icarus.2018.09.004

Abbey, W. J., Bhartia, R., Beegle, L. W., DeFlores, L., Paez, V., Sijapati, K., Sijapati, S., Williford, K., Tuite, M., Hug, W., & Reid, R. (2017). Deep UV Raman spectroscopy for planetary exploration: The search for in situ organics. *Icarus*, 290, 201–214. https://doi.org/10.1016/j.icarus.2017.01.039

Abell, P. A., Vilas, F., Jarvis, K. S., Gaffey, M. J., & Kelley, M. S. (2010). Mineralogical composition of (25143) Itokawa 1998 SF36 from visible and near-infrared reflectance spectroscopy: Evidence for partial melting. *Meteoritics and Planetary Science*, 42, 2165–2177.

Acuña, M. H., Connerney, J. E. P., Ness, N. F., Lin, R. P., Mitchell, D., Carlson, C. W., McFadden, J., Anderson, K. A., Rème, H., Mazelle, C., Vignes, D., Wasilewski, P., & Cloutier, P. (1999). Global distribution of crustal magnetization discovered by the Mars Global Surveyor MAG/ER experiment. *Science*, 284(5415), 790–793. https://doi.org/10.1126/science.284.5415.790

Adams, J. & McCord, T. (1970). Remote sensing of lunar surface mineralogy: Implications from visible and near-infrared reflectivity of Apollo 11 samples. In: *Proceedings of the Apollo 11 Lunar Science Conference*. Geochimica eta Cosmochimica Acta Supplement, vol. 3. Pergamon, New York, pp. 1937–1945.

Adams, J. B., Charette, M. P., & Rhodes, J. M. (1975). Chemical fractionation of the lunar regolith by impact melting. *Science*, 190(4212), 380–381.

Agee, C. B., N. V. Wilson, F. M. McCubbin, K. Ziegler, V. J. Polyak, Z. D. Sharp, Y. Asmerom, M. H. Nunn, R. Shaheen, M. H. Thiemens et al. (2013). Unique meteorite from Early Amazonian Mars: Water-rich basaltic breccia Northwest Africa 7034. *Science*, 339(6121), 780–785.

Akhmanova, M., Dement'ev, B., & Markov, M. (1978). Possible water in Luna 24 Regolith from the Sea of Crises. *Geochemistry International*, 15(166).

Allton, J. H., and Waltz, S. R. (1980). Depth scales for Apollo 15, 16, and 17 drill cores. *Proceedings from the 11th Lunar and Planetary Science Conference*, 1463–1477.

Altan, H. (1973). Effects of heavy ions on bacteria. *Life Sciences in Space Research*, 11, 273–280.

Anderson, R. C., Jandura, L., Okon, A. B., Sunshine, D., Roumeliotis, C., Beegle, L. W., Hurowitz, J., Kennedy, B., Limonadi, D., McCloskey, S., Robinson, M., Seybold, C., & Brown, K. (2012). Collecting samples in Gale Crater, Mars; an overview of the Mars Science Laboratory Sample Acquisition, Sample Processing and Handling System. *Space Science Reviews*, 170(1–4), 57–75. https://doi.org/10.1007/s11214-012-9898-9

Archer, S. D. J., de los Ríos, A., Lee, K. C., Niederberger, T. S., Cary, S. C., Coyne, K. J., Douglas, S., Lacap-Bugler, D. C., & Pointing, S. B. (2017). Endolithic microbial diversity in sandstone and granite from the McMurdo Dry Valleys, Antarctica. *Polar Biology*, 40(5), 997–1006. https://doi.org/10.1007/s00300-016-2024-9

Arnold, J. R. (1979). Ice in the lunar polar regions. *Journal of Geophysical Research*, 84(B10), 5659–5668.

Arp, G., Reimer, A., Simon, K., Sturm, S., Wilk, J., Kruppa, C., Hecht, L., Hansen, B. T., Pohl, J., Reimold, W. U., Kenkmann, T., & Jung, D. (2019a). The Erbisberg drilling 2011: Implications for the structure and postimpact evolution of the inner ring of the Ries impact crater. *Meteoritics and Planetary Science*, 54(10), 2448–2482. https://doi.org/10.1111/maps.13293

Arp, G., Schultz, S., Karius, V., & Head, J. W. (2019b). Ries impact crater sedimentary conglomerates: Sedimentary particle "impact pre-processing", transport distances and provenance, and implications for Gale crater conglomerates, Mars. *Icarus*, 321, 531–549. https://doi.org/10.1016/j.icarus.2018.12.003

Baker, D. M. H., Head, J. W., Fassett, C. I., Kadish, S. J., Smith, D. E., Zuber, M. T., & Neumann, G. A. (2011). The transition from complex crater to peak-ring basin on the moon: New observations from the Lunar Orbiter Laser Altimeter (LOLA) instrument. *Icarus*, 214(2), 377–393. https://doi.org/10.1016/j.icarus.2011.05.030

Baker, V. R., Carr, M. H., Gulick, V. C., Williams, C. R., & Marley, M. S. (1992). Channels and valley networks. *Mars*, 493–522.

Baldwin, R. B. (1964). Lunar crater counts. *The Astronomical Journal*, 69(5), 377–392.

Bar-On, Y. M., Phillips, R., & Milo, R. (2018). The biomass distribution on Earth. *Proceedings of the National Academy of Sciences*, 115(25), 6506–6511. https://doi.org/10.1073/pnas.1711842115

Barsukov, V. L. (1977). Preliminary data for the regolith core brought to earth by the automatic lunar station Luna 24. *Proceedings from the 8th Lunar Science Conference*, 3, 3303–3318.

Basu, A., & Bower, J. F. (1977). Provenance of Apollo 15 deep drill core sediments. *Proceedings from the 8th Lunar Science Conference*, 3, 2841–2867.

Basu, A., McKay, D. S., & Fruland, R. M. (1978). Origin and modal petrography of Luna 24 soils. In R. B. Merrill & J. J. Papike (Eds.), *Mare Crisium: The View from Luna 24 (pp. 321–337)*. Pergamon Press.

Beaty, D. W., Grady, M. M., McSween, H. Y., Sefton-Nash, E., Carrier, B. L., Altieri, F., Amelin, Y., Ammannito, E., Anand, M., Benning, L. G. et al. (2019). The potential science and engineering value of samples delivered to Earth by Mars Sample Return: International MSR Objectives and Samples Team (iMOST). *Meteoritics & Planetary Science*, 54, S3–S152

Beegle, L. W., Feldman, S., Johnson, P., & Dreyer, C. B. (2009). Instruments for in-situ sample analysis. In: *Drilling in Extreme Environments: Penetration and Sampling on Earth and Other Planets* (Y. Bar-Cohen and K. Zacny editors). John Wiley & Sons, pp. 643–706. ISBN: 978-3-527-40852-8.

Bell, E. A., Boehnke, P., Harrison, T. M., & Mao, W. L. (2015). Potentially biogenic carbon preserved in a 4.1 billion-year-old zircon. *Proceedings of the National Academy of Sciences of the United States of America*, 112(47), 14518–14521. https://doi.org/10.1073/pnas.1517557112

Bibring, J.-P., Langevin, Y., Gendrin, A., Gondet, B., Poulet, F., Berthé, M., Soufflot, A., Arvidson, R., Mangold, N., Mustard, J., et al. (2005). Mars surface diversity as revealed by the OMEGA/Mars Express observations. *Science*, 307(5715), 1576–1581.

Biemann, K. & Lavoie, J. M. (1979). Some final conclusions and supporting experiments related to the search for organic-compounds on the surface of mars. *Journal of Geophysical Research*, 84, 8385–8390.

Biemann, K. et al. (1976). Search for organic and volatile inorganic-compounds in 2 surface samples from Chryse-Planitia region of Mars. *Science*, 194(4260), 72–76.

Bierhaus, E. B., Chapman, C. R., & Merline, W. J. (2005). Secondary craters on Europa and implications for cratered surfaces. *Nature*, 437(7062), 1125.

Bierhaus, E. B., Clark, B. C., Harris, J. W., Payne, K. S., Dubisher, R. D., Wurts, D. W., Hund, R. A., Kuhns, R. M., Linn, T. M., Wood, J. L., May, A. J., Dworkin, J. P., Beshore, E., Lauretta, D. S., & the OSIRIS-REx Team. (2018). The OSIRIS-REx Spacecraft and the Touch-and-Go Sample Acquisition Mechanism (TAGSAM). *Space Science Reviews*, 214, 107.

Binder, A. B., Arvidson, R. E., Guinness, E. A., Jones, K. L., Morris, E. C., Mutch, T. A., Pieri, D. C., & Sagan, C. (1977). The geology of the Viking Lander 1 site. *Journal of Geophysical Research*, 82(28), 4439–4451.

Binzel, R. P., Rivkin, A. S., Bus, S. J., & Burbine, T. H. (2010). MUSES-C target asteroid (25143) 1998 SF36: A reddened ordinary chondrite. *Meteoritics and Planetary Science*, 36, 1167–1172.

Bishop, J. L., Fairén, A. G., Michalski, J. R., Gago-Duport, L., Baker, L. L., Velbel, M. A., Gross, C., & Rampe, E. B. (2018). Surface clay formation during short-term warmer and wetter conditions on a largely cold ancient Mars. *Nature Astronomy*. https://doi.org/10.1038/s41550-017-0377-9

Boehnke, P., & Harrison, T. M. (2016). Illusory late heavy bombardments. *Proceedings of the National Academy of Sciences*, 113(39), 10802–10806. https://doi.org/10.1073/pnas.1611535113

Boston, P. J., Ivanov, M. V., & McKay, C. P. (1992). On the possibility of chemosynthetic ecosystems in subsurface habitats on Mars. *Icarus*. doi:10.1016/0019-1035(92)90045-9.

Bramble, M. S., Mustard, J. F., & Salvatore, M. R. (2017). The geological history of Northeast Syrtis Major, Mars. *Icarus*, doi:10.1016/j.icarus.2017.03.030.

Bramson, A. M., Byrne, S., Putzig, N. E., Sutton, S., Plaut, J. J., Brothers, T. C., & Holt, J.W. (2015). Widespread excess ice in Arcadia Planitia, Mars. *Geophysical Research Letters*, 42(16), 6566–6574.

Bridges, J. C. & Warren, P. (2006). The SNC meteorites: Basaltic igneous processes on Mars. *Journal of the Geological Society*, 163(2), 229–251.

Bristow, T. F., Haberle, R. M., Blake, D. F., Marais, D. J. D., Eigenbrode, J. L., Fairén, A. G., Grotzinger, J. P., Stack, K. M., Mischna, M. A., Rampe, E. B., Siebach, K. L., Sutter, B., Vaniman, D. T., & Vasavada, A. R. (2017). Low Hesperian PCO2 constrained from in situ mineralogical analysis at Gale Crater, Mars. *Proceedings of the National Academy of Sciences*, 114(9), 2166–2170. https://doi.org/10.1073/pnas.1616649114

Bristow, T. F. et al. (2018). Clay mineral diversity and abundance in sedimentary rocks of Gale crater, Mars. *Science Advances*, doi:10.1126/sciadv.aar3330.

Britt, D. T. & Pieters, C. M . (1998). Bidirectional reflectance properties of iron-nickel meteorites. *Proceedings of the Lunar and Planetary Science Conference*, 18, 503–512.

Britt, D. T., Yeomans, D., Housen, K., & Consolmagno, G. (2002). Asteroid density, porosity, and structure. In: *Asteroids III* (W. Bottke, A. Cellino, P. Paolicchi, and R.P. Binzel editors). University of Arizona Press, Tucson, AZ, pp. 485–500.

Burns, R. G. (1993). *Mineralogical Applications of Crystal Field Theory*, 2nd edn. Cambridge University Press, Cambridge.

Cabrol, N., Grin, E., Landheim, R., Kuzmin, R., & Greeley, R. (1998). Duration of the Ma'adim Vallis/Gusev crater hydrogeologic system, Mars. *Icarus*, 133(1), 98–108.

Caprarelli, G., & Orosei, R. (2015). Probing the hidden geology of Isidis Planitia (Mars) with impact craters. *Geosciences (Switzerland)*, doi:10.3390/geosciences5010030.

Carr, M. H. (1995). The Martian drainage system and the origin of valley networks and fretted channels. *Journal of Geophysical Research: Planets*, 100(E4), 7479–7507.

Carr, M. H. (1996). *Water on Mars*. University of Oxford Publishing, Oxford.

Chamberlain, T. P. & Hunten, D. M. (1990). *Theory of Planetary Atmospheres: An Introduction to Their Physics and Chemistry*. Academic Press, p. 36.

Chapman, C. R., Cohen, B. A., & Grinspoon, D. H. (2007). What are the real constraints on the existence and magnitude of the late heavy bombardment? *Icarus*, 189(1), 233–245. https://doi.org/10.1016/j.icarus.2006.12.020

Cheng, A. F., Izenberg, N., Chapman, C. R., & Zuber, N. T. (2002). Ponded deposits on asteroid 433 Eros. *Meteoritics and Planetary Science*, 37, 1095–1105.

Chou, C. L., Boynton, W. V., Sundberg, L. L., & Wasson, J. T. (1975). Volatiles on the surface of Apollo 15 green glass and volatile-element systematics in soils from edges of Mare Basins. *Lunar and Planetary Science Conference*, 6, 137.

Clark, B. E., Hapke, B., Pieters, C., & Britt, D. (2002). Asteroid space weathering and regolith evolution. In: *Asteroids III* (W. Bottke, A. Cellino, P. Paolicchi, and R.P. Binzel editors). University of Arizona Press, Tucson, AZ, pp. 585–599.

Clifford, S., Bianchi, R., De Sanctis, M., Duke, M., Kim, S., Mancinelli, R., Ming, D., Passey, Q., Smrekar, S., & Beaty, D. (2001). *Science Rationale and Priorities for Subsurface Drilling: Final Report*.

Clifford, S. M., & Parker, T. J. (2001). *The evolution of the Martian hydrosphere: Implications for the fate of a primordial ocean and the current state of the Northern plains: Icarus*, doi:10.1006/icar.2001.6671.

Cloutis, E., Szymanski, P., Applin, D., & Goltz, D. (2016). Identification and discrimination of polycyclic aromatic hydrocarbons using Raman spectroscopy. *Icarus*, 274, 211–230.

Cloutis, E. A., Hudon, P., Hiroi, T., & Gaffey, M. J. (2012). Spectral reflectance properties of carbonaceous chondrites 4: Aqueously altered and thermally metamorphosed meteorites. *Icarus*, 220, 586–617.

Cockell, C. S. (2014). Trajectories of Martian habitability. *Astrobiology*, 14(2), 182–203. https://doi.org/10.1089/ast.2013.1106

Cockell, C. S., Catling, D. C., Davis, W. L., Snook, K., Kepner, R. L., Lee, P., & McKay, C. P. (2000). The ultraviolet environment of Mars: Biological implications past, present, and future. *Icarus*, 146(2), 343–359. https://doi.org/10.1006/icar.2000.6393

Cohen, B. A., Swindle, T. D., & Kring, D. A. (2000). Support for the lunar cataclysm hypothesis from lunar meteorite impact melt ages. *Science*, 290(5497), 1754–1756. https://doi.org/10.1126/science.290.5497.1754

Committee on the Planetary Science Decadal Survey Space Studies Board Division on Engineering and Physical Sciences (2001). *VISION and VOYAGES for Planetary Science in the Decade 2013–2022*. The National Academies Press, Washington, DC.

Consolmagno, G., Britt, D. T., & Stoll, C. P. (2010). The porosity of ordinary chondrites: Models and interpretation. *Meteoritics and Planetary Science*, 33, 1221–1229.

Cottin, H., Kotler, J. M., Bartik, K., Cleave, H. J., Cockell, C. S., de Vera, J.-P. P., Ehrenfreund, P., Leuko, S., Ten Kate, I. L., Martins, Z., Pascal, R., Quinn, R., Rettberg, P., & Westall, F. (2015). Astrobiology and the possibility of life on earth and elsewhere... *Space Science Reviews*, 209, 1–42. doi: 10.1007/s11214-015-0196-1.

Crawford, I. A. (2006). The astrobiological case for renewed robotic and human exploration of the moon. *International Journal of Astrobiology*, 5(3), 191–197. https://doi.org/10.1017/S1473550406002990

Crawford, I. A., Anand, M., Cockell, C. S., Falcke, H., Green, D. A., Jaumann, R., & Wieczorek, M. A. (2012). Back to the moon: The scientific rationale for resuming lunar surface exploration. *Planetary and Space Science*, 74(1), 3–14. https://doi.org/10.1016/j.pss.2012.06.002

Crawford, I. A., Fagents, S. A., Joy, K. H., & Rumpf, M. E. (2010). Lunar palaeoregolith deposits as recorders of the galactic environment of the solar system and implications for astrobiology. *Earth, Moon and Planets*, 107(1), 75–85. https://doi.org/10.1007/s11038-010-9358-z

Crawford, Ian A., & Joy, K. H. (2014). Lunar exploration: Opening a window into the history and evolution of the inner solar system. *Philosophical Transactions of the Royal Society A: Mathematical, Physical and Engineering Sciences*, 372(2024), 1–21. https://doi.org/10.1098/rsta.2013.0315

Crotts, A. (2011). Water on the Moon, I. Historical overview. *Astronomical Review*, 6(7), 4–20. https://doi.org/10.1080/21672857.2011.11519687

Dartnell, L. R., Desorgher, L., Ward, J. M., & Coates, A. J. (2007). Modelling the surface and subsurface Martian radiation environment: Implications for astrobiology. *Geophysical Research Letters*, 34(2). https://doi.org/10.1029/2006GL027494

Dartnell, L. R., Patel, M. R., Storrie-Lombardi, M. C., Ward, J. M., & Muller, J.-P. (2012). Experimental determination of photostability and fluorescence-based detection of PAHs on the Martian surface. *Meteoritics & Planetary Science*, 47(5), 806–819. https://doi.org/10.1111/j.1945-5100.2012.01351.x

Darwin, C. 1859, *On the Origin of Species* Published by John Murray: London.

Davidsson, B. J. R., Sierks, H., Güttler, C., Marzari, F., Pajola, M., Rickman, H., A'Hearn, M. F., Auger, A.-T., El-Maarry, M. R., Fornasier, S., Gutiérrez, P. J., Keller, H. U., Massironi, M., Snodgrass, C., Vincent, J.-B., Barbieri, C., Lamy, P. L., Rodrigo, R., Koschny D. ... Tubiana, C. (2016). The primordial nucleus of comet 67P/Churyumov-Gerasimenko. *Astronomy and Astrophysics*, 562, A63 (30 pp.).

De Sanctis, M. C., Altieri, F., Ammannito, E., Biondi, D., De Angelis, S., Meini, M., Mondello, G., Novi, S., Paolinetti, R., Soldani, M., Mugnuolo, R., Pirrotta, S., Vago, J. L., & the Ma_MISS Team (2017). Ma_MISS on ExoMars: Mineralogical characterization of the Martian subsurface. *Astrobiology*, 17, 612–620.

Delbo, M. & Michel, P. (2011). Temperature history and dynamical evolution of (101955) 1999 RQ36: A potential target for sample return from a primitive asteroid. *Astrophysical Journal Letters*, 782, L42 (5 pp.).

DellaGiustina, D. N., Emery, J. P., Golish, D. R., Rozitis, B., Bennett, C. A., Burke, K. N., Ballouz, R.-L., Becker, K. J., Christensen, P. R., Drouet d'Aubigny, C. Y., Hamilton, V. E., Reuter, D. C., Rizk, B., Simon, A. A., Asphaug, E., Bandfield, J. L., Barnouin, O. S., Barucci, M. A., Bierhaus E. B. ... the OSIRIS-REx Team (2019). Properties of rubble-pile asteroid (101955) Bennu from OSIRIS-REx imaging and thermal analysis. *Nature Astronomy*, 3, 341–351.

DeMeo, F. E., Binzel, R. P., Slivan, S. M., & Bus, S. J. (2009). An extension of the Bus asteroid taxonomy into the near-infrared. *Icarus*, 202, 160–180.

Des Marais, D. J. (2010). Exploring Mars for evidence of habitable environments and life. *Proc Am Philos Soc*, 154, 402–421.

Dismukes, G. C., Klimov, V. V., Baranov, S. V., Kozlov, Y. N., DasGupta, J., & Tyryshkin, A. (2001). The origin of atmospheric oxygen on Earth: The innovation of oxygenic photosynthesis. *Proceedings of the National Academy of Sciences*, 98(5), 2170–2175. https://doi.org/10.1073/pnas.061514798

Duke, M. B., Gaddis, L. R., Taylor, G. J., & Schmitt, H. H. (2006). Development of the Moon. In B. L. Jolliff, M. A. Wieczorek, C. K. Shearer, & C. R. Neal (Eds.), *Reviews in Mineralogy and Geochemistry*, 60, 597–656. Mineralogical Society of America. https://doi.org/10.2138/rmg.2006.60.6

Duke, M. B., & Nagle, J. S. (1976). *Lunar core catalog* (Publicatio). NASA.

Dzaugis, M., Spivack, A. J., & Hondt, S. (2018). Radiolytic H2 production in Martian environments. *Astrobiology*, 18(9), 1137–1146. https://doi.org/10.1089/ast.2017.1654

Edwards, C. S., & Ehlmann, B. L. (2015). Carbon sequestration on Mars. *Geology*, 43(10), 863–866. https://doi.org/10.1130/G36983.1

El-Maarry, M. R., Groussin, O., Thomas, N., Pajola, M., Auger, A.-T., Davidsson, B., Hu, X., Hviid, S. F., Knollenberg, J., Güttler, C., Tubiana, C., Fornasier, S., Feller, C., Hasselmann, P., Vincent, J.-B., Sierks, H., Barbieri, C., Lamy, P., Rodrigo, R. ... Shi, X. (2017). Surface changes on comet 67P/Churyumov-Gerasimenko suggest a more active past. *Science*, 355, 1392–1395.

Eshelman, E. J., Malaska, M. J., Manatt, K. S., Doloboff, I. J., Wanger, G., Willis, M. C., Abbey, W. J., Beegle, L. W., Priscu, J. C., & Bhartia, R. (2019). WATSON: In situ organic detection in subsurface ice using deep-UV fluorescence spectroscopy. *Astrobiology*, 19(6) 771–784. doi: 10.1089/ast.2018.1925.

Fa, W., Zhu, M.-H., Liu, T., & Plescia, J. B. (2015). Regolith stratigraphy at the Chang'E-3 landing site as seen by lunar penetrating radar. *Geophysical Research Letters*, 42(23), 10179–10187. https://doi.org/10.1002/2015GL066537

Fairén, A. G., Davila, A. F., Lim, D., Bramall, N., Bonaccorsi, R., Zavaleta, J., Uceda, E. R., Stoker, C., Wierzchos, J., Dohm, J. M., Amils, R., Andersen, D., & McKay, C. P. (2010). Astrobiology through the ages of Mars: The study of terrestrial analogues to understand the habitability of mars. *Astrobiology*, 10, 821–843.

Farley, K., Stack Morgan, K., & Williford, K. (2018). *Jezero-Midway interellipse traverse mission concept*. Fourth landing site selection workshop for Mars 2020. Pasadena, California.

Farley, K. A., Malespin, C., Mahaffy, P., Grotzinger, J. P., Vasconcelos, P. M., Milliken, R. E., Malin, M., Edgett, K. S., Pavlov, A. A., Hurowitz, J. A., Grant, J. A., Miller, H. B., Arvidson, R., Beegle, L., Calef, F., Conrad, P. G., Dietrich, W. E., Eigenbrode, J., Gellert, R., . . . Team, the M. S. (2014). In situ radiometric and exposure age dating of the Martian surface. *Science*, 343(6169). https://doi.org/10.1126/science.1247166

Farmer, V. C. (1974). *Infrared Spectra of Minerals*. Mineralogical Society of Great Britain and Ireland, London.

Fassett, C. I. & Head, J. W., III. (2005). Fluvial sedimentary deposits on Mars: Ancient deltas in a crater lake in the Nili Fossae region. *Geophysical Research Letters*, 32(14), L14201.

Fassett, C. I. & Head, J. W., III. (2008). The timing of Martian valley network activity: Constraints from buffered crater counting. *Icarus*, 195(1), 61–89.

Fassett, C. I. & Head, J. W. (2011). Sequence and timing of conditions on early Mars. *Icarus*, 211, 1204–1214.

Feldman, W. C., Maurice, S., Lawrence, D. J., Little, R. C., Lawson, S. L., Gasnault, O., Wiens, R. C., Barraclough, B. L., Elphic, R. C., Prettyman, T. H., Steinberg, J. T., & Binder, A. B. (2001). Evidence for water ice near the lunar poles. *Journal of Geophysical Research E: Planets*, 106(E10), 23231–23251. https://doi.org/10.1029/2000JE001444

Fisk, M. R., & Giovannoni, S. J. (1999). Sources of nutrients and energy for a deep biosphere on Mars. *Journal of Geophysical Research E: Planets*, doi:10.1029/1999JE900010.

Flahaut, J., Blanchette-Guertin, J. F., Jilly, C., Sharma, P., Souchon, A., Van Westrenen, W., & Kring, D. A. (2012). Identification and characterization of science-rich landing sites for lunar lander missions using integrated remote sensing observations. *Advances in Space Research*, 50(12), 1647–1665. https://doi.org/10.1016/j.asr.2012.05.020

Florensky, C. P., Basilevsky, A. T., Ivanov, A. V., Pronin, A. A., & Rode, O. D. (1977). Luna 24: Geologic Setting of landing site and characteristic of sample core (preliminary data). *Proceedings from the 8th Lunar Science Conference*, 3, 3257–3279.

Gaffey, M. J. (2010). Space weathering and the interpretation of asteroid reflectance spectra. *Icarus*, 209, 564–574.

Garenne, A., Beck, P., Montes-Hernandez, G., Brissaud, O., Schmitt, B., Quirico, E., Bonal, L., Beck, C., & Howard, K. T. (2016). Bidirectional reflectance spectroscopy of carbonaceous chondrites: Implications for water quantification and primary composition. *Icarus*, 264, 172–183.

Garrick-Bethell, I. & Carr, C. E. (2007). Working and walking on small asteroids with circumferential ropes. *Acta Astronautica*, 61, 1130–1135.

Garrick-Bethell, I., Head, J. W., & Pieters, C. M. (2011). Spectral properties, magnetic fields, and dust transport at lunar swirls. *Icarus*, 212(2), 480–492. https://doi.org/10.1016/j.icarus.2010.11.036

Gellert, R., Campbell, J., King, P., Leshin, L., Lugmair, G., Spray, J., Squyres, S., & Yen, A. (2009). The Alpha-Particle-X-ray-Spectrometer (APXS) for the Mars Science Laboratory (MSL) Rover mission. 2364.

Gibson, M. A., & Knudsen, C. W. (1985). Lunar oxygen production from Ilmenite. In W. W. Mendell (Ed.), *Lunar Bases and Space Activities of the 21st Century* (p. 543).

Gillis-Davis, J. J., Lucey, P. G., Bradley, J. P., Ishii, H. A., Kaluna, H. M., Misra, A., & Connolly, H. C., Jr. (2017). Incremental laser space weathering of Allende reveals non-lunar like space weathering effects. *Icarus*, 286, 1–14.

Gohn, G. S., Koeberl, C., Miller, K. G., Reimold, W. U., Browning, J. V., Cockell, C. S., Horton, J. W., Kenkmann, T., Kulpecz, A. A., Powars, D. S., Sanford, W. E., & Voytek, M. A. (2008). Deep drilling into the Chesapeake Bay impact structure. *Science*, 320(5884), 1740–1745. https://doi.org/10.1126/science.1158708

Gohn, G. S., Koeberl, C., Miller, K. G., Reimold, W. U., Cockell, C. S., Horton, J. W., Sanford, W. E., & Voytek, M. A. (2006). Chesapeake Bay impact structure drilled. *Eos*, 87(35), 349–351. https://doi.org/10.1029/2006EO350001

Gorevan, S. P., Myrick, T., Davis, K., Chau, J. J., Bartlett, P., Mukherjee, S., Anderson, R., Squyres, S. W., Arvidson, R. E., Madsen, M. B., Bertelsen, P., Goetz, W., Binau, C. S., & Richter, L. (2003). Rock Abrasion Tool: Mars Exploration Rover mission. *Journal of Geophysical Research: Planets*, 108(E12). https://doi.org/10.1029/2003JE002061

Goudge, T. A., Milliken, R. E., Head, J. W., Mustard, J. F., & Fassett, C. I. (2017). Sedimentological evidence for a deltaic origin of the western fan deposit in Jezero crater, Mars and implications for future exploration. *Earth and Planetary Science Letters*, 458, 357–365. https://doi.org/10.1016/j.epsl.2016.10.056

Grotzinger, J. P., Gupta, S., Malin, M. C., Rubin, D. M., Schieber, J., Siebach, K., Sumner, D. Y., Stack, K. M., Vasavada, A. R., Arvidson, R. E. & Calef, F. (2015). Deposition, exhumation, and paleoclimate of an ancient lake deposit, Gale crater, Mars. *Science*, 350(6257).

Gulick, S., Morgan, J., Mellet, C. L., Lofi, J., Chenot, E., Christeson, G., Claeys, P., Cockell, C., Coolen, M. J. L., Ferrière, L., Gebhardt, C., Goto, K., Jones, H., Kring, D. A., Lowery, C. M., Ocampo-Torres, R., Perez-Cruz, L., Pickersgill, A. E., Poelchau, M., . . . Bralower, T. J. (2017). Expedition 364 preliminary report: Chicxulub: Drilling the K-Pg impact crater. *In International Ocean Discovery Program*. https://doi.org/10.14379/iodp.sp.364.2016

Hamilton, V. E., Simon, A. A., Christensen, P. R., Reuter, D. C., Clark, B. E., Barucci, M. A., Bowles, N. E., Boynton, W. V., Brucato, J. R., Cloutis, E. A., Connolly H. C., Jr., Donaldson Hanna, K. L., Emery, J. P., Enos, H. L., Fornasier, S., Haberle, C. W., Hanna, R. D., Howell, E. S., Kaplan, H. H. ... the OSIRIS-REx Team. (2019). Evidence for widespread hydrated minerals on asteroid (101955) Bennu. *Nature Astronomy*, 3, 332–340.

Hapke, B. (2001). Space weathering from Mercury to the asteroid belt. *Journal of Geophysical Research*, 106, 10,039–10,073.

Hartmann, W. K. (1965). Terrestrial and lunar flux of large meteorites in the last two billion years. *Icarus*, 4(2), 157–165. https://doi.org/10.1016/0019-1035(65)90057-6

Hartmann W. K. (1973). Ancient lunar mega-regolith and subsurface structure. *Icarus*, 18, 634–636.

Hartmann, W. K. (2003). Megaregolith evolution and cratering cataclysm models—Lunar cataclysm as a misconception (28 years later). *Meteoritics & Planetary Science*, 38(4), 579–593. https://doi.org/10.1111/j.1945-5100.2003.tb00028.x

Hartmann, W. K., Tholen, D. J., Meech, K. J., & Cruikshank, D. P. (1990). 2060 Chiron: Colorimetry and cometary behavior. *Icarus*, 83, 1–15.

Haskin, L. A., Gillis J., Korotev R. L., & Jolliff, L. (2000). The materials of the lunar Procellarum KREEP Terrane: A synthesis of data from geomorphological mapping, remote sensing, and sample analyses. *Journal of Geophysical Research*, 105, 20403–20415.

Head, J.W. (1974). Orientale multi-ringed basin interior and implications for the petrogenesis of lunar highland samples. *The Moon*, 11(3–4), 327–356.

Head, J. W. (1976). Lunar volcanism in space and time. *Reviews of Geophysics*, 14(2), 265–300.

Head, J. W., & Wilson, L. (1992). Lunar mare volcanism: Stratigraphy, eruption conditions, and the evolution of secondary crusts. *Geochimica et Cosmochimica Acta*, 56(6), 2155–2175. https://doi.org/10.1016/0016-7037(92)90183-J

Hecht, M. H. (2002). Metastability of liquid water on Mars. *Icarus*, 156(2), 373–386.

Hecht, M. H. et al. (2009). Detection of perchlorate and the soluble chemistry of Martian soil at the Phoenix Lander site. *Science*, 325(5936), 64–67. doi: 10.1126/science.1172466.

Heiken, G. H., Vaniman, D. T., & French, B. M. (Eds.) (1991). *Lunar Sourcebook: A User's Guide to the Moon*. Cambridge University Press, Cambridge, UK.

Heiken, G. (1975). Petrology of lunar soils. *Reviews of Geophysics*, 13(4), 567–587. https://doi.org/10.1029/RG013i004p00567

Heiken, G. H., & McKay, D. S. (1974). Petrography of Apollo 17 soils. *Proceedings of the 5th Lunar Science Conference*, 1, 843–860.

Heiken, G. H., McKay, D. S., & Brown, R. W. (1974). Lunar deposits of possible pyroclastic origin. *Geochimica et Cosmochimica Acta*, 38(11), 1703–1718. https://doi.org/10.1016/0016-7037(74)90187-2

Heiken, G. H., McKay, D. S., & Fruland, R. M. (1973). Apollo 16 soils: Grain size analyses and petrography. *Proceedings of the 4th Lunar Science Conference*, 1, 251–265.

Heiken, G. H., Morris, R. V., & McKay, D. S. (1976). Petrographic and ferromagnetic resonance studies of the Apollo 15 deep drill core. *Abstract of the Lunar and Planetary Science Conference*, 7, 361–363.

Hiesinger, H., Head, J. W., Wolf, U., Jaumann, R., & Neukum, G. (2003). Ages and stratigraphy of mare basalts in Oceanus Procellarum, Mare Nubium, Mare Cognitum, and Mare Insularum. *Journal of Geophysical Research E: Planets*, 108(E7), 1–27. https://doi.org/10.1029/2002je001985

Hiesinger, H., Head, J. W., Wolf, U., Jaumann, R., & Neukum, G. (2011). Ages and stratigraphy of lunar mare basalts: A synthesis. *In Special Paper of the Geological Society of America* (Vol. 477). https://doi.org/10.1130/2011.2477(01)

Hirata, N. & Ishiguro, M. (2012). *Properties and possible origin of black boulders on the asteroid Itokawa*. In: *Lunar and Planetary Science Conference*, vol. 42, abstract #1821. The Woodlands, TX, USA.

Hiroi, T., Pieters, C. M., Zolensky, M. E., & Lipschutz, E. (1993). Evidence of thermal metamorphism on the C, G, B, and F asteroids. *Science*, 261, 1016–1018.

Hoehler, T. M., & Jørgensen, B. B. (2013). Microbial life under extreme energy limitation. *Nature Reviews Microbiology*, 11(2), 83–94. https://doi.org/10.1038/nrmicro2939

Hood, L., P. Coleman, and D. Wilhelms (1979). The Moon: Sources of the crustal magnetic anomalies. *Science*, 204(4388), 53–57.

Hu, R., Kass, D. M., Ehlmann, B. L., & Yung, Y. L. (2015). Tracing the fate of carbon and the atmospheric evolution of Mars. *Nature Communications*, 6. https://doi.org/10.1038/ncomms10003

Hughes, D. W. (1975). Cometary outburst, a brief survey. *Quarterly Journal of the Royal Astronomical Society*, 16, 410–427.

Hutton, J. (1788). Theory of the Earth; or an investigation of the laws observable in the composition, dissolution, and restoration of land upon the Globe. *Transactions of the Royal Society of Edinburgh*, vol. 1, Part 2, pp. 209–304.

Hynek, B. M., M. Beach, and M. R. Hoke (2010). Updated global map of Martian valley networks and implications for climate and hydrologic processes. *Journal of Geophysical Research: Planets*, 115(E9), E09008.

Jaumann, R., Hiesinger, H., Anand, M., Crawford, I. A., Wagner, R., Sohl, F., Jolliff, B. L., Scholten, F., Knapmeyer, M., Hoffmann, H., Hussmann, H., Grott, M., Hempel, S., Köhler, U., Krohn, K., Schmitz, N., Carpenter, J., Wieczorek, M., Spohn, T., . . . Oberst, J. (2012). Geology, geochemistry, and geophysics of the Moon: Status of current understanding. *Planetary and Space Science*, 74(1), 15–41. https://doi.org/10.1016/j.pss.2012.08.019

Jaumann, R., Schmitz, N., Ho, T.-M., Schröder, S. E., Otto, K. A., Stephan, K., Elgner, S., Krohn, K., Preusker, F., Scholten, F., Biele, J., Ulamec, S., Krause, C., Sugita, S., Matz, K.-D., Roatsch, T., Parekh, R., Mottola, S., Grott, M. . . . Kouyama, T. (2019). Images from the surface of asteroid Ryugu show rocks similar to carbonaceous chondrite meteorites. *Science*, 365, 817–820.

Jin, W., Li, F., Yan, J., Yang, X., Ye, M., Andert, T. P., & Peytav, G. (2019). Simulation of global GM estimate of Asteroid (469219) 2016 HO3 for China's future asteroid mission. In: 2019 EPSC-DPS Joint Meeting 2019, abstract #1485-3. Geneva, Switzerland.

Jolliff, B. L., Gillis, J. J., Haskin, L. A., Korotev, R. L., & Wieczorek, M. A. (2000). Major lunar crustal terranes: Surface expressions and crust-mantle origins. *Journal of Geophysical Research*, 105(E2), 4197–4216. https://doi.org/10.1029/1999JE001103

Karachevtseva, I., Kozlova, N., Nadezhdina, I., Zubarev, A., Abdrakhimov, A., Basilevsky, A., & Oberst, J. (2013). New processing of Luna archive panoramas and geologic assessment of the Lunokhod landing sites. European Planetary Science Congress Abstracts, 8, EPSC2013-532.

Karr, C., Jr. (1975). *Infrared and Raman Spectroscopy of Lunar and Terrestrial Minerals*. Academic Press, New York.

Kasting, J. F., Kopparapu, R., Ramirez, R. M., & Harman, C. E. (2014). Remote life-detection criteria, habitable zone boundaries, and the frequency of Earth-like planets around M and late K stars. *Proceedings of the National Academy of Sciences*, 111(35), 12641–12646.

Kaula, W. M., Schubert, G., Lingenfelter, R. E., Sjogren, W. L., & Wollenhaupt, W. R. (1972). Analysis and interpretation of lunar laser altimetry. In A. E. Metzger, J. I. Trombka, L. E. Peterson, R. C. Reedy, & J. R. Arnold (Eds.), *Lunar and Planetary Science Conference Proceedings* (Vol. 3, p. 2189).

Kaula, W. M., Schubert, G., Lingenfelter, R. E., Sjogren, W. L., & Wollenhaupt, W. R. (1974). Apollo laser altimetry and inferences as to lunar structure. *Lunar and Planetary Science Conference Proceedings*, 5, 3049–3058.

Keller, L. P., & McKay, D. S. (1997). The nature and origin of rims on lunar soil grains. *Geochimica et Cosmochimica Acta*, 61(11), 2331–2341.

Kita, I., Matsuo, S., & Wakita, H. (1982). H2 generation by reaction between H2O and crushed rock: An experimental study on H2 degassing from the active fault zone. *Journal of Geophysical Research: Solid Earth*, 87(B13), 10789–10795. https://doi.org/10.1029/JB087iB13p10789

Kitazato, K., Milliken, R. E., Iwata, T., Abe, M., Ohtake, M., Matsuura, S., Arai, T., Nakauchi, Y., Nakamura, T., Matsuoka, M., Senshu, H., Hirata, N., Hiroi T., Pilorget, C., Brunetto, R., Poulet, F., Riu, L., Bibring, J.-P., Takir D. . . . Tsuda, Y. (2019). The surface composition of asteroid 162173 Ryugu from Hayabusa-2 near-infrared spectroscopy. *Science*, 364, 272–275.

Klein, H. P., Horowita, N. H., & Biemann, K. (1992). The search for extant life on Mars. In: *Mars* (H.H. Kieffer, et al. editors). University of Arizona Press, Tucson, AZ, pp. 1221–1233.

Kminek, G., & Bada, J. L. (2006). The effect of ionizing radiation on the preservation of amino acids on Mars. *Earth and Planetary Science Letters*, 245(1), 1–5. https://doi.org/10.1016/j.epsl.2006.03.008

Konopliva, A. S., Park, R. S., Vaughan, A. T., Bills, B. G., Asmar, S. W., Ermakova, A. I., Rambaux, N., Raymond, C. A., Castillo-Rogez, J. C., Russell, C. T., Smith, D. E., & Zuber, M. T. (2018). The Ceres gravity field, spin pole, rotation period and orbit from the Dawn radiometric tracking and optical data. *Icarus*, 299, 411–429.

Kramer, G. Y., Besse, S., Dhingra, D., Nettles, J., Klima, R., Garrick-Bethell, I., Clark, R. N., Combe, J. P., Head, J. W., Taylor, L. A., Pieters, C. M., Boardman, J., & McCord, T. B. (2011). M3 spectral analysis of lunar swirls and the link between optical maturation and surface hydroxyl formation at magnetic anomalies. *Journal of Geophysical Research E: Planets*, 116(9), 1–20. https://doi.org/10.1029/2010JE003729

Kribs, D. A. (1969). *Familiarization and Support Manual for Apollo Lunar Surface Drill (ALSD)* (p. 270).

Kring, David A., & Cohen, B. A. (2002). Cataclysmic bombardment throughout the inner solar system 3.9–4.0 Ga. *Journal of Geophysical Research: Planets, 107*(E2), 4-1–4-6. https://doi.org/10.1029/2001JE001529.

Kring D. A., & Durda D. D. (2012). *A Global Lunar Landing Site Study to Provide the Scientific Context for Exploration of the Moon.* Edited by Lunar and Planetary Institute.

Kring, D. A., Kramer, G. Y., Collins, G. S., Potter, R. W. K., & Chandnani, M. (2016). Peak-ring structure and kinematics from a multi-disciplinary study of the Schrödinger impact basin. *Nature Communications, 7*, 1–10. https://doi.org/10.1038/ncomms13161

Küppers, M., Keller, H. U., Kührt, E., A'Hearn, M. F., Altwegg, K., Bertrand, R., Busemann, H., Capria, M. T., Colangeli, L., Davidsson, B., Ehrenfreund, P., Knollenberg, J., Mottola, S., Rathke, A., Weiss, P., Zolensky, M., Akim, E., Basilevsky, A., Galimov, E. . . . Zarnecki, J. C. (2009). Triple F—A comet nucleus sample return mission. *Experimental Astronomy, 23*(3), 809–847. doi: 10.1007/s10686-008-9115-8.

Lantz, C., Brunetto, R., Barucci, M. A., Fornasier, S., Baklouti, D., Bourcois, J., & Godard, M. (2017). Ion irradiation of carbonaceous chondrites: A new view of space weathering on primitive asteroids. *Icarus, 285*, 43–57.

Lapen, T., M. Righter, A. Brandon, V. Debaille, B. Beard, J. Shafer, & A. Peslier (2010). A younger age for ALH84001 and its geochemical link to shergottite sources in Mars. *Science, 328*(5976), 347–351.

LaRowe, D. E., & Amend, J. P. (2015). Power limits for microbial life. *Frontiers in Microbiology, 6.* https://doi.org/10.3389/fmicb.2015.00718

Lauretta, D. S., Balram-Knutson, S. S., Beshore, E., Boynton, W. V., Drouet d'Aubigny, C., DellaGiustina, D. N., Enos, H. L., Golish, D. R., Hergenrother, C. W., Howell, E. S., Bennett, C. A., Morton, E. T., Nolan, M. C., Rizk, B., Roper, H. L., Bartels, A. E., Bos, B. J., Dworkin, J. P., Highsmith, D. E. . . . Sandford, S. A. (2017). OSIRIS-REx: Sample return from asteroid (101955) Bennu. *Space Science Reviews, 212*, 925–984.

Lauretta, D. S., Squyres, S. W., Messenger, S., Nakamura-Messenger, K., Nakamura, T., Glavin, D. P., Dworkin, J. P., Nguyen, A., Clemett, S., Furukawa, Y., Kimura, Y., Takigawa, A., Blake, G., Zega, T. J., Mumma, M., Milam, S., Herd, C. D. K., & the CAESAR Project Team (2018). *The CAESAR New Frontiers mission: 2. Sample science.* In: *Lunar and Planetary Science Conference, 49,* abstract #1334. The Woodlands, TX, USA.

Lefticariu, L., Pratt, L. M., & Ripley, E. M. (2006). Mineralogic and sulfur isotopic effects accompanying oxidation of pyrite in millimolar solutions of hydrogen peroxide at temperatures from 4 to 150°C. *Geochimica et Cosmochimica Acta, 70*(19), 4889–4905. https://doi.org/10.1016/j.gca.2006.07.026

Li, J. & Jewitt, D. (2013). Recurrent perihelion activity in (3200) Phaethon. *Astronomical Journal, 145,* 154 (9pp).

Li, L., Wing, B. A., Bui, T. H., McDermott, J. M., Slater, G. F., Wei, S., Lacrampe-Couloume, G., & Lollar, B. S. (2016). Sulfur mass-independent fractionation in subsurface fracture waters indicates a long-standing sulfur cycle in Precambrian rocks. *Nature Communications, 7*(1), 13252. https://doi.org/10.1038/ncomms13252

Li, S., Lucey, P. G., Milliken, R. E., Hayne, P. O., Fisher, E., Williams, J. P., Hurley, D. M., & Elphic, R. C. (2018). Direct evidence of surface exposed water ice in the lunar polar regions. *Proceedings of the National Academy of Sciences of the United States of America, 115*(36), 8907–8912. https://doi.org/10.1073/pnas.1802345115

Lin, H., He, Z., Yang, W., Lin, Y., Xu, R., Zhang, C., Zhu, M., Chang, R., Zhang, J., Li, C., Lin, H., Liu, Y., Gou, S., Wei, Y., Hu, S., Xue, C., Yang, J., Zhong, J., Fu, X., . . . Zou, Y. (2019). Olivine-norite rock detected by the lunar rover Yutu-2 likely crystallized from the SPA impact melt pool. *National Science Review,* 1–15.

Lin, L. H., Wang, P.-L., Rumble, D., Lippmann-Pipke, J., Boice, E., Pratt, L. M., Sherwood Lollar, B., Brodie, E., Hazen, T., Andersen, G., DeSantis, T., Moser, D. P., Kershaw, D., & Onstott, T. C. (2006). Long term biosustainability in a high energy, low diversity crustal biome. *Science, 314,* 479–482.

Lissauer, J. J., Dawson, R. I., & Tremaine, S. (2014). Advances in exoplanet science from Kepler. *Nature, 513*(7518), 336.

LoCoco, J. (2018). Advances in slimline borehole geophysical logging. *SAGEEP,* 2018, 1.

Lollar, B. S., Voglesonger, K., Lin, L. H., Lacrampe-Couloume, G., Telling, J., Abrajano, T. A., Onstott, T. C., & Pratt, L. M. (2007). Hydrogeologic controls on episodic H2 release from Precambrian fractured rocks— Energy for deep subsurface life on Earth and Mars. *Astrobiology.* https://doi.org/10.1089/ast.2006.0096

Lucey, P., Korotev, R. L., Gillis, J. J., Taylor, L. A., Lawrence, D., Campbell, B. A., Elphic, R., Feldman, B., Hood, L. L., Hunten, D., Mendillo, M., Noble, S., Papike, J. J., Reedy, R. C., Lawson, S., Prettyman, T., Gasnault, O., & Maurice, S. (2006). Understanding the lunar surface and space-moon interactions. In B. L. Jolliff, M. A. Wieczorek, C. K. Shearer, & C. R. Neal (Eds.), *New Views of the Moon, 60,* 83–219. Chantilly, VA: Mineralogical Society of America, pp. 83–219. https://doi.org/10.2138/rmg.2006.60.2

Lyell, C. (1830). *Principles of Geology, Being an Attempt to Explain the Former Changes of the Earth's Surface, by Reference to Causes Now in Operation.* London: John Murray. Volume 1.

Lyell, C. (1832). *Principles of Geology, Being an Attempt to Explain the Former Changes of the Earth's Surface, by Reference to Causes Now in Operation.* London: John Murray. Volume 2.

Lyell, C. (1833). *Principles of Geology, Being an Attempt to Explain the Former Changes of the Earth's Surface, by Reference to Causes Now in Operation.* London: John Murray. Volume 3.

Lyons, J. R., Manning, C., & Nimmo, F. (2005). Formation of methane on Mars by fluid-rock interaction in the crust: *Geophysical Research Letters*, doi:10.1029/2004GL022161.

Magnani, P., Re, E., Senese, S., Rizzi, F., Gily, A., & Baglioni, P. (2010). *The drill and sampling system for the ExoMars Rover.* International Symposium on Artificial Intelligence, Robotics and Automation in Space (i-SAIRAS). Sapporo, Japan.

Mayhew, L. E., Ellison, E. T., McCollom, T. M., Trainor, T. P., & Templeton, A. S. (2013). Hydrogen generation from low-temperature water–rock reactions. *Nature Geoscience*, 6(6), 478–484. https://doi.org/10.1038/ngeo1825

Mazrouei, S., Daly, M. G., Barnouin, O. S., Ernst, C. M., & DeSouza, I. (2014). Block distribution on Itokawa. *Icarus*, 229, 181–189.

McEwen, A. S. et al. (2007a). A closer look at water-related geologic activity on Mars. *Science*, 317, doi:10.1126/science.1143987

McEwen, A. S. et al. (2007b). Mars reconnaissance orbiter's High Resolution Imaging Science Experiment (HiRISE). *Journal of Geophysical Research E: Planets*, doi:10.1029/2005JE002605.

McEwen, A. S. et al. (2010). The High Resolution Imaging Science Experiment (HiRISE) during MRO's Primary Science Phase (PSP). *Icarus*, 205, doi:10.1016/j.icarus.2009.04.023.

McEwen, A. S., Dundas, C. M., Mattson, S. S., Toigo, A. D., Ojha, L., Wray, J. J., Chojnacki, M., Byrne, S., Murchie, S.L., & Thomas, N. (2014). Recurring slope lineae in equatorial regions of Mars. *Nature Geoscience*, doi:10.1038/ngeo2014.

McKay, C. P. (1997). The search for life on Mars. *Origins of Life and Evolution of the Biosphere*, 27(1–3), 263–289.

McKay, C. P. (2004). What is life—And how do we search for it in other worlds? *Plos Biology*, 2(9), 1260–1263.

McKay, D. S., Basu, A., & Nace, G. (1980). Lunar core 15010/11: Grain size, petrology, and implications for regolith dynamics. *Proceedings from the 11th Lunar and Planetary Science Conference*, 2, 1531–1550.

McKay, D. S., Dungan, M. A., Morris, R. V., & Fruland, R. M. (1977). Grain size, petrographic, and FMR studies of the double core 60009/10: A study of soil evolution. *Proceedings from the 8th Lunar Science Conference*, 3, 2929–2952.

McKay, D. S., Fruland, R. M., & Heiken, G. H. (1974). Grain size and evolution of lunar soils. *Proceedings of the 5th Lunar Science Conference*, 1, 887–906.

McKay, D. S., Gibson, E. K., Thomas-Keprta, K. L., Vali, H., Romanek, C. S., Clemett, S. J., Chillier, X. D., Maechling, C. R., & Zare, R. N. (1996). Search for past life on Mars: Possible relic biogenic activity in Martian meteorite ALH84001. *Science*, 273(5277), 924–930.

McKay, D. S., Heiken, G. H., Basu, A., Blanford, G., Steven, S., Reedy, R., French, B. M., & Papike, J. J. (1991). The Lunar Regolith. In G. H. Heiken, D. T. Vaniman, & B. M. French (Eds.), *Lunar Sourcebook: A User's Guide to the Moon* (pp. 285–356). Cambridge: Cambridge University Press.

McLennan, S. M. et al. (2011). Planning for Mars Returned Sample Science: Final report of the MSR End-to-End International Science Analysis Group (E2E-iSAG).

McSween, H. Y., Jr. (1985). SNC meteorites: Clues to Martian petrologic evolution? *Reviews of Geophysics*, 23(4), 391–416.

McSween, H. Y., Jr., Binzel, R. P., De Sanctis, M. C., Ammannito, E., Prettyman, T. H., Beck, A. W., Reddy, V., Le Corre, L., Gaffey, M. J., McCord, T. B., Raylond, C. A., Russell, C. T., & the Dawn Science Team. (2013). Dawn; the Vesta–HED connection; and the geologic context for eucrites, diogenites, and howardites. *Meteoritics and Planetary Science*, 48, 2090–2104.

MEPAG (2020). Mars scientific goals, objectives, investigations, and priorities: 2020 (D. Banfield editor), 89p. White Paper posted March, 2020 by the Mars Exploration Program Analysis Group (MEPAG) at https://mepag.jpl.nasa.gov/reports.cfm.

Meslier, V., Casero, M. C., Dailey, M., Wierzchos, J., Ascaso, C., Artieda, O., McCullough, P. R., & DiRuggiero, J. (2018). Fundamental drivers for endolithic microbial community assemblies in the hyperarid Atacama Desert. *Environmental Microbiology*, 20(5), 1765–1781. https://doi.org/10.1111/1462-2920.14106

Meyer, C. (2007a). *Lunar Sample Compendium: Synopsis of Deep Lunar Drill Strings* (pp. 1–8). NASA.

Meyer, C. (2007b). *Lunar Sample Compendium: Summary of Apollo Drive Tubes* (pp. 1–13). NASA.

Meyer, C. (2007c). Lunar Sample Compendium: A15 Deep Drill 15001–15006 (pp. 1–19). NASA.

Meyer, C. (2016). The lunar sample compendium. Astromaterials Research & Exploration Science (ARES). https://curator.jsc.nasa.gov/

Meyer, H. O. A., & McCallister, R. H. (1977). The Apollo 16 deep drill core. *Proceedings from the 8th Lunar Science Conference*, 3, 2889–2907.

Michalski, J. R., Cuadros, J., Niles, P. B., Parnell, J., Deanne Rogers, A., and Wright, S. P. (2013). Groundwater activity on Mars and implications for a deep biosphere. *Nature Geoscience*, doi:10.1038/ngeo1706.

Michalski, J. R., Onstott, T. C., Mojzsis, S. J., Mustard, J., Chan, Q. H. S., Niles, P. B., & Johnson, S. S. (2018). The Martian subsurface as a potential window into the origin of life. *Nature Geoscience*, 11(1), 21–26. https://doi.org/10.1038/s41561-017-0015-2

Michel, P. & Delbo, M. (2010). Orbital and thermal evolutions of four potential targets for a sample return space mission to a primitive near-Earth asteroid. *Icarus*, 209, 520–534.

Mitrofanov, I. G., Sanin, A. B., Boynton, W. V., Chin, G., Garvin, J. B., Golovin, D., Evans, L. G., Harshman, K., Kozyrev, A. S., Litvak, M. L., Malakhov, A., Mazarico, E., McClanahan, T., Milikh, G., Mokrousov, M., Nandikotkur, G., Neumann, G. A., Nuzhdin, I., Sagdeev, R., . . . Zuber, M. T. (2010). Hydrogen mapping of the lunar south pole using the LRO neutron detector experiment LEND. *Science*, 330(6003), 483–486. https://doi.org/10.1126/science.1185696

Miyamoto, H., Yano, H., Scheeres, D. J., Abe, S., Barnouin-Jha, O., Cheng, A. F., Demura, H., Gaskell, R. W., Hirata, N., Ishiguro, M., Michikami, T., Nakamura, A. M., Nakamura, R., Saito, J., & Sasaki, S. (2007). Regolith migration and sorting on asteroid Itokawa. *Science*, 316, 1011–1014.

Mojzsis, S. J., Arrhenius, G., McKeegan, K. D., Harrison, T. M., Nutman, A. P., & Friend, C. R. L. (1996). Evidence for life on Earth before 3,800 million years ago. *Nature*, 384(6604), 55–59. https://doi.org/10.1038/384055a0

Moreschini, P., Rutberg, M., Zacny, K., & Paulsen, G. (2010). *Dihedral: Downhole regolith interrogation with helium-assisted drill and laser induced breakdown spectroscopy system.* In: *Lunar and Planetary Science Conference*, vol. 41, abstract #1722. The Woodlands, TX, USA.

Morgan, J. V., Gulick, S. P. S., Bralower, T., Chenot, E., Christeson, G., Claeys, P., Cockell, C., Collins, G. S., Coolen, M. J. L., Ferrière, L., Gebhardt, C., Goto, K., Jones, H., Kring, D. A., Le Ber, E., Lofi, J., Long, X., Lowery, C., Mellett, C., . . . Zylberman, W. (2016). The formation of peak rings in large impact craters. *Science*, 354(6314), 878–882. https://doi.org/10.1126/science.aah6561

Morris, R. V. (1978). The surface exposure (maturity) of lunar soils: Some concepts and I/FeO compilation. *Lunar and Planetary Science Conference Proceedings*, 2, 2287–2297.

Morris, R. V. (1980). Origins and size distribution of metallic iron particles in the lunar regolith. *Proceedings from the 11th Lunar and Planetary Science Conference*, 2, 1697–1712.

Mustard, J. F., Brinckerhoff, W. B., Carr, M., Marais, D. J. D., Drake, B., Edgett, K. S., Elkins-Tanton, L. T., Grant, J. A., Milkovich, S. M., Ming, D., Murchie, S., Onstott, T. C., Ruff, S. W., Sephton, M. A., & Steele, A. (n.d.). *Report of the Mars 2020 Science Definition Team.* 52.

Mustard, J. F., Poulet, F., Head, J. W., Mangold, N., Bibring, J. P., Pelkey, S. M., Fassett, C. I., Langevin, Y., & Neukum, G. (2007). Mineralogy of the Nili Fossae region with OMEGA/Mars Express data: 1. Ancient impact melt in the Isidis basin and implications for the transition from the Noachian to Hesperian. *Journal of Geophysical Research E: Planets*, 112, doi:10.1029/2006JE002834

NASA. (1968). Apollo Lunar Surface Drill (ALSD) Final Report. Report number NASA CR 92412. https://ntrs.nasa.gov/archive/nasa/casi.ntrs.nasa.gov/19690002958.pdf

NASA. (April 1975). Apollo Program Summary Report (PDF) (Report). JSC-09423. Retrieved September 29, 2017. https://www.hq.nasa.gov/alsj/APSR-JSC-09423-OCR.pdf

NASA. (2013). *Vision and Voyages for Planetary Science in the Decade 2013–2022.* The National Academies Press, Washington, DC.

National Academies of Sciences, Engineering, and Medicine (2018). *Space Studies Board Annual Report 2017*, The National Academies Press, Washington, DC. https://doi.org/10.17226/25146.

National Research Council. 2011. *Vision and Voyages for Planetary Science in the Decade 2013–2022.* Washington, DC: The National Academies Press. https://doi.org/10.17226/13117.

Neal, C. R. (2009). The Moon 35 years after Apollo: What's left to learn? *Chemie Der Erde - Geochemistry*, 69(1), 3–43. https://doi.org/10.1016/j.chemer.2008.07.002

Nealson, K. H., Inagaki, F., & Takai, K. (2005). Hydrogen-driven subsurface lithoautotrophic microbial ecosystems (SLiMEs): Do they exist and why should we care? *Trends in Microbiology*, 13(9), 405–410. https://doi.org/10.1016/j.tim.2005.07.010

Neukum, G., & Wise, D. U. (1976). Mars: A standard crater curve and possible new time scale. *Science*, 194(4272), 1381–1387. https://doi.org/10.1126/science.194.4272.1381

Nienow, J. A., McKay, C. P., & Friedmann, E. I. (1988). The cryptoendolithic microbial environment in the Ross Desert of Antarctica: Light in the photosynthetically active region. *Microbial Ecology*, 16(3), 271–289. https://doi.org/10.1007/BF02011700

Nisbet, E. G. & Sleep, N. H. (2001). The habitat and nature of early life. *Nature*, 409, 1083–1091.

Nittler, L. R., Starr, R. D., Lim, L., McCoy, T. J., Burbine, T. H., Reedy, R. C., Trombka, J. I., Gorenstein, P., Squyres, S. W., Boynton, W. V., McClanahan, T. P., Bhangoo, J. S., Clark, P. E., Murphy, M. E., & Killen R. (2001). X-ray fluorescence measurements of the surface elemental composition of asteroid 433 Eros. *Meteoritics and Planetary Science*, 36, 1673–1695.

Noguchi, T., Nakamura, T., Kimura, M., Zolensky, M. E., Tanaka, M., Hashimoto, T., Konno, M., Nakato, A., Ogami, T., Fujimura, A., Abe, M., Yada, T., Mukai, T., Ueno, M., Okada, T., Shirai, K., Ishibashi, Y., & Okazaki, R . (2011). Incipient space weathering observed on the surface of Itokawa dust particles. *Science*, 333, 1211–1215.

Norman, M. D. (2009). The lunar cataclysm: Reality or "mythconception"? *Elements*, 5(1), 23–28. https://doi.org/10.2113/gselements.5.1.23

Norman, R. B., Gronoff, G., & Mertens, C. J. (2014). Influence of dust loading on atmospheric ionizing radiation on Mars. *Journal of Geophysical Research: Space Physics*, doi:10.1002/2013JA019351.

Nutman, A. P., Bennett, V. C., Friend, C. R. L., Kranendonk, M. J. V., & Chivas, A. R. (2016). Rapid emergence of life shown by discovery of 3,700-million-year-old microbial structures. *Nature*, 537(7621), 535–538. https://doi.org/10.1038/nature19355

Ogawa, N., Kawakatsu, Y., Ohtake, H., Imada, T., Shimada, T., Tkaji, A., & Ikeda, H. (2018). *Overview of MMX mission and its planetary protection plan*. In: *42nd COSPAR Scientific Assembly*; Pasadena, CA, July 14–22, 2018, Abstract id. PPP.2-11-18.

Okada, T., Kebukawa, Y., Aoki, J., Matsumoto, J., Yano, H., Iwata, T., Mori, O., Bibring, J.-P., Ulame, S., Jaumann, R., & Solar Poser Sail Science Team (2018). Science exploration and instrumentation of the OKEANOS mission to a Jupiter Trojan asteroid using the solar power sail. *Planetary and Space Science*, 161, 99–106.

Onstott, T. C., Ehlmann, B. L., Sapers, H., Coleman, M., Ivarsson, M., Marlow, J. J., Neubeck, A., & Niles, P. (2019). Paleo-rock-hosted life on Earth and the search on Mars: A review and strategy for exploration. *Astrobiology*, 19(10), 1230–1262. https://doi.org/10.1089/ast.2018.1960

Onstott, T. C., McGown, D., Kessler, J., Lollar, B. S., Lehmann, K. K., & Clifford, S. M. (2006). Martian CH_4: Sources, flux, and detection. *Astrobiology*, 6(2), 377–395. https://doi.org/10.1089/ast.2006.6.377

Orosei, R. et al. (2018), Radar evidence of subglacial liquid water on Mars: *Science*, doi:10.1126/science.aar7268.

Osburn, M. R., LaRowe, D. E., Momper, L. M., & Amend, J. P. (2014). Chemolithotrophy in the continental deep subsurface: Sanford Underground Research Facility (SURF), USA. *Frontiers in Microbiology*, 5. https://doi.org/10.3389/fmicb.2014.00610

Oze, C., & Sharma, M. (2005). *Have olivine, will gas: Serpentinization and the abiogenic production of methane on Mars: Geophysical Research Letters*, doi:10.1029/2005GL022691.

Papanastassiou, D. A., & Wasserburg, G. J. (1971). Lunar chronology and evolution from RbSr studies of Apollo 11 and 12 samples. *Earth and Planetary Science Letters*, 11(1–5), 37–62. https://doi.org/10.1016/0012-821X(71)90139-7

Papike, J., J. Karner, C. Shearer, & P. Burger (2009). Silicate mineralogy of Martian meteorites. *Geochimica et Cosmochimica Acta*, 73(24), 7443–7485.

Papike, J. J., Simon, S. B., White, C., & Laul, J. C. (1981). The relationship of the lunar regolith <10 μm fraction and agglutinates. Part I: A model for agglutinate formation and some indirect supportive evidence. *Proceedings from the 12th Lunar and Planetary Science Conference*, 409–420.

Parnell, J., & McMahon, S. (2016). Physical and chemical controls on habitats for life in the deep subsurface beneath continents and ice. *Philosophical Transactions of the Royal Society A: Mathematical, Physical and Engineering Sciences*, 374(2059), https://doi.org/10.1098/rsta.2014.0293

Pavlov, A. A., Vasilyev, G., Ostryakov, V. M., Pavlov, A. K., & Mahaffy, P. (2012). Degradation of the organic molecules in the shallow subsurface of Mars due to irradiation by cosmic rays. *Geophysical Research Letters*, 39(13). https://doi.org/10.1029/2012GL052166

Pavlov, A. K., Blinov, A. V., and Konstantinov, A. N. (2002). Sterilization of Martian surface by cosmic radiation: *Planetary and Space Science*, doi:10.1016/S0032-0633(01)00113-1.

Peters, G. H., Carey, E. M., Anderson, R. C., Abbey, W. J., Kinnett, R., Watkins, J. A., Schemel, M., Lashore, M. O., Chasek, M. D., Green, W., Beegle, L. W., & Vasavada, A. R. (2018). Uniaxial compressive strengths of rocks drilled at Gale Crater, Mars. *Geophysical Research Letters*, 45(1), 108–116. https://doi.org/10.1002/2017GL075965

Picardi, G. (2005). Radar soundings of the subsurface of Mars: *Science*, doi:10.1126/science.1122165.

Pieri, D. (1976). Distribution of small channels on the Martian surface. *Icarus*, 27(1), 25–50.

Pieri, D. C. (1980). Martian valleys: Morphology, distribution, age, and origin. *Science*, 210(4472), 895–897.

Pieters, C. M., Ammannito, E., Blewett, D. T., Denevi, B. W., De Sanctis, M. C., Gaffey, M. J., Le Corre, L., Li, J.-Y., Marchi, S., McCord, T. B., McFadden, L. A., Mittlefehldt, D. W., Nathues, A., Palmer, E., Reddy, V., Raymond, C. A., & Russell, C. T. (2012). Distinctive space weathering on Vesta from regolith mixing processes. *Nature*, 491, 79–82.

Pieters, C. M. & Noble, S. K . (2016). Space weathering on airless bodies. *Journal of Geophysical Research*, 121, 1865–1884.

Pillinger, C. T., Gowar, A. P., & Massey, H. S. W. (1977). The separation and subdivision of two 0.5g samples of lunar soil collected by the Luna 16 and 20 missions. *Philosophical Transactions of the Royal Society of London . Series A, Mathematical and Physical Sciences*, 284(1319), 137–143. https://doi.org/10.1098/rsta.1977.0003

Poch, O., Noblet, A., Stalport, F., Correia, J. J., Grand, N., Szopa, C., & Coll, P. (2013). Chemical evolution of organic molecules under Mars-like UV radiation conditions simulated in the laboratory with the "Mars organic molecule irradiation and evolution" (MOMIE) setup. *Planetary and Space Science*, 85, 188–197. https://doi.org/10.1016/j.pss.2013.06.013

Qian, Y., Xiao, L., Yin, S., Zhang, M., Zhao, S., Pang, Y., Wang, J., Wang, G., & Head, J. W. (2020). The regolith properties of the Chang'e-5 landing region and the ground drilling experiments using lunar regolith simulants. *Icarus*, 337. https://doi.org/10.1016/j.icarus.2019.113508

Qian, Y. Q., Xiao, L., Zhao, S. Y., Zhao, J. N., Huang, J., Flahaut, J., Martinot, M., Head, J. W., Hiesinger, H., & Wang, G. X. (2018). Geology and scientific significance of the Rümker Region in Northern Oceanus Procellarum: China's Chang'E-5 Landing Region. *Journal of Geophysical Research: Planets*, 123(6), 1407–1430. https://doi.org/10.1029/2018JE005595

Quinn, D. P., & Ehlmann, B. L. (2019). The deposition and alteration history of the Northeast Syrtis Major layered sulfates. *Journal of Geophysical Research: Planets*. https://doi.org/10.1029/2018JE005706

Razzell Hollis, J., Rheingold, D., Bhartia, R., & Beegle, L. (2020). An optical model for quantitative Raman micro-spectroscopy. *Applied Spectroscopy*. Article number 0003702819895299. doi: 10.1177/0003702819895299

Rhodes, J. M., Adams, J. B., Blanchard, D. P., Charette, M. P., Rodgers, K. V., Jacobs, J. W., Brannon, J. C., & Haskin, L. A. (1975). Chemistry of agglutinate fraction in lunar soils. *Lunar and Planetary Science Conference Proceedings*, 2, 2291–2307.

Rivkin, A. S., Howell, E. S., Lebofsky, L. A., Clark, B. E., & Britt, D. T . (2000). The nature of M-class asteroids from 3-μm observations. *Icarus*, 145, 351–368.

Roberts, J. H., Kahn, E. G., Barnouin, O. S., Ernst, C. M., Proctor, L. M., & Gaskell, R. W. (2014). Origin and flatness of ponds on asteroid 433 Eros. *Meteoritics and Planetary Science*, 49, 1735–1748.

Rothschild, L. J. (2008). The evolution of photosynthesis . . . again? *Philosophical Transactions of the Royal Society B: Biological Sciences*, 363(1504), 2787–2801. https://doi.org/10.1098/rstb.2008.0056

Rothschild, L. J., & DesMarais, D. (1989). Stable carbon isotope fractionation in the search for life on early Mars. *Advances in Space Research*, 9(6), 159–165. https://doi.org/10.1016/0273-1177(89)90223-8

Rummel, J. D., Beaty, D. W., Jones, M. A., Bakermans, C., Barlow, N. G., Boston, P. J., Chevrier, V. F., Clark, B. C., De Vera, J. P. P., Gough, V. R., Hallsworth, J. E., Head, J. W., Hipkin, V. J., Kieft, T. L., Mcewen, A. S., Mellon, M. T., Mikucki, J. A., Nicholson, W. L., Omelon, C. R., . . . Wray, J. J. (2014). A new analysis of Mars "Special Regions": Findings of the second MEPAG special regions science analysis group (SR-SAG2). *Astrobiology*. https://doi.org/10.1089/ast.2014.1227

Saal, A. E., Hauri, E. H., Cascio, M. L., Van Orman, J. A., Rutherford, M. C., & Cooper, R. F. (2008). Volatile content of lunar volcanic glasses and the presence of water in the moon's interior. *Nature*, 454(7201), 192–195. https://doi.org/10.1038/nature07047

Sawaryn, S. J., Bustin, P., Cain, M. G., Crawford, I. A., Lim, S., Linossier, A., & Smith, D. J. (2018). *Lunar drilling—Challenges and opportunities*. In *SPE Annual Technical Conference and Exhibition*. Society of Petroleum Engineers. https://doi.org/10.2118/191624-MS

Scheeres, D. J., French, A. S., McMahon, J. W., Davis, A. B., Brack, D. N., Leonard, J. M., Antreasian, P., Brozovic, M., Chesley, S. R., Farnocchia, D., Jacobson, R. A., Takahashi, Y., Mazarico, E. M., Liounis, A., Rowlands, D., Highsmith, D. E., Getzandanner, K., Moreau, M., Tricarico, P. ... the OSIRIS-REx Team. (2019). *OSIRIS-REx gravity field estimates for Bennu using spacecraft and natural particle tracking data*. In: *EPSC-DPS 2019 Joint Meeting*, abstract EPSC-DPS2019-106-2. Geneva, Switzerland.

Schmidt, M. E., Herd, C. D. K., McCoy, T. J., McSween, H. Y., Rogers, A. D., & Treiman, A. H. (2019). *Igneous Mars: Crust and mantle evolution as seen by rover geochemistry, Martian meteorites, and remote sensing, in 50th Lunar and Planetary Science Conference*, p. 2419.

Schott, W. (1955). Rate of sedimentation of recent deep-sea sediments. In *Recent Marine Sediments*. SEPM (Society for Sedimentary Geology). https://doi.org/10.2110/pec.55.04

Schwandt, C., Hamilton, J. A., Fray, D. J., & Crawford, I. A. (2012). The production of oxygen and metal from lunar regolith. *Planetary and Space Science*, 74(1), 49–56. https://doi.org/10.1016/j.pss.2012.06.011

Sefton-Nash, E., Williams, J. P., Greenhagen, B. T., Warren, T. J., Bandfield, J. L., Aye, K. M., Leader, F., Siegler, M. A., Hayne, P. O., Bowles, N., & Paige, D. A. (2019). Evidence for ultra-cold traps and surface water ice in the lunar south polar crater Amundsen. *Icarus*, 332, 1–13. https://doi.org/10.1016/j.icarus.2019.06.002

Shestopalov, D. I., Golubeva, L. F., & Cloutis, E. A . (2013). Optical maturation of asteroid surfaces. *Icarus*, 225, 781–793.

Shevchenko, V., Rodionova, Z., & Michael, G. (2015). *Lunar and Planetary Cartography in Russia*. Springer.

Shoemaker, E. M. (1961). Interplanetary correlation of geologic time: ABSTRACT. *AAPG Bulletin*, 45(1), 130–130.

Siddiqi, A. A. (2018). *Beyond Earth: A Chronicle of Deep Space Exploration, 1958–2016*. The NASA History Series (2nd ed.). NASA History Program Office, Washington, DC, p. 1. ISBN: 9781626830424. LCCN 2017059404. SP2018-4041.

Smith, D. E., Zuber, M. T., Solomon, S. C., Phillips, R. J., Head, J. W., Garvin, J. B., Banerdt, W. B., Muhleman, D. O., Pettengill, G. H., Neumann, G. A., Lemoine, F. G., Abshire, J. B., Aharonson, O., Brown, C. D., Hauck, S. A., Ivanov, A. B., McGovern, P. J., Zwally, H. J., Duxbury, T. C. (1999). The global topography of Mars and implications for surface evolution. *Science*, 284(5419), 1495–1503.

Smith, I. B., and Holt, J. W. (2010). Onset and migration of spiral troughs on Mars revealed by orbital radar. *Nature*, doi:10.1038/nature09049.

Soderblom, L. A., Condit, C. D., West, R. A., Herman, B. M., & Kreidler, T. J. (1974). Martian planetwide crater distributions: Implications for geologic history and surface processes. *Icarus*, 22(3), 239–263. https://doi.org/10.1016/0019-1035(74)90175-4

Solomon, S. C. (1978). On volcanism and thermal tectonics on one-plate planets. *Geophysical Research Letters*, 5(6), 461–464.

Soo, R. M., Hemp, J., Parks, D. H., Fischer, W. W., & Hugenholtz, P. (2017). On the origins of oxygenic photosynthesis and aerobic respiration in cyanobacteria. *Science*, 355(6332), 1436–1440. https://doi.org/10.1126/science.aal3794

Space Science Board, N.R.C. (1978). Strategy for Exploration of the Inner Planets 1978–1987.

Squyres, S. W., Arvidson, R. E., Bell, J. F., Brückner, J., Cabrol, N. A., Calvin, W., Carr, M. H., Christensen, P. R., Clark, B. C., Crumpler, L., Des Marais, D. J., D'Uston, C., Economou, T., Farmer, J., Farrand, W., Folkner, W., Golombek, M., Gorevan, S., Grant, J. A. . . . Yen, A. (2004). The Spirit Rover's Athena science investigation at Gusev crater, Mars. *Science*, 305(5685), 794–799.

Stamenković, V., Beegle, L. W., Zacny, K., Arumugam, D. D., Baglioni, P., Barba, N., Baross, J., Bell, M. S., Bhartia, R., Blank, J. G., Boston, P. J., Breuer, D., Brinckerhoff, W., Burgin, M. S., Cooper, I., Cormakovic, V., Davila, A., Davis, R. M., Edwards, C., . . . Woolley, R. (2019). The next frontier for planetary and human exploration. *Nature Astronomy*, 3(2), 116–120. https://doi.org/10.1038/s41550-018-0676-9

Starukhina, L. V., & Shkuratov, Y. G. (2000). The lunar poles: Water ice or chemically trapped hydrogen? *Icarus*, 147(2), 585–587. https://doi.org/10.1006/icar.2000.6476

Steenstra, E. S., Martin, D. J. P., McDonald, F. E., Paisarnsombat, S., Venturino, C., O'Hara, S., Calzada-Diaz, A., Bottoms, S., Leader, M. K., Klaus, K. K., van Westrenen, W., Needham, D. H., & Kring, D. A. (2016). Analyses of robotic traverses and sample sites in the Schrödinger basin for the HERACLES human-assisted sample return mission concept. *Advances in Space Research*, 58(6), 1050–1065. https://doi.org/10.1016/j.asr.2016.05.041

Steno, N., Ferdinando Grand-Duke of Tuscany, & Accademia della Crusca. (1669). *Nicolai Stenonis De solido intra solidum naturaliter contento dissertationis prodromus*. Ex typographia sub signo Stellae,. https://www.biodiversitylibrary.org/item/250134

Stern, S. A., Weaver, H. A., Spencer, J. R., Olkin, C. B., Gladstone, G. R., Grundy, W. M., Moore, J. M., Cruikshank, D. P., Elliott, H. A., McKinnon, W. B., Parker, J. W., Verbiscer, A. J., Young, L. A., Aguilar, D. A., Albers, J. M., Andert, T., Andrews, J. P., Bagenal, F., Banks, M. E. . . . Zurbuchen, T. H. (2019). Initial results from the New Horizons exploration of 2014 MU$_{69}$, a small Kuiper Belt object. *Science*, 364. doi: 10.1126/science.aaw9771.

Stevens, T. O., & McKinley, J. P. (1995). Lithoautotrophic microbial ecosystems in deep basalt aquifers. *Science*, 270(5235), 450–455. https://doi.org/10.1126/science.270.5235.450

Stöffler, D. (1977). Research drilling, Nördlingen 1973: Polymict breccias, crater basement and cratering model of the Ries impact structure. *Geologica Bavarica*, 75(January 1977), 443–458.

Stöffler, D., & Ryder, G. (2001a). Stratigraphy and isotope ages of lunar geologic units: Chronological standard for the inner solar system. *Space Science Reviews*, 96(1), 9–54. https://doi.org/10.1023/A:1011937020193

Stöffler, D., & Ryder, G. (2001b). Stratigraphy and isotope ages of lunar geologic units: Chronological standard for the inner solar system. In R. Kallenbach, J. Geiss, & W. K. Hartmann (Eds.), *Chronology and Evolution of Mars* (pp. 9–54). Springer Netherlands.

Stöffler, D., Ryder, G., Ivanov, B. A., Artemieva, N. A., Cintala, M. J., & Grieve, R. A. F. (2006). Cratering History and Lunar Chronology. In Bradley L. Jolliff, M. A. Wieczorek, C. K. Shearer, & C. R. Neal (Eds.), *New Views of the Moon* (Vol. 60, pp. 519–596). Mineralogical Society of America. https://doi.org/10.2138/rmg.2006.60.05

Stoker, C. R., Lemke, L. G., & Gonzalez, A. A. (2006). *Application of burrowing moles for planetary and lunar subsurface access.* In: *Lunar and Planetary Science Conference*, vol. 37, abstract #1542. League City, Texas, USA.

Strazzulla, G., Dotto, E., Binzel, R., Brunetto, R., Barucci, M. A., Blanco, A., & Orofino, V. (2005). Spectral alteration of the Meteorite Epinal (H5) induced by heavy ion irradiation: A simulation of space weathering effects on near-Earth asteroids. *Icarus*, 174, 31–35.

Stuurman, C. M., Osinski, G. R., Holt, J. W., Levy, J. S., Brothers, T. C., Kerrigan, M., & Campbell, B. A. (2016). SHARAD detection and characterization of subsurface water ice deposits in Utopia Planitia. *Mars : Geophysical Research Letters*, doi:10.1002/2016GL070138.

Sugita, S., Honda, R., Morota, T., Kameda, S., Sawada, H., Tatsumi, E., Yamada, M., Honda, C., Yokota, Y., Kouyama, T., Sakatani, N., Ogawa, K., Suzuki, H., Okada, T., Namiki, N., Tanaka, S., Iijima, Y., Yoshioka, K., Hayakawa, M., Cho, Y. . . . Tsuda, Y. (2019). The geomorphology, color, and thermal properties of Ryugu: Implications for parent-body processes. *Science*, 364. doi: 10.1126/science.aaw0422.

Surkov, Y. A., Barsukov, V. L., Moskalyeva, L. P., Kharyukova, V. P., & Kemurdzhian, A. L. (1984). New data on the composition, structure, and properties of Venus rock obtained by Venera 13 and Venera 14. *Journal of Geophysical Research: Solid Earth*, 89(S02), B393–B402. https://doi.org/10.1029/JB089iS02p0B393

Takir, D., Stockstill-Cahill, K. R., Hibbitts, C. A., & Nakauchi, Y. (2019). 3-μm reflectance spectroscopy of carbonaceous chondrites under asteroid-like conditions. *Icarus*, 333, 243–251.

Tanaka, K. L. (1986). The stratigraphy of Mars. *Journal of Geophysical Research: Solid Earth*, 91(B13), E139–E158.

Tanaka, K. L., Robbins, S. J., Fortezzo, C. M., Skinner, J. A., & Hare, T. M. (2014). The digital global geologic map of Mars: Chronostratigraphic ages, topographic and crater morphologic characteristics, and updated resurfacing history. *Planetary and Space Science*, 95, 11–24. https://doi.org/10.1016/j.pss.2013.03.006

Tarnas, J. D., Mustard, J. F., Sherwood Lollar, B., Bramble, M. S., Cannon, K. M., Palumbo, A. M., & Plesa, A.-C. (2018). Radiolytic H2 production on Noachian Mars: Implications for habitability and atmospheric warming. *Earth and Planetary Science Letters*, 502, 133–145. https://doi.org/10.1016/j.epsl.2018.09.001

Taylor, F. W. (2011). Comparative planetology, climatology and biology of Venus, Earth and Mars. *Planetary and Space Science*, 59(10), 889–899. https://doi.org/10.1016/j.pss.2010.11.009

Taylor, G. J., Warner, R. D., & Keil, K. (1979). Stratigraphy and depositional history of the Apollo 17 drill core. *Proceedings from the 10th Lunar and Planetary Science Conference*, 2, 1159–1184.

Tera, F., Papanastassiou, D. A., & Wasserburg, G. J. (1974). Isotopic evidence for a terminal lunar cataclysm. *Earth and Planetary Science Letters*, 22(1), 1–21. https://doi.org/10.1016/0012-821X(74)90059-4

Thaisen, K. G., Head, J. W., Taylor, L. A., Kramer, G. Y., Isaacson, P., Nettles, J., Petro, N., & Pieters, C. M. (2011). Geology of the Moscoviense Basin. *Journal of Geophysical Research E: Planets*, 116(4), 1–14. https://doi.org/10.1029/2010JE003732

The United States. (2018). Space Policy Directive-1: Reinvigorating America's human space exploration program. *Federal Register*, 82, 1–2.

Toksöz, M. N., Anton, M. D., Solomon, S. C., & Anderson, K. R. (1974). Structure of the Moon. *Reviews of Geophysics and Space Physics*, 12(4), 539–567. https://doi.org/10.1029/RG012i004p00539

Torre, J. R. de la, Goebel, B. M., Friedmann, E. I., & Pace, N. R. (2003). Microbial diversity of cryptoendolithic communities from the McMurdo Dry Valleys, Antarctica. *Applied and Environmental Microbiology*, 69(7), 3858–3867. https://doi.org/10.1128/AEM.69.7.3858-3867.2003

Vago, J., Witasse, O., Svedhem, H., Baglioni, P., Haldemann, A., Gianfiglio, G., Blancquaert, T., McCoy, D., & de Groot, R. (2015). ESA ExoMars program: The next step in exploring Mars. *Solar System Research*, 49(7), 518–528. https://doi.org/10.1134/S0038094615070199

Vago, J. L., Westall, F., Pasteur Instrument Teams, Landing Site Selection Working Group, and Other Contributors, & the ExoMars Project Team. (2017). Habitability on early Mars and the search for biosignatures with the ExoMars Rover. *Astrobiology*, 17, 471–510.

Van Thienen, P., Vlaar, N. J., & Van den Berg, A. P. (2004). Plate tectonics on the terrestrial planets. *Physics of the Earth Planetary Interior*, 142(1–2), 61–74.

Vaniman, D. T., Labotka, T. C., & Papike, J.J. (1979). The Apollo 17 drill core: Petrologic systematics and the identification of a possible Tycho component. *Proceedings from the 10th Lunar and Planetary Science Conference*, 2, 1185–1227.

Vernazza, P., Brunetto, R., Strazzulla, G., Fulchignoni, M., Rochette, P., Meyer-Vernet, N., & Zouganelis, I. (2006). Asteroid colors: A novel tool for magnetic field detection? The case of Vesta. *Astronomy and Astrophysics*, 451, L43–L46.

Vine, F. J. & Matthews, D. H. (1964). Magnetic anomalies over oceanic ridges. *Nature*, 199(4897), 947–949.

Vinogradov, A. P. (1971). Preliminary data on lunar ground brought to Earth by automatic probe "Luna-16". *Proceedings of the 2nd Lunar Science Conference*, 1, 1–16.

Vinogradov, A. P. (1972). Preliminary data on the lunar samples returned by the automatic station Luna 20. In *Pravada*.

Vinogradov et al. (Eds.) (1971). *Lunokhod 1—Mobile laboratory on the moon*, Vol. 1. Nauka, Moscow (in Russian).

Walsh, K. J., Jawin, E. R., Ballouz, R.-L., Barnouin, O. S., Bierhaus, E. B., Connolly, H. C., Jr., Molaro, J. L., McCoy, T. J., Delbo', M., Hartzell, C. M., Pajola, M., Schwartz, S. R., Trang, D., Asphaug, E., Becker, K. J., Beddingfield, C. B., Bennett, C. A., Bottke, W. F., Burke, K. N. … the OSIRIS-REx Team. (2019). Craters, boulders and regolith of (101955) Bennu indicative of an old and dynamic surface. *Nature Geoscience*, 12, 242–246.

Wang, A. et al. (2004). Mineralogy of a Martian meteorite as determined by Raman spectroscopy. *Journal of Raman Spectroscopy*, 35(6), 504–514.

Watanabe, S.-I., Tsuda, Y., Yoshikawa, M., Tanaka, S., Saiki, T., & Nakazawa, S. (2017). Hayabusa-2 mission overview. *Space Science Reviews*, 208, 3–16.

Watson, K., Murray, B. C., & Brown, H. (1961). The behavior of volatiles on the lunar surface. *Journal of Geophysical Research*, 66(9), 3033–3045. https://doi.org/10.1111/j.1749-6632.1965.tb20380.x

Westall, F. (2012). The early Earth. In: *Frontiers of Astrobiology* (C. Impey, J. Lunine, and J. Funes editors). Cambridge University Press, Cambridge, p. 331.

Westall, F., de Vries, S. T., Nijman, W., Rouchon, V., Orberger, B., Pearson, V., Watson, J., Verchovsky, A., Wright, I., Rouzaud, J.-N., Marchesini, D., & Severine, A. (2006). The 3.466 Ga "Kitty's Gap Chert", an early Archean microbial ecosystem. In Special Paper 405: Processes on the early Earth. *Geological Society of America*, pp. 105–131.

Wilde, S. A., Valley, J. W., Peck, W. H., & Graham, C. M. (2001). Evidence from detrital zircons for the existence of continental crust and oceans on the Earth 4.4 Gyr ago. *Nature*, 409, 175–178.

Wilhelms, D. (1976). Secondary impact craters of lunar basins. *Lunar and Planetary Science Conference Proceedings*, 7, 2883–2901.

Williford, K. H., Farley, K. A., Stack, K. M., Allwood, A. C., Beaty, D., Beegle, L. W., Bhartia, R., Brown, A. J., de la Torre Juarez, M., Hamran, S.-E., Hecht, M. H., Hurowitz, J. A., Rodriguez-Manfredi, J. A., Maurice, S., Milkovich, S., & Wiens, R. C . (2018). Chapter 11—The NASA Mars 2020 Rover mission and the search for extraterrestrial life. In: *From Habitability to Life on Mars*, pp. 275–308. Elsevier.

Wordsworth, R., Kalugina, Y., Lokshtanov, S., Vigasin, A., Ehlmann, B., Head, J., Sanders, C., & Wang, H. (2017). Transient reducing greenhouse warming on early Mars. *Geophysical Research Letters*, 44(2), 665–671. https://doi.org/10.1002/2016GL071766

Wordsworth, R. D., Kerber, L., Pierrehumbert, R. T., Forget, F., & Head, J. W. (2015). Comparison of "warm and wet" and "cold and icy" scenarios for early Mars in a 3-D climate model. *Journal of Geophysical Research: Planets*, 120(6), 1201–1219. https://doi.org/10.1002/2015JE004787

Xiao, L., Zhu, P., Fang, G., Xiao, Z., Zou, Y., Zhao, J., Zhao, N., Yuan, Y., Qiao, L., Zhang, X., Zhang, H., Wang, J., Huang, J., Huang, Q., He, Q., Zhou, B., Ji, Y., Zhang, Q., Shen, S., . . . Gao, Y. (2015). A young multilayered terrane of the northern Mare Imbrium revealed by Chang'E-3 mission. *Science*, 347(6227), 1226–1229. https://doi.org/10.1126/science.1259866

Zacny, K., Bar-Cohen, Y., Brennan, M., Briggs, G., Cooper, G., Davis, K., Dolgin, B., Glaser, D., Glass, B., Gorevan, S., Guerrero, J., McKay, C., Paulsen, G., Stanley, S., & Stoker, C. (2008). Drilling systems for extraterrestrial subsurface exploration. *Astrobiology*, 8(3), 665–706. https://doi.org/10.1089/ast.2007.0179

Zacny, K., Chu, P., Davis, K., Paulsen, G., & Craft, J. (2014). Mars 2020 sample acquisition and caching technologies and architectures, in *IEEE Aerospace Conference Proceedings*, doi:10.1109/AERO.2014.6836211.

Zacny, K., Mueller, J., Costa, T., Cwik, T., Zimmerman, W., Gray, A., Chow, P., Rehnmark, F., & Adams, G . (2018). *SLUSH: Europa hybrid deep drill*. In: *2018 IEEE Aerospace Conference*. doi:10.1109/AERO.2018.8396596, Big Sky, Montana, USA.

Zahnle, K., Arndt, N., Cockell, C., Halliday, A., Nisbet, E., Selsis, F., & Sleep N. H. (2007). Emergence of a habitable planet. *Space Science Reviews*, 129, 35–78.

Zeki, S., Nealson, K. H., & Conrad, P. G. (1999). Life: Past, present and future. *Philosophical Transactions of the Royal Society of London. Series B: Biological Sciences*, 354(1392), 1923–1939. https://doi.org/10.1098/rstb.1999.0532

Zellner, N. E. B. (2017). Cataclysm no more: new views on the timing and delivery of lunar impactors. *Origins of Life and Evolution of Biospheres*, 47(3), 261–280. https://doi.org/10.1007/s11084-017-9536-3

Zhang, T., Zhang, W., Wang, K., Gao, S., Hou, L., Ji, J., & Ding, X. (2017). Drilling, sampling, and sample-handling system for China's asteroid exploration mission. *Acta Astronautica*, 137, 192–204.

Zheng, Y., Ouyang, Z., Li, C., Liu, J., & Zou, Y. (2008). China's lunar exploration program: Present and future. *Planetary and Space Science*, 56(7), 881–886. https://doi.org/10.1016/j.pss.2008.01.002

Zolensky, M. E., K. Nakamura, A. F. Cheng, M. J. Cintala, F. Hörz, R. V. Morris, & D. Criswell (2002). *Meteoritic evidence for the mechanism of pond formation on asteroid Eros*. In: *Lunar and Planetary Science Conference, 33*, abstract #1593. League City, Texas, USA.

4 Biological Contamination Control and Planetary Protection Measures as Applied to Sample Acquisition

James N. Benardini, Moogega Stricker, and Kasthuri J. Venkateswaran

Jet Propulsion Laboratory (JPL)/California Institute of Technology (Caltech), Pasadena, CA

CONTENTS

4.1 BIOLOGICAL CONTAMINATION CONTROL AND PLANETARY PROTECTION

For the terrestrial analysis of extraterrestrial samples or investigations of life on a celestial body of biological interest, such as Mars or Europa, a biological contamination control program may be required to meet both planetary protection and/or to ensure project science objectives can be met to avoid false positives. Planetary protection background, requirements, and standards will be detailed and contrasted with a project-developed science requirements implementation approach. Biological contamination sources and potential biological cleanliness assessment and verification methods are presented. Finally, engineering end-to-end considerations will be discussed with mitigations and case studies covering a mission's early design phase, manufacturing, integration, test, launch operations, through sample acquisition. The utilization of an integrated biological contamination control program into the hardware engineering life cycle is essential for ensuring that the samples acquired for study meet planetary protection requirements and enable meaningful downstream science investigations.

4.1.1 PLANETARY PROTECTION POLICY

Planetary protection policy is an international policy stemming from the 1967 Outer Space Treaty requiring each space-faring nation to maintain a PP policy, develop a set of requirements, and implement a series of standards for spacecraft destined to biologically sensitive planetary bodies and to minimize harmful contamination. Planetary protection policy aims to reduce the risk of harmful contamination from Earth-based organisms to a target body (referred to as forward planetary protection) and the inadvertent contamination of the Earth's biosphere from extraterrestrial material (referred to as backward planetary protection). Although planetary protection focuses on both forward and backward planetary protection, the focus of this chapter will be forward planetary protection. In addition to the planetary protection requirements, a mission's science requirements can also impact a spacecraft's biological contamination requirements. Based on the biological contamination requirements, the mission approach can vary from just simple documentation to a comprehensive microbial reduction and verification implementation campaign. NASA has developed a set of technical standards for microbial reduction and recontamination prevention as guidance for missions in developing their implementation approach. In addition to this guidance the mission can also propose to develop alternative approaches or leverage industrial approaches from other environmental microbiology communities (e.g., medical device industry or other governmental agencies).

4.1.1.1 Background and Establishment of Mission Planetary Protection Requirements

After the formation of NASA in 1958, it was recognized that biological sciences in planetary exploration needed to have "adequate emphasis" as a discipline in order to not lose valuable science opportunities for future missions (Sagan and Coleman, 1966). The foundation of the current planetary protection guidelines stem from the United Nations 1967 "Treaty on Principles Governing the Activities of States in the Exploration and Use of Outer Space, Including the Moon and Other Bodies" (Outer Space Treaty). The UN 1967 Treaty states that signatory countries

> shall pursue studies of outer space, including the moon and other celestial bodies, and conduct exploration of them so as to avoid their harmful contamination, and also adverse changes in the environment of Earth resulting from the introduction of extraterrestrial matter.

From this treaty, the international community established a panel on Planetary Protection (PP) at the Committee on Space Research (COSPAR) which established and currently maintains PP policy and approaches. COSPAR policy then serves as a reference standard for spacefaring nations and in guiding compliance with article IC of the Outer Space Treaty (Kminek et al., 2017). For example, in the United States, NASA has created NPR 8020.12 Planetary Protection Provisions for Robotic Extraterrestrial Missions (2011), and in Europe the European Space Agency has created ESSB-ST-U-001 (2013) ESA Planetary Protection Requirements. To establish mission-specific requirements, each mission is first assigned a PP mission category based on the target body and the mission type (Table 4.1).

The PP categories are established based on the probability of the target body harboring life, the target body containing evidence for past life and/or chemical evolution of the solar system, and that a contamination event from the spacecraft could cause an adverse impact on current or future investigations. A Category I mission is defined as one that will not study life and the target body has the lowest likelihood for harboring life, thus, any spacecraft contamination level would be acceptable. If the mission plans to conduct scientific studies in regard to chemical evolution and the origin of life but the target body has an extremely low chance of harboring life, then the mission would be characterized as a PP Category II mission. PP Category II missions would also require spacecraft cleanliness that would ensure current or future investigations would not be impacted by spacecraft biological contamination. When the target body has both a possibility to harbor life and could provide evidence for past life, missions exploring it are designated as either PP Category III or IV missions. The distinction between a PP Category III and IV mission is defined by the chance of a spacecraft causing an inadvertent contamination release to a target body. If the spacecraft is an orbiter or flyby mission with a lower probability of contamination of the target body, then the mission is categorized as a PP Category III mission. Consequently, if the spacecraft is an orbiter or flyby mission with a high probability of impacting, for example, a lander or a probe that comes in direct contact with the target body, then it would be a categorized as a PP Category IV mission. Notably, the PP category of a given mission is defined by the highest category target body encountered for a given spacecraft. For example, a spacecraft going to an undifferentiated asteroid that requires a Mars gravity assist would be considered a PP Category III mission because the flight path would bring the spacecraft close to Mars. Sample return missions have a unique PP categorization, Category V, that is determined by whether the return sample has any potential to contain viable indigenous life that could pose a threat to the terrestrial biosphere. If the sample is unlikely to contain indigenous life, then it would be an "unrestricted Earth-return" mission. Whereas, if the sample could have life then it would be a "restricted Earth-return" mission.

After the mission PP category is assigned by the nation's space agency, then the associated PP requirements are derived from agency-level PP requirements documents. In general, the greater the biological interest in the target body and the more potential for the spacecraft to cause a contamination event, the more requirements are imposed on a mission to minimize biological contamination. For missions going to target bodies thought to have a potential to harbor life, the PP policies and additional biological requirements increase in stringency. As shown in Table 4.2, these range from missions providing documentation as a minimum to hardware biological control and cleanroom

TABLE 4.1
The Planetary Protection Mission Categories Based on Target Bodies and Spacecraft Type as per NASA NPR 8020.12

Types of Planetary Bodies	Mission Type	Mission Category	Target Body
A	Bodies not of direct interest for understanding the process of chemical evolution or the origin of life	Any	I	Undifferentiated, metamorphosed asteroids, Io, others TBD
B	Bodies of significant interest relative to the process of chemical evolution and the origin of life, but where only a remote chance that contamination carried by a spacecraft could compromise future investigations.	Any	II	Venus, Earth's Moon, Comets; non-Category I Asteroids, Jupiter, Jovian Satellites (except Io and Europa), Saturn, Saturnian Satellites (except Titan and Enceladus), Uranus, Uranian Satellites, Neptune, Neptunian Satellites (except Triton), Kuiper-Belt Objects (<1/2 the size of Pluto), others TBD
			II*	Icy Satellites where there is a "remote" potential for contamination of the liquid-water environments that may exist beneath their surfaces, addressing both the existence of such environments and the prospects of accessing them. Ganymede, Titan, Triton, Pluton, and Charon (Neptune), others TBD
C	Bodies of significant interest to the process of chemical evolution and/or the origin of life and where scientific opinion provides a significant chance that contamination could compromise future investigations	Flyby, Orbiter	III	Mars, Europa, Enceladus, others TBD
		Lander, Probe	IV	
All	Earth-return missions from bodies deemed by scientific opinion to have no indigenous life forms	Unrestricted Earth-Return	V (unrestricted)	Venus, Moon, others TBD
	Earth-return missions from bodies deemed by scientific opinion to be of significant interest to the process of chemical evolution and/or the origin of life	Restricted Earth-Return	V (restricted)	Mars, Europa, Enceladus, others TBD

Source: NASA (2005).

TABLE 4.2

High-level Planetary Protection Requirements and Mission Examples for each PP Mission by Category

Mission Category	Mission Type	Implementation Requirements	Mission Example
I	Any	Documentation	
II	Any	Documentation	Galileo, JUNO, (Voyager), GRAIL, (Ranger), (Surveyor)
III	Flyby, Orbiter	Impact avoidance and/or contamination control including: cleanroom assembly, microbial reduction, and trajectory biasing	(Viking Orbiter), Dawn, JUNO, Europa Clipper, Mars Odyssey, MAVEN
IV	Lander, Probe	Impact avoidance and contamination control including: cleanroom assembly, microbial reduction, trajectory biasing, organics archiving.	(Viking Lander), Mars Pathfinder, Mars Science Laboratory, Mars Exploration Rover, InSight
V Unrestricted Earth Return	Lander	As appropriate for the specified category of the outbound mission. No inbound PP requirement	Hayabusa, OSIRIS-REx, Stardust, Genesis, (Apollo 15–17)
V Restricted Earth Return	Lander	Impact avoidance and contamination control including: cleanroom assembly, microbial containment of sample, breaking chain of contact with target planet, sample containment and biohazard testing in receiving laboratory (continuing monitoring of project activities, pre-project advanced studies and research, as needed).	(Apollo 11, 12, 14)

Source: NASA (2005). Note that the PP Categorizations have changed since 1983 and missions flown before 1983, as denoted by the parentheses, would be classified differently than the current approach. The PP Categories reflected in this table represent the PP Categories these missions would receive using the current classification system.

assembly. All missions are required to address agency-level PP requirements, but the extent of the documentation and implementation varies depending on the target body, mission design, and mission science objective. PP Category I and II missions which have the lowest possibility of discovering life and pose no risk for spacecraft contamination events only have to document the mission design and provide a description of why the target body does not pose a PP risk. PP Category III missions are a common PP Category due to the prevalence of Mars orbiters, including Mars Odyssey and the Mars Atmosphere and Volatile Evolution (MAVEN) missions. PP Category III missions are also assigned when Mars is used for gravity assists for flyby missions to the outer solar system such as the Dawn mission. For these missions, cleanroom assembly, trajectory biasing, and impact avoidance or biological constraints are required. Cleanroom assembly in an ISO 8 cleanroom is required to reduce the contamination potential. To ensure that the spacecraft is not launched on a direct course with the body of interest the mission design is required to perform trajectory biasing. In addition to cleanroom assembly and trajectory biasing, the mission design should avoid impact of the body of interest with a high enough certainty or to account for the possibility of a direct contamination event, with cleanliness levels akin to landing a surface mission. For landed vehicles on target bodies with biological exploration PP Category IV, requirements include trajectory biasing, impact avoidance, microbial reduction, cleanroom assembly as a prevention measure, an organics inventory, and an archive of organics in bulk. To date, the missions that have undergone these requirements have all been Mars-based to include Viking, Mars Exploration Rover, Mars Pathfinder, Mars Science Laboratory, and InSight.

For Mars lander and probe missions, there are three sub-categories that also drive PP requirements. Category IVa is the lowest risk category and is designated when mission objectives do not seek to perform life detection or traverse into a region of Mars where life is possible. Category IVb is a risk category where bioburden requirements are more stringent but are isolated to a subsystem that is accessing a special region, collecting samples, or performing a life detection study. Lastly, Category IVc is the categorization where the entire spacecraft undergoes increased microbial reduction levels. This category is applied when the entire spacecraft is expected to interact with a special region, collect samples, or where the sensitivity of a life detection instrument drives the cleanliness level. There are no additional PP requirements for Category V "unrestricted Earth-return" missions, but for a "restricted Earth-return" the sample must be contained, the chain of contact with the target body should be broken, and sample containment should be in place until a biohazard assessment determines the sample is safe and poses no risk to life on Earth or the sample itself is sterilized. Note that breaking the chain of contact, also referred to as break-the-chain, is defined by isolating the hardware that contacted the extraterrestrial body, directly or indirectly, through the utilization of a containment system such that it will not come in contact with the Earth's biosphere until the container is either sterilized or secured within the sample handling facility.

To establish the PP category and associated PP requirements for missions involving drilling, excavation, and sample acquisition, the following should be considered. First begin to establish the target body, understand the trajectory design space, and evaluate all the celestial bodies encountered by the mission. Next, perform mission design analysis to understand the probability of impact of any encountered bodies to determine if the spacecraft will pose a threat of contamination in an off-nominal situation. Finally, assess the in situ science goals for biological or chemical evolution study, establish the need for sample collection, and determine if mission plans include Earth return.

4.1.1.2 Planetary Protection Technical Standards

Technical standards and specifications exist for PP spacecraft implementation and bioburden sampling and processing for biological cleanliness verification. For spacecraft implementation, a given mission can utilize a variety of approaches described in the standards. Should the mission choose to not use the agency PP technical standard, they can also perform hardware destructive testing to establish a hardware actual bioburden level. This is performed in conjunction with approval from the specific agency's PP officer.

4.1.1.2.1 *Spore Bioburden Surface and Non-Metallic Density Specification Values*

All hardware surfaces and densities have to be accounted for in a biologically sensitive mission for bioburden accounting purposes. The process of accounting for hardware bioburden is an extensive process that tracks each component's total surface area and non-metallic volume (e.g., adhesives, lubrication, electronics, composites, etc.) similar to that of a mission's mass equipment list. As such, each component must be evaluated for bioburden density utilizing specification values and/or test values. The source specific microbial specification values for cleanroom assembly and encapsulated bioburden density are conservative values set forth by NASA (Table 4.3). The surface burden density specification value ranges from 1×10^5 spores/m^2 in an uncontrolled manufacturing environment to 50 spores/m^2 in an ISO Class 7 cleanroom with stringent controls (e.g., established biological control areas if multiple missions are utilizing the same facility, biological monitoring assessments of the facility, full coverall gowning to include surgical facemasks, etc.). NASA selected the bacterial spore as a proxy for total microbiological cleanliness on Mars-bound spacecraft given its robust ability to (high level of resistant potential) resist harsh environmental conditions and biological reduction processes. In addition to surface bioburden, planetary protection is also concerned about

TABLE 4.3
Source Specific Microbial Specifications as per NASA NPR 8020.12 for Cleanroom Assembly and Encapsulated Bioburden Density

Parameter Specification	Value
Surface burden density: uncontrolled manufacturing	1×10^5 spores/m^2
Surface burden density: ISO Class 8 cleanroom with normal controls	1×10^4 spores/m^2
Surface burden density: ISO Class 8 cleanroom with stringent controls	1×10^3 spores/m^2
Surface burden density: ISO Class 7 cleanroom with normal controls	5×10^2 spores/m^2
Surface burden density: ISO Class 7 cleanroom with stringent controls	50 spores/m^2
Encapsulated burden density (in non-metallic materials), average	130 spores/cm^3
Encapsulated burden density specific to electronic piece parts	3–150 spores/cm^3
Encapsulated burden density specific to non-electronic non-metallic materials	1–30 spores/cm^3

Source: NASA (2005).

bioburden density inside of non-metallic materials (e.g. adhesives, epoxies, paints, and composite materials). These non-metallic materials are tracked and of concern where hardware can impact a biological body of interest.

4.1.1.2.2 Heat Microbial Reduction

Heat microbial reduction is the most widely used and the most well understood of all the hardware microbial reduction methods. The heat microbial specifications were originally developed from the food industry and further applied to various microbial reduction and sterilization applications in not only the food industry but the medical device industry as well (Otterbein & Pflug, 2010). For applicability to spacecraft use, NASA conducted an extensive heat microbial reduction materials compatibility assessment for the Viking lander program, establishing *Bacillus pumilus* var niger (now *B. atrophaeus*) as the standard biological challenge organism. As part of the Viking studies, a series of low-abundance environmental organisms that were even more "hardy" than the standard were identified (Puleo et al., 1978). In the early 2000s NASA and ESA undertook a study on a new "hardy" biological challenge standard, *Bacillus* strain ATCC 29669, as this was a spacecraft assembly facility-associated isolate that exhibited a hardier heat resistance profile as compared to the industry standard of *Bacillus atrophaeus* ATCC 9372 (Schubert and Beaudet, 2001). The hardier isolate testing revised the NASA and ESA microbial heat standards and are currently captured in the ESA PP standards (ECSS-Q-ST-70-57C, 2013). NASA has also adopted these standards and is currently updating documentation. InSight, Mars 2020, and the Europa Clipper missions utilize the new heat microbial reduction specifications. For the application of heat microbial reduction, there are multiple curves that take into account hardy and non-hardy microbial populations. The implementation of heat microbial reduction includes heat in the range of 110°C–200°C and specification factors to account for free surfaces, mated surfaces, and embedded microorganisms in non-metallic materials.

4.1.1.2.3 Heat Microbial Reduction for Absolute Sterility

Absolute sterility for a given spacecraft component can be accounted when the hardware reaches 500°C for 0.5 seconds. Utilization of this specification occurs as part of a mission's entry heating analysis and credit is taken in the bioburden accounting. When this specification is applied this is an exception to the launch bioburden requirements, as the entry heating occurs typically as the spacecraft enters the target body's atmosphere.

4.1.1.2.4   *Vapor Hydrogen Peroxide (VHP)*

Vapor phase bioburden reduction can be either utilized in a parametric microbial reduction modality or for complete sterilization as per the joint ESA- and NASA-approved microbial reduction method. Both ESA and NASA developed these specifications in the early 2000s, developing both an ambient and a vacuum-based VHP delivery system (Chung et al., 2009; ECSS-Q-ST-70-56C, 2013). Due to the variety of hardware configurations, material types, and possible sterilization equipment setup, the processing credit is only allowed using a biological challenge study to verify the efficacy of the test process on a case-by-case basis. For 2- to 6-orders-of-magnitude microbial credit, parametric microbial reduction can be performed. For each decimal reduction value (i.e., D-value defined as what is needed to reduce the microbial population by 90%) of surface bioburden reduction desired at the end of the process, a time integrated concentration of 200 (mg/L) sec is added to the process duration. This can be in either an ambient or vacuum chamber with a required surface hydrogen peroxide concentration of 0.5–1.1 mg/L and 1.1 mg/L, respectively. The complete sterilization specification requires a concentration of 6 mg/L–8.6 g/L in an ambient chamber with a final time integrated concentration of 14,000 (mg/L) sec, equivalent to a medical "overkill" sterilization process.

4.1.1.3   Microbial Reduction Methodologies Not Captured by Agency Standards

The current NASA-approved microbial reduction processes have extensive engineering heritage from a system engineering approach in that the exposure conditions, hardware compatibility, post-reduction hardware performance, and contamination control aspects are well characterized. Hardware is qualified to these conditions and have often flown on multiple missions with success. Despite the heritage approaches, as missions become more biologically sensitive and hardware components cannot withstand heat microbial reduction or VHP, the engineering community has to conduct hardware sterilization trades using industry-based sterilization guidelines and requirements to find acceptable technologies. Non-standard approaches need project-specific testing, protocol development and verification to the sponsoring agency to demonstrate the proposed efficacy. This can be done using industry-based sterilization guidelines that include the "Guidelines for Disinfection and Sterilization in Healthcare Facilities" by the Center for Disease Control (CDC) and ISO 14937:2009 standards (Rutala et al., 2008; ISO, 2009). A variety of industry techniques have been considered over the years for alternative spacecraft use, but some of the recent methods considered for flight hardware microbial reduction include chemical (e.g., solvent based, ethylene oxide), plasma, and radiation (e.g., gamma, ultraviolet (UV)).

4.1.1.3.1   *Precision and Solvent-Based Cleaning*

Critical spacecraft fluid handling systems such as sampling, environmental life support systems, and propulsion systems typically undergo a precision cleaning process to ensure they meet particulate and molecular contamination control levels. Precision cleaning can vary vastly from vendor to vendor, but the overall concept that is employed for robotic spacecraft and associated launch hardware usually involves the following steps: pre-cleaning visual inspection, coarse mechanical scrubbing in a detergent or initial bath, ultra-pure water rinse, series of solvent and ultra-pure water ultrasonic baths, high-pressure solvent or ultra-pure water rinse, and post-cleaning visual inspection (could include various particulate, non-volatile residue or biological testing). Activities for cleaning usually occur in a specialized cleanroom and all cleaned components are tracked and contained in a protective enclosure (e.g., bagging, totes, boxes) to prevent recontamination prior to integration into the engineering assembly. For the Mars Science Laboratory and InSight missions, the propulsion systems and launch vehicle environmental control systems that passed non-volatile residue and particulate tests also met biological cleanliness requirements typically ranging <30 spores/m^2 (Benardini et al., 2014, 2014); Hendrickson et al., 2020, 2020).

Typical daily recontamination prevention for cleanroom, ground support equipment, and hardware surfaces involve solvent cleaning with cleanroom wipes or swabs wicked with high-performance liquid chromatography (HPLC)-grade isopropanol or ethanol. As a best practice cleaning technique, each damp wipe or swab is passed in a single motion over the surface to be cleaned. During cleaning, the wipe or swab is lifted between each pass and each pass is overlapped by 50% first in a horizontal, then vertical, and finally a diagonal direction. For hardware that cannot undergo alcohol cleaning, either methyl ethyl ketone or molecular-grade sterile water can also be utilized as a cleaning agent for mechanical systems. For all precision cleaning and solvent cleaning methods, bioburden verification has to be performed to understand the biological burden present on each surface after cleaning.

4.1.1.3.2 Chemical—Ethylene Oxide

Ethylene oxide was first employed for the Ranger Project as a recommendation from NASA Headquarters to the NASA Centers in October 1959: "of the several means of sterilization proposed, NASA considers the use of ethylene oxide in its gaseous phase as the most feasible agent at this time" (Hall, 2010). A combination of ethylene oxide and heat microbial reduction was employed on lunar missions for sterilization processes. After a series of mission mishaps on the Ranger Project, heat microbial reduction of spacecraft components and ethylene oxide was discontinued as a terminal sterilization step. After the 1960s, NASA abandoned the use of ethylene oxide and ceased the development of material testing and implementation advancements. Given that ethylene oxide continues to be a widely used industrial sterilization technology in the medical device industry, it is still considered as a viable surface sterilization treatment for spacecraft. It does not require that a cell be metabolically active for ethylene oxide to be an effective sterilization mechanism as it relies on its high reactivity to alkylate cellular components to inactive organisms. Gas concentration, exposure time, temperature, and humidity are the four factors that impact ethylene oxide sterilization (Mendes et al., 2007) and thus further materials and processing testing will need to factor into any parametric or lethality testing for spacecraft material use. Other microbial reduction gases such as chlorine, chlorine dioxide, or fluorine gas may certainly be valid for consideration with the appropriate materials and process testing.

4.1.1.3.3 Plasma

Plasma ionizes the molecules in the surrounding gas to generate active species such as ions, free radicals, and unstable molecules. These active species can directly disrupt the cell membranes and impact cellular macromolecules and membrane lipids leading to microbial lethality. The plasma system can be further tuned by altering the reactant carrier gas. The plasma sterilization technology has been selected by NASA as a potential supplement to containment and sterilization technologies under study for a possible Mars Sample Return Campaign. This system has the capability to provide microbial reduction and surface etching to indiscriminately remove all forms of contamination at the target location. Plasma systems have been widely and reliably used across several disciplines, including the semiconductor industry and ceramic etching (An, 2008). As a surface-only treatment, plasma does not penetrate non-metallic materials to reduce embedded bioburden within spacecraft, and as such is focused for use on cleaning hardware surfaces prior to critical assembly, final surface cleaning, or as an in-flight cleaning technology for Mars Sample Return.

Jet Propulsion Laboratory's preliminary, unpublished testing has demonstrated at least a 6-log reduction of heat-resistant bacterial endospores in vacuum conditions. Control studies were also performed to demonstrate that the vacuum alone was not sufficient to inactivate the microorganisms, and the resulting effect was due to plasma exposure. Both oxygen and fluorine were tested as carrier gasses. Oxygen is relatively safe to work with; however, fluorine has a lower recombination rate, which prolongs the effect of plasma by allowing radical species to penetrate further in a volume.

The key challenge in the plasma sterilization system is to develop the same breadth of knowledge as traditional heat and chemical sterilization technologies. The plasma technology continues to prove its value as a system that not only inactivates but destroys the possible life forms, which makes it an attractive candidate for sterilizing microbial life both known and unknown.

4.1.1.3.4   Radiation—Ionizing and Non-Ionizing

Gamma and ultraviolet (UV) space radiation environments are well understood during spaceflight and flight operations and as such, have a high potential as a microbial reduction alternative. UV has been studied, and using Earth-based experimental test setups, demonstrated complete microbial reduction on the simulated surfaces of Mars or on the Moon (Schuerger et al., 2006, 2019). In-flight space environmental studies have also shown significant microbial reduction using the International Space Station as an experimental platform to demonstrate the impact of solar UV under various conditions of space vacuum and Mars gas mixes (Rabbow et al., 2012; Vaishampayan et al., 2012). UV is being considered as an in-orbit active process through active engineering systems (e.g. light-emitting diode bulbs), passive in-space environmental processes, and operational scenarios have also been considered for on Mars sterilization of rover surfaces using the Mars environmental conditions. UV is a surface-only treatment and microbial reduction radiation modeling is particularly sensitive to shadowing effects (caused by hardware configurations or dust). NASA has not utilized UV to date as a microbial reduction technique on an interplanetary mission. With more advances in modeling and a developed understanding through lab tests on the impacts of indirect UV, the potential utilization of UV as a mission-approved microbial reduction technology is feasible. Notably, UV has also been implemented in overhead lighting for the facility bioburden control of the ExoMars Mars Organic Molecule Analyzer (MOMA) aseptic assembly cleanroom (Lalime et al., 2018).

Gamma sterilization can also be conducted on Earth prior to launch, but currently is in the development phase for characterizing microbial reduction as well as material compatibility of spacecraft power system components. Gamma sterilization has been employed on the Ranger spacecraft series, USSR/Russian missions and Beagle 2 as a microbial reduction protocol (Meltzer, 2012; Calvin and Gazenko, 1975; National Academies, 2018). To implement gamma sterilization both personnel safety and hardware dosage considerations must be addressed. Personnel safety concerns have to be addressed due to the potential deleterious impacts to human health of ionizing radiation. Hardware dosage considerations are key to ensure proper microbial reduction, but the hardware has to have adequate support equipment. While gamma radiation has some personnel safety issues to overcome, it may prove effective for large system sterilization with minimal electronics installed (e.g. parachutes, blanketing material, aeroshells).

The majority of the radiation in the Jovian environment is trapped electrons. Since they are able to penetrate through hardware surfaces, they are being utilized for in-flight sterilization for NASA missions and contributed missions to the outer planets such as the JUNO, JUICE, and the Europa Clipper missions. From a mission standpoint, this is quite effective as exposed hardware can experience upwards of >10 Mrad of radiation during the lifetime of the mission, effectively rendering complete microbial reduction of the hardware surface. Flight hardware is designed with shielding to protect critical avionics and scientific equipment, and as a result, differential microbial reduction of a spacecraft is calculated using zones based on environmental radiation gradients.

4.1.2   Mission Science Requirements Driving the Need for Biological Contamination Control

While planetary protection requirements impose biological cleanliness levels on spacecraft to meet space agency and international COSPAR policy requirements, biological cleanliness requirements may also be driven by a mission's science requirements. Given that science instrumentation is increasing in the level of detection sensitivity required for science missions, the typical contamination control requirements and implementation methods will inevitably start to need more stringent biological cleanliness requirements to ensure that project science goals can be met. These science sensitivity requirements would be negotiated by each project's science team similar to a project contamination control requirement. Biological contamination on spacecraft could include viable microbial cells that could replicate, non-viable cells that could interfere with microscopy techniques as they may appear to be intact cells, and non-viable cells that could be a source of abiotic chemicals (e.g. adenosine tri-phosphate (ATP), proteins, deoxyribonucleic acid (DNA), ribonucleic acid (RNA), or other secondary metabolites). The biological verification technique may also need to be expanded as the valid verification technique using infrared (IR) and Raman spectroscopy may not be sensitive enough to detect the single cell to tens of cells limit of modern microbiological and molecular methods.

4.2 BIOLOGICAL CONTAMINATION TYPES AND ASSESSMENT METHODS

Having a developed and clear biological contamination requirement that identifies the type of contamination, threshold of contamination, and appropriate biological cleanliness method is paramount to the foundation of a biological contamination control program. The biological contamination types include both cells and cellular debris from cell remnants. These types could encompass a wide variety of microbial populations from space-relevant extreme environments to include spore forming, radiation-resistant (both UV and ionizing), and cold-tolerant biological materials. While NASA only has an approved cultivation-based biological cleanliness verification standard, it also leverages biochemical methods to provide a more rapid assessment of cleanliness. Additional biochemical and molecular biology-based methods are being considered as supplements to the NASA Standard Assay (NSA).

4.2.1 BIOLOGICAL CONTAMINATION TYPES

The NSA focuses on heat-tolerant, spore-forming organisms as a proxy for overall microbial cleanliness for Mars-destined spacecraft. Although the spore has been a key for Mars-based requirements, as missions expand to sample return, outer planets, and life detection, broader classes of biological life have to be considered. Organisms that are likely to survive the space environment and celestial bodies of interest are considered important when factoring in the total viable population for probabilistic-based requirements. Along with meeting these probabilistic-based requirements post-microbial reduction, cellular debris may be present on the hardware surfaces. Thus, these remnants may pose as a contamination source for sensitive science investigations looking into biotic compounds.

4.2.1.1 Microbial Types

The Space Studies Board (NRC, 2000) task group on Europa suggested NASA should consider the following four groupings of microorganisms when determining the biological contamination risk:

- **Type A**: Typical, common microorganisms of all types (bacteria, archaea, fungi, etc.).
- **Type B**: Spores of microorganisms that are known to be resistant to environmental insults (such as desiccation, heat, and radiation; e.g., *Bacillus subtilis*).
- **Type C**: Spores that are especially radiation-resistant (e.g. *Bacillus pumilus* SAFR-032).
- **Type D**: Rare but highly radiation-resistant, non-spore microorganisms (e.g., *Deinococcus radiodurans*).

The basis for this categorization is radiation sensitivity with respect to the Jovian radiation environment. Although each species will have a somewhat different survival response to ionizing radiation, these are the four general categories that can be readily distinguished by straightforward assay protocols. A majority of microbial diversity protocols, as detailed later, do not test the organisms for environmental tolerance or spore formation but infer the observed NRC type of the microbe based on peer-reviewed published results of similar genera and species having those characteristics.

During microbial diversity surveys performed over the last decade, several hundred aerobic microbial strains have been isolated from spacecraft-associated surfaces and characterized for their taxonomic affiliations (Ghosh et al., 2010; Kempf et al., 2005; La Duc et al., 2003, 2004, 2007).

When Phoenix mission samples were analyzed (Ghosh et al., 2010), microorganisms grown under mesophilic conditions (Type A) yielded a total of 21 species, 5 of which were novel species. Spore formers (Type B) demonstrated diverse metabolic capabilities: 11 species were detected from the spore-favoring heat shock assay, 7 were obtained from high-alkaline media, 7 were cultured under anaerobic conditions, 5 were isolated on neutral pH R2A under mesophilic conditions, 2 were isolated under psychrotolerant conditions, 4 survived high UV exposure, and 1 was determined to be both hydrogen peroxide resistant and thermophilic.

It should be noted that a single species, *B. pumilus*, was found to tolerate nearly all experimental stresses (Type B), including growth at high temperature (~65°C) and exposure to hydrogen peroxide conditions. Eight thermophilic bacterial strains were obtained from the Phoenix mission samples; in addition to a single isolate of *B. pumilus*, 7 strains of *Geobacillus stearothermophilus* were collected.

Along with spore-forming bacteria (Type B), known to be the most common microbial inhabitant of spacecraft assembly facilities, several non-spore-forming bacteria (Type D) have been found to be tolerant to extreme environmental conditions (Ghosh et al., 2010; La Duc et al., 2007). Non-spore-forming bacteria of the genus *Deinococcus* have been reported to survive acute exposures to ionizing radiation, UV light (UV), and prolonged desiccation (Cox and Battista, 2005; Mattimore and Battista, 1996; Rainey et al., 2005). These Type D organisms are more likely to escape standard planetary protection assays, since such protocols account only for the presence of spores. Hence, without the appropriate cleaning/sterilization protocols, the occurrence of an extremely radiation-resistant strain of a species of the genus *Deinococcus* in the cleanroom facility could pose a serious risk to the integrity of future explorations of extraterrestrial life and sample return missions.

Characterization of alkali-tolerant and psychrotolerant strains (extremophiles; belonging to Type A) further demonstrated physiological diversity present in the NASA cleanrooms. Direct cultivation in high-alkaline medium yielded a total of 23 species, 2 of which were novel; low-temperature conditions resulted in the isolation of 18 species, 6 of which were novel. *Acinetobacter johnsonii* and *B. diminuta* dominated the mesophilic and alkali-tolerant bacteria recovered. Psychrotolerant bacteria predominantly belonged to Proteobacteria and Actinobacteria, with no single species dominating the isolates. Of the 6 UV-resistant bacterial species analyzed (Type D), 4 belonged to the Firmicutes. In addition to aerobic microbes, anaerobic organisms (Type A) were also isolated in these environments (Probst et al., 2010) which were not specifically called out, but by definition binned into Type A in the Europa Forward Contamination SSB report.

When next-generation DNA sequencing methods are employed, the spacecraft assembly facility environmental samples exhibited >1000 different kinds of microbial types which require careful consideration when categorizing them (Cooper et al., 2011; La Duc et al., 2012; Mahnert et al., 2015; Minich et al., 2018; Vaishampayan et al., 2010, 2013; Weinmaier et al., 2015). Most of these studies reported they were dominated by *Acinetobacter lwoffii*, *Paracoccus marcusii*, *Mycobacterium* sp., and *Novosphingobium*, which were reported to be present in >75% of the samples (~200+ samples tested). According to microbial spatial topography, the most abundant cleanroom contaminant, *A. lwoffii*, is related to human foot traffic exposure and belongs to Type A microorganisms as defined by the Europa SSB report (NRC, 2000). This begs a question about the incidence of other types of microorganisms. Cultivation-based analyses conducted during the Phoenix mission revealed 0.3%–11% of Type B microorganisms are forming spores and known to be resistant to environmental insults (Ghosh et al., 2010). Several novel spore-forming microorganisms belong to Type B microorganisms also reported in samples associated with the Phoenix mission (Benardini et al., 2011). A systematic study is needed to measure the incidence of other types of microorganisms (Type C and D) categorized by the Europa SSB report.

4.2.1.2 Cells and Cellular Debris—Microbial Reduction and Cell Remnants

The advent of molecular technologies has dramatically increased the resolution and accuracy of detection of distinct microbial lineages in mixed microbial assemblages. Despite an expanding array of approaches for detecting microbes in a given sample, a rapid and robust means of assessing the differential viability of these cells, as a function of phylogenetic lineage, remains elusive. In several studies, pre-PCR propidium monoazide (PMA) treatment was coupled with downstream molecular (Illumina sequencing, pyrosequencing, and PhyloChip DNA microarray) analyses to better understand the frequency, diversity, and distribution of viable bacteria in spacecraft assembly cleanrooms. This also enables removal of dead bug body husks from the analysis. If dead bug body information is required, the difference between PMA-treated and non-PMA-treated samples will describe what percentage of microbial contamination was attributed to the dead bug bodies. Sample fractions not treated with PMA, which were indicative of the presence of both live and dead cells, yielded a great abundance of highly diverse microbial DNA sequences. In contrast, only 1%–10% of all of the sequencing reads, arising from a few robust microbial lineages, originated from sample fractions that had been pre-treated with PMA. The results of PhyloChip analyses of PMA-treated and -untreated sample fractions broadly agreed with those of DNA sequencing (Vaishampayan et al., 2013).

4.2.1.3 Other Biotic Compounds—ATP and DNA

Nucleic acid-based microbial detection is described earlier. Detection of other contaminates and current analysis methods for their detection is included here. A firefly luciferase bioluminescence assay method differentiates free extracellular ATP (dead cells, etc.) from intracellular ATP (viable microbes). This can be used to determine the viable microbial cleanliness of various cleanroom facilities (Venkateswaran et al., 2003). For comparison, samples were taken from cleanrooms, where the air was filtered to remove particles >0.5 μm, and from ordinary rooms with unfiltered air. The intracellular ATP was determined after enzymatically degrading the sample's free ATP. ATP-based determinations indicate that the microbial burden was lower in cleanroom facilities than in ordinary rooms. Comparisons can be made with microbial burden cultivated on agar (a much slower method); however, there was no direct correlation between the two sets of measurements because the two approaches measured very different sub-populations.

4.2.1.4 Biotic Secondary Metabolites to Include Amino Acids and Lipids

University of California San Diego researchers (Prof. Pieter Dorrestein and Prof. Rob Knight) are currently developing joint-targeted metabolomics using a triple quadrupole mass spectrometer, including molecules such as bacterial-associated molecules such as phenyl sulfate, indole, cresols, kynurenine, serotonin amongst others (Melnik et al., 2017). This pipeline also targets amino acids and lipids, and is complemented by untargeted metabolomics analysis performed on an Q Exactive mass spectrometer (QE) for discovery of additional molecules that may be involved in skin host-microbe interactions and/or involved in triggering inflammatory responses. This procedure yields the maximum amount of knowledge that is currently available for any mass spectrometry run in terms of annotations, and will continue to grow. The mass spectrometers used for the metabolomics analysis are quadrupole time of flight (Q-TOF), QE, matrix-assisted laser desorption/ionization-time of flight (MALDI-TOF), and ion traps. The Q-TOF and QE spectrometers are used as workhorse instruments for untargeted analysis. Microbial imaging mass spectrometry is conducted by the MALDI-TOF. Ion traps are used for testing ionization methods before implementing them on the workhorse instruments and for mass spectrometry-based purification workflows to obtain pure samples of unknown molecules for nuclear magnetic resonance analysis.

4.2.2 Biological Assessment Methods

The NASA standard assay is an approved cultivation technique to estimate the biological cleanliness of spacecraft hardware. In addition to the NASA Standard Assay, NASA also recognizes the use of the ATP assay as a rapid means to estimate microbial cleanliness, but not to be used to verify cleanliness levels. While these are approved approaches, other microbiological, biochemical, and molecular biology techniques may be useful to assess the overall biological cleanliness of a sample acquisition system. Tailored cultivation approaches to specific classes of organisms such as with media and conditions to promote radiation-resistant organisms or organisms resistant to varying temperatures or pH ranges could be leveraged. To garner a larger breadth of microbial populations, direct microscopic or molecular biology-based methods can be explored. NASA is currently evaluating these biochemical, direct microscopic, and molecular biology methods as supplemental means to discern biological contamination for future missions.

4.2.2.1 Cultivation-Based Microbial Population Estimation

The enumeration of spores (NASA, 2005), mesophilic heterotrophic bacteria (NASA, 2005), archaea (Moissl et al., 2002), fungi (Anastasi et al., 2005), and anaerobes (Bryant, 1972; Hespell et al., 1987) can be carried out according to established protocols. In order to estimate the physiological breadth of the bacterial population present in an environment, several cultivation and analysis strategies can be employed (La Duc et al., 2003, 2004). This includes, but is not limited to media and conditions promoting the growth of alkaliphiles (pH 10.0), acidophiles (pH 3.0), thermophiles (65°C), psychrophiles (4°C), halophiles (NaCl 25%), and UV- (1,000 J/m^2), gamma- (1 Mrad) (La Duc et al., 2007), and electron-beam-radiation- (Pillai et al., 2006) resistant microbes. For an example, an in-depth description of methods aimed at the selective isolation of oligotrophs and psychrophiles is presented next. In addition, several molecular and visualization-based analyses, such as epifluorescence microscopy, ATP-assays (biochemical-based (La Duc et al., 2007; Venkateswaran et al., 2003)), and quantitative polymerase chain reactions (qPCR) can be performed to derive estimates of total and viable microbial numbers (Nocker et al., 2009), as well as specific populations (bacteria (Suzuki et al., 2000, 2001), archaea (Suzuki et al., 2000), and eukaryotes (Haugland et al., 2002)).

4.2.2.2 Isolation and Enumeration of Oligotrophic Microorganisms

Oligotrophs are organisms capable of surviving and proliferating in extremely nutrient-deprived conditions (total organic carbon level <2 mg/L). Species inhabiting extremely oligotrophic environments on Earth have been shown to display an array of distinctive metabolic activities for community growth, including atmospheric nitrogen and organic carbon fixation, methane, hydrogen and iron oxidation, the breakdown of complex organic molecules, and the mobilization of inorganic phosphorous (Barton and Northup, 2007; Barton et al., 2004; Chelius and Moore, 2004). As with all environments, an understanding of the potential energy sources (both electron donors/acceptors) and nutrients that support microbial growth and propagation is critical to understanding microbial adaptation, physiology, colonization, and propagation within oligotrophic environments.

The researchers need to identify: (1) microbial species capable of growing on identified nutrient sources, and (2) microbial consortia/biofilms capable of growing on identified nutrients sources. For example, samples collected can be placed in appropriate aliquots of sterile buffer. Suitable volumes of each sample are distributed into media plates augmented with various nutrient additives, based on the nutrient and energy sources identified in the sample. To obtain microbial consortia (rather than individual species) for analyzing biofilm growth, samples are placed in a flask containing a suitable volume of sample of filter-sterilized site water sample.

4.2.2.3 Isolation and Enumeration of Psychrophilic Microorganisms

Psychrophiles are microbes that exhibit growth optima below 15°C (Bakermans and Nealson, 2004; Junge et al., 2002). Deming and co-workers (Deming, 2002) investigated the lower temperature range of the obligate psychrophilic and salt-requiring Arctic marine bacterium, *Colwellia psychrerythraea* and reported metabolic activity at −20°C. Additionally, cell replication has been observed at −10°C in microbes isolated from sea ice off the Alaskan coast (Breezee et al., 2004) and from Siberian permafrost (Bakermans et al., 2003; Jakosky et al., 2003). To this end, the researchers attempts (1) examination of the lower temperature limits of the oligotrophic microbes isolated from nutrient-deprived environments and (2) experiments promoting the selection of cold-adapted microbes within environments (Deming, 2002).

For example, using a refrigerated water bath, psychrophiles were grown under different temperature-controlled conditions to investigate growth and viability. This includes "warm" conditions (10°C), growth permissive sub-zero conditions (−2°C), and slowly freezing conditions (~ −5°C to −10°C). Growth is assessed by viable counting using petrifilm, which is added to growth media, and by microscopy (Marx et al., 2009). The effect of different growth conditions is investigated using micro-ion flux estimation techniques developed at the University of Tasmania. This method utilizes a pair of aligned microelectrodes containing an ion-selective solution to measure electrochemical gradients of different ions (H^+, Na^+, K^+, etc.) near cells (Shabala et al., 2009). This allows estimation of the net flux of ions across cell walls. Probes are positioned close to cells using a micromanipulator and a microscope. For bacteria, it is necessary to test them as a monolayer on poly-lysine treated glass slides. The cell layers can be directly exposed to different conditions including ionic solutions, and/or temperature (the system includes a temperature-controlled stage that can be stably lowered to <0°C). It is possible to conduct experiments that measure levels of cell energy and activity in real time (on the order of seconds to minutes). ATP estimation via luminometry and ion flux data (H^+, K^+, Na^+) is useful for interpretation of the physiological consequences of strains under osmotic stress as well as providing insights into energy generation and conservation. The approaches described here have been previously demonstrated in studying the effect of different osmotic pressures on various microorganisms (Shabala et al., 2009). In addition, an epifluorescence microscopy and/or fluorometry method could be performed to assess metabolic activity of the psychrophiles as described elsewhere (Martin et al., 2011).

4.2.2.4 Isolation and Enumeration of Radiation Resistant Microorganisms

4.2.2.4.1 UV-Radiation Resistance

Microbial cultures, consortia, or raw samples can be exposed to UVA (315–400 nm), UVA+B (280–400 nm), and total UV (200–400 nm) at the Mars-simulated solar constant of 590 W m^{-2} total irradiance (190 nm to 3 µm) (Appelbaum and Flood, 1990) using an X-25 solar simulator (Spectrolab Inc.) equipped with a Xenon Arc-lamp (Newcombe et al., 2005). The intensity of UV irradiation is approximately 10% (59 W m^{-2}) of the full spectrum and the total irradiance level in our simulations will be close to the actual Mars solar constant of 590 W m^{-2} (Appelbaum and Flood, 1990). The various bandwidths are generated by placing Corning glass (UVA) or plastic petri lids (UVA+B, Fisher Scientific) as filters in front of the sample cuvette and the spectrometer. Irradiation intensity can be constantly adjusted by fine-tuning the lamp power to achieve ± 5% variability as indicated by an Optronics Laboratories OL754 UV-VIS spectrometer. After exposure, appropriate aliquots of samples can be placed into suitable media that promotes growth and cultivation conditions of the test extremophilic microorganisms, and colony-forming units will be counted. For environmental samples, DNA are extracted and subjected to molecular methods to measure microbial diversity (see the next section). When viability measurement is required, the propidium monoazide methodology was implemented (Venkateswaran et al., 2014).

4.2.2.4.2 Gamma-Radiation Resistance

For gamma radiation-tolerance studies, *Dienococcus radiodurans* is used as a biological challenge organism as it exhibits high tolerance to gamma radiation that is well characterized. The concentrations and composition of nutrients used are those that limit the survival of *D. radiodurans* exposed to radiation (Venkateswaran et al., 2000). In addition, although extremely resistant to γ-radiation and UV-C (100–295 nm), *D. radiodurans* is more sensitive to UV-A (320–400 nm) than *E. coli* is (Slade and Radman, 2011). While *D. radiodurans* is very efficient at removing reactive oxygen species generated by ionizing radiation, during desiccation, and following exposure to UV-C (Daly et al., 2010; Fredrickson et al., 2008; Krisko and Radman, 2010), *D. radiodurans* appears highly sensitive to singlet oxygen (Slade and Radman, 2011). For the purpose of this research communication, radiation-resistant extremophiles are defined as the microbes that exhibit normal growth over 100 population doublings at the rate of 0.2 rad/sec (2 mGy/sec) under the nutrient conditions of the environment from where samples are collected (Daly et al., 2010). To this end, it is possible to (1) examine the radiation dosage limits of the oligotrophic microbes isolated from nutrient-deprived environments and (2) conduct experiments promoting the selection of cold-adapted radiation-resistant microbes within the samples.

To determine radiation response in previous work, desiccated samples were loaded into sealed containers and then placed in a dry cell or in water-tight tubes and positioned near enough to the reactor (or lanthanum source) to achieve the desired dose rate 7.2–72 Gy/hr (from 0.2 to 2 rad/s) for high total doses up to the Mrad (kGy) range. Additionally, continuously growing cultures located in an incubator containing growing samples at the appropriate temperature are placed in the dry cell and exposed at the 0.2 rad/sec trial dose rate. Samples are retrieved periodically for assessment of viability at various dose points. Total doses of up to 5 Mrad (50 kGy) can be attained over the course of one month of operation using these dose rates. Total dose should be monitored, e.g., by using a combination of detectors and chemical dosimeters.

4.2.2.5   Cultivation-Independent Microbial Diversity Approaches to Estimate Microbial Populations

Several state-of-the art molecular methods as well as non-destructive techniques have been tested and validated to elucidate microbial breadth present in environmental samples. Some of them are:

1. Targeted amplicon sequencing using next-generation sequencing (NGS) methods to render a comprehensive, in-depth catalog of microbial diversity associated with environmental samples (Checinska et al., 2015; La Duc et al., 2012; Petrosino et al., 2009; Vaishampayan et al., 2013; Venkateswaran et al., 2012).
2. Shotgun metagenomic approaches to assess both microbial diversity and functional capabilities of the community as a whole by exploring its entire gene content (Demaere et al., 2013; Singh et al., 2018).
3. A novel cell-sorting technology, *polymerase chain reaction (PCR)-activated cell sorting (PACS)* (Eastburn et al., 2013; Lim and Abate, 2013), that allows cells to be sorted based on nucleic acid biomarkers. The PACS technology works by encapsulating each cell in a microfluidic drop, lysing the cell, and performing PCR on the lysate to detect nucleic acids of interest.
4. *Microbial Community Screening and Profiling (MCSP)* (Probst et al., 2013) is a rapid and non-destructive molecular spectroscopic method capable of obtaining microbial diversity information directly from a sample without intermediate preparation. An appropriately calibrated MCSP platform can enable rapid microbial identification at the class and phylum levels, can quantify microbial abundance within microbial communities, and can also elucidate entire community functional relationships at a chemical level (Probst et al., 2013).

In utilizing such capabilities, substantial amounts of empirical data describing the full spectrum of phylogenetic and physiological breadth of the diverse consortia of microbes present in a given environment sample can be generated.

4.2.2.6   ATP Assay to Estimate Microbial Cleanliness

NASA measures and validates the biological cleanliness of spacecraft surfaces by counting endospores using the NASA Standard Assay (NSA). NASA has also approved the previously mentioned ATP-based detection methodology as a means to prescreen surfaces for the presence of microbial contamination, prior to the spore assay. During Mars Science Laboratory (MSL) spacecraft assembly, test, and launch operations, 4853 surface samples were collected to verify compliance with the bioburden requirement at launch. A subset of these samples was measured for microbial cleanliness using both the NSA (n = 272) and ATP assay (n = 249). NSA results revealed that ~8% (22/272) of the samples showed the presence of at least one spore, whereas ATP assay measurements indicated that ~15% (35/249) of samples exceeded the "threshold cleanliness limit" of 2.3×10^{-11} mmol ATP per 25 cm^2 used by MSL. Of the 22 NSA samples with a spore, 18% (4/22) were considered above the level of acceptance by both techniques. Based on post-launch data analysis presented here, it was determined that this threshold cleanliness limit of 2.3×10^{-11} mmol ATP per 25 cm^2 could be adopted as a benchmark for assessing spacecraft surface cleanliness. This study clearly demonstrates the value of using alternative methods to rapidly assess spacecraft cleanliness, and provides useful information regarding the process.

A large fraction of the samples collected from spacecraft and spacecraft-associated surfaces have no colony formers on TSA, but were positive for intracellular ATP. Subsequently, genomic DNA was isolated directly from selected samples and 16S rDNA fragments were cloned and sequenced, identifying nearest neighbors, many of which are known to be non-cultivable in the media employed. It was concluded that viable microbial contamination can be reliably monitored by measurement of intracellular ATP, and that this method may be considered superior to cultivable colony counts due to its speed and its ability to report the presence of viable but non-cultivable organisms. When the detection of nonviable microbes is of interest, the ATP assay can be supplemented with DNA analysis.

4.3 SAMPLE ACQUISITION FOR BIOLOGICAL CLEANLINESS

Understanding whether or not biological cleanliness applies to a particular mission is essential to establish early on in the mission life cycle. Note that not all missions will have these exact contraints as detailed herein, but these contraints and implementation approaches have been captured from Mars-based exploration and would have to be evaluated on a mission-by-mission basis. Developing a cleanliness plan for a Mars sampling mission may include: starting to work with engineering during the mechanical design phase, selecting material, designing features for ease of cleaning and recontamination prevention, contamination modeling, establishing a cleaning method for hardware assembly and testing, developing an implementation and monitoring program during assembly and testing, and managing operations.

4.3.1 MECHANICAL DESIGN FEATURES FOR BIOLOGICAL CLEANLINESS IN SAMPLE ACQUISITION

A successful planetary protection and contamination control campaign starts long before a sample is acquired. The most seamless examples of integrating planetary protection goals within a mission commonly tackle the issue in the design phase using a system's engineering approach. It is important to discuss possible solutions in the early stages of a project, as hardware designs can be rigid to future material or configuration changes. Selected parameters that have an impact on both planetary protection and contamination control are material choice, cleanability of materials, and design features.

4.3.1.1 Material Choice

Material selection plays a direct role in the scientific value of the acquired samples. Carbon dating, for example, will preclude the use of metal alloys that contain significant levels of lead that pose a contamination transfer risk. It is inevitable that the acquisition tool will impart trace amounts of the tool itself to a sample; therefore, the material can be strategically chosen to minimize, or at least maintain cognizance of the background levels contributed by the device. Characteristics to target for material selection include:

- **Low outgassing materials**: choosing low outgassing materials aids in reducing the overall contamination imparted by the material. Planning a preliminary outgassing bakeout of the hardware can also help reduce the risk of not meeting the overall outgassing levels for the hardware.
- Low particulate-generating materials (sealed or controlled edges, minimize the use of Velcro or felt seals).
- **Organic composition**: minimize the use of organic containing parts, lubricants, or epoxies. These materials can pose both an organic molecular and biological contamination source, especially if wear-products are generated through its use. These products may spread embedded, unsterilized microbial contaminants.

Strategic material selection will allow for a balance in meeting the scientific goals and overcoming the engineering obstacles to design a sample acquisition system.

4.3.1.2   Ability for Materials to be Cleaned

The selection of a material and surface coating type plays an important role in how easy or difficult the subsequent cleaning process will be. Materials such as paint can be cleaned, but have the potential to flake or experience degraded thermal properties depending on the cleaning method selected and degree of application. An additional factor to consider is the point of the hardware build schedule at which the selected cleaning process is implemented. In the case of ultrasonic cleaning for example, the energy profile of a sonicator is much different depending on the position of the hardware within the bath and the geometry of the part. Ultrasonic wave propagation is affected by reflections from the wall of the sonicator, bottom surface, and secondary reflections at the interface of the part (Gogate et al., 2002; Niemczewski, 2007; Pasumarthi, 2016). Therefore, cleaning efficiencies may lead to the selection of a process flow where sonication occurs at a piece-part level prior to assembly coupled with an environmentally controlled assembly process to maintain the level of cleanliness, rather than at a higher level of assembly. Alternatively, tests using the assembled geometry of interest can be conducted to demonstrate that the desired cleaning levels are achieved.

4.3.1.3   Design Features

Hardware design features focus on designing and the operation of hardware to protect and minimize the potential for biological contamination. This includes both active and passive methods that are verified through a series of hardware contamination modeling and testing. During this project phase the hardware undergoes an extensive contamination vector input and output analysis to ensure that contamination requirements are well understood through the operations phase for sample acquisition. An appropriate test and monitoring program is then devised and implemented for that hardware to monitor key steps in the hardware testing, assembly, and operations.

4.3.1.3.1   Design

Features that protect or isolate the sample acquisition tool can minimize transport of contaminants by direct transfer. Design tools such as barriers or doors can provide temporary protection until the point of use. These options can be one-time use or reusable. The reuse of barriers or doors pose a significant engineering challenge to mitigate the failure of deploying the mechanism, which poses a critical risk of achieving the prime goal—to acquire the sample. The Phoenix arm is an example of designing the system to meet planetary protection needs from the early stages of development. The choice of biobarrier material considered material compatibility with two separate options of microbial reduction (Salinas et al., 2007). The Phoenix arm consisted of a fully assembled mechanical arm that was encased with a bagging material, heat microbial reduced, and then the entire assembly was mounted on the surface of the spacecraft. The bag was preferentially vented through high-efficiency particulate air filters preventing recontamination and then the bag was opened with a mechanism on Mars right before surface operations commenced. Implementation of the subsystem biobarrier ensured that the protected subsystem remained in a microbial-reduced state until operational use as not to introduce inadvertent contamination on the Martian surface.

Molecular absorbers, or "getters," provide a design feature that allows for an added level of protection on top of selecting low outgassing materials. Getters come in a variety of compositions (Giorgi et al., 1985), but depending on the use case, there can be only limited options available to suit the spacecraft applications. For example, space applications dramatically limit the availability of materials certified for use. Abraham et al. utilized zeolites to address molecular contamination sources during their thermal vacuum tests in order to protect the contamination critical laser optics of the Advanced Topographic Laser Altimeter System (ATLAS) instrument (Abraham and Jallice, 2018).

Thermal control is also an option to minimize sources of contaminants, in combination with other methods. Sources that are kept at lower temperatures outgas at lower rates. The concept employing lower temperatures takes into account chemical species migrating from a warmer to colder surface. Similar to a cold trap or cold finger being used that is utilized in the lab or chamber, portions of the spacecraft can be designed with thermal gradients in mind to preferentially ensure that surfaces that are most sensitive (e.g., sampling systems or optical surfaces) are thermally regulated at higher temperatures. The temperature regulation can either be active, like a heater, or passive, considering mission design through navigation such as sun pointing compared to moving a critical system adjacent to a warm electronic box.

4.3.1.3.2 *Configuration*

The configuration of the sample acquisition tool with respect to the surrounding hardware and environment is equally as important, as the surrounding environment poses a contamination threat. Strategies in designing the configuration of the hardware include controlling contaminants that are within line of sight and isolating the contamination hazards completely.

Line of sight refers to having a straight line or direct path from the source to the target. There are design options that allow for sources of contamination to be configured in a place where there is no direct line of sight to the target. Conversely, there are also ways to isolate the target so that the line of sight is constrained to a very small window.

An effective way of controlling the configuration to limit its ability to contamination to the target is to isolate the source completely. This can be achieved with doors, covers, and filtered vents that point to less sensitive areas. The venting strategy is important, as it can be used to direct effluent strategically away from key critical areas to include the target as well as other inlets or apertures.

4.3.2 Contamination/Recontamination Vector Assessments and Modeling

Dust and contamination transport physics has been leveraged for past missions to include Earth-orbiting satellites (Chang, 1999), lunar missions (Hess et al., 2015; Wang et al., 2007), missions to Mars (ten Kate et al., 2008; Beaudet, 2000), and missions en route to other solar bodies (Borson, 2002; Barengoltz and Edgars, 1975). As transport models continue to be produced to support the science and engineering needs of the mission, a fundamental shift has occurred from a spacecraft-centric model to a sample-centric model. This shift in perspective gives further credibility to the cleanliness implementation approach for the mission in terms of all relevant contamination vectors to the acquired sample. For example, the Phoenix mission sample acquisition system was extraordinarily clean and sufficiently protected with a biobarrier in the spacecraft-centric view. From the sample-centric approach, models were also developed to determine the transport of contaminants from the spacecraft top deck to the soil at the target location of sample acquisition. This sample-centric view also revealed that the spacecraft itself was clean enough to meet the scientific needs of the mission. This sample-centric approach continues to be utilized in the development of the contamination control approach for the Mars Science Laboratory and the Mars 2020 missions. This approach has driven been used to justify reduced cleaning methods over more stringent cleaning and control methods (Figure 4.1). This will be discussed further in the Mars 2020 case study section.

Particulate and molecular contamination transport models affect the following measures in practice:

- **Hardware cleaning levels**: Transport models provide insight into appropriate cleaning levels required for the hardware or can verify that the chosen approach meets the needs of the mission. Models may drive an otherwise inconspicuous piece of hardware to greater cleaning lengths considering the environmental factors and contamination transport potential for that hardware. This also ties into hardware reliability, as there is a great dependence in the transport model for particular components to be isolated throughout the entire duration of the operation or mission. Any change to these initial assumptions will have an impact on the results of the model.

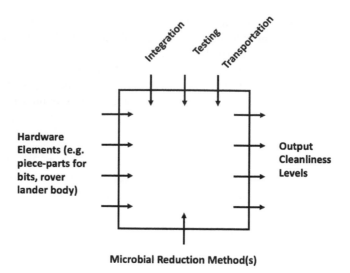

FIGURE 4.1 Process controls that can be modified to achieve the desired cleanliness goal.

- **Hardware integration flow**: Hardware integration flow can be affected by the results of the model. For example, the consideration of when to perform thermal vacuum bakeouts can be critical to achieving the end hardware cleanliness. Baking out at the piece part level allows for inspection and verification leading up to bakeout at the subassembly or assembly level. This may also result in the piece part being baked out multiple times to ensure that the component can meet the desired cleanliness at the end subsystem bakeout. In particular, if hardware is nested inside a complicated assembly, it would have to be baked out on its own to avoid contaminating the entire assembly. Outgassing models help define these lower-level piece part bakeouts.
- **Experimental validation**: Transport models are effective in gaining an understanding of the wide variety of possible contamination vectors to the targeted region. The model itself can be validated in part or end-to-end using testbeds to simulate key operations of concern in the relevant environment. The development of these models also helps to identify knowledge gaps in contamination vectors that can be addressed experimentally using flight-like hardware in relevant environmental conditions.
- **Hardware Transportation**: The packaging used to protect a component to preserve the cleanliness levels becomes more important if the transport models determine that the initial conditions must be maintained at the same or low levels of re-accumulation. This may dictate hardware storage and transportation constraints, such as use of a nitrogen-purged container or a hermetically sealed vessel.

4.3.3 CLEANING AND RECONTAMINATION PREVENTION STRATEGIES

The approach to developing a plan to clean and prevent recontamination is highly coupled with understanding the properties of the source material, known contamination vectors, and most importantly through a strong relationship with the design, assembly, integration, and test engineers throughout the entire hardware life cycle. An example flow is illustrated in Figure 4.2, which outlines the process of developing an approach in the early phases of the mission. There are several factors to consider and the process is iterative with new information and design changes. In the planning phase of the mission, the materials and manufacturing processes are reviewed and a risk analysis is conducted to establish the implementation approach for microbial reduction and organic

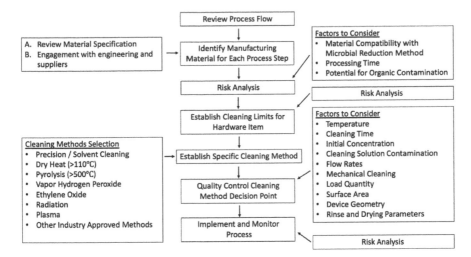

FIGURE 4.2 Hardware cleaning implementation approach to achieve biological, organic, and particulate cleanliness. Integrated tools, particularly in the form of risk assessments, can be utilized to aid in process decisions.

contamination introduction through microbial processing. Once the methodology of cleaning is finalized, then the cleaning limit is established and then various cleaning methodologies such as solvent cleaning or heat microbial reduction is established. The process is then conducted with appropriate quality control measures and oversight to monitor the process and verify intermediate or final hardware processing steps.

4.3.3.1 Consulting with Manufacturers and Engineers

An integral approach with the engineering team and an understanding of the manufacturing processes at suppliers should be adopted at the earliest stages of the mission to facilitate proper flow-down of requirements and cost-effective mission planning. Communicating with suppliers can provide additional information on implemented processes and controls throughout the manufacturing flow. For example, there may be processes that can be tailored to meet a mission-unique cleanliness requirement. Utilizing existing manufacturing steps and standards, such as United States Military Standard specification parts, can also be used for microbial reduction or cleaning process credits. For example, elevated burn-in temperatures can provide up to a 6-log reduction in surface and embedded microbial contamination for electronic circuit boards.

Reviewing the material specifications can ultimately improve the breadth and range of cleaning methods and parameters available for use. Paints and other coatings may change color, flake, or experience degraded thermal properties as a result of one microbial reduction technique.

4.3.3.2 Establish Specific Cleaning Methods

It is important to establish specific cleaning methods and guidelines, as there are material-specific nuances that may affect the cleanliness of the final product. For example, variants of a cleaning protocol may be applied to accommodate certain parts with protective coatings due to hardware sensitivities. This ties heavily to the open dialogue with the engineering team to make sure the cleaning process does not inadvertently damage or degrade the part.

There are also material types that are easy to clean, but may not be able to be sampled. Properties such as surface roughness, surface particulate cleanliness, and maintaining explicit optical finishes are typically surfaces that cannot be directly sampled to establish microbial cleanliness. Identifying such handling constraints and locations where witness coupons will be used, as an alternative mode to assessing cleanliness, is also part of the developmental process.

4.3.3.3　Implement and Monitor Processes

Cleaning methods and recontamination prevention strategies may seem effective and even simple on paper, but keen involvement of the process developer is necessary to identify and mitigate potential gaps and inefficiencies. For example, donning sterile garmenting prior to entering a cleanroom is simple and effective with proper training. However, practitioners may not be aware of how to open doors into a cleanroom facility with sterile garmenting and gloves. Details such as these are key to document and communicate consistently with all operators that will utilize these processes. Another example of process controls that affect cleanliness is a process control during application of lubricants. Lubricants are known to migrate when applied to a hardware component. Including handling and glove-change control processes would mitigate the risk of spread of the lubricant as a potential contaminant.

4.3.4　Operational Constraints for Sample Site Selection

For planetary protection Category IV missions, site selection is an important consideration due to the potential contamination risk to life detection or sample return missions. For these biologically significant missions (e.g. to Mars or outer planet icy worlds), planetary protection issues are typically addressed as part of the mission's landing site certification process (Golombek et al., 2017). The features of a landing site that planetary protection is concerned about include water and ice availability, geological surface features, subsurface discontinuities, concentration of water-bearing minerals, location and temperature of the landing site, and the spacecraft surface operations. The process for planetary protection certification for such missions is directly coupled to the landing site review and down-select process for the mission. Typically, there can be a planetary protection set of questions that is answered for each landing site during the site selection presentations and/or a splinter planetary protection review to present each site. For example, in the case of the InSight mission, the landing site scientists were able to demonstrate through a detailed analysis of the terrain with orbital assets that there was no evidence for geological features of PP relevance (e.g., recurring slope lineae, gullies, ground ice, etc.) (Golombek et al., 2017) in candidate landing sites. The mission could land as a Category IVa planetary protection mission because no special region was present. A *special region* at Mars is defined by NASA to be either a naturally occurring or spacecraft-induced region on Mars with physical parameters defined to meet both the <0.5 availability of water and >−25°C for a time period of 500 years (Rummel et al., 2014; NASA NPR 8020.12). As far as operation constraints at the landing site for InSight, there was also an operational limit to how far the (non-sterile) mole could penetrate the surface of Mars; as it was not allowed to go below 5 m of the surface, where the water environment is less well understood. The mole is a subsurface hammering mechanism intended to bore down in the Martian subsurface up to 5 m conducting heat transfer experiments throughout the mole operations. Hence, it is key for any such mission to understand the risk of contamination from landing the spacecraft at the surface as well as understanding the type of operation that will horizontally or vertically interact with the site of interest.

4.3.5　Operational Constraints for Drilling for Samples

The concept of protecting hardware from Earth-sourced contaminants does not stop at launch. Although considerable efforts go into cleaning and protecting the hardware from recontamination, there are still sources of microbial and organic contamination present, inherent to the spacecraft. For this reason, there may be operational constraints that may be required during sample drilling. Key phases to consider when determining if operational constraints are required are:

1. *Commissioning activities upon landing*—These activities may be required to shed additional protective layers that may limit the diffusion of organic molecular or biological contamination.

An example of a commissioning activity constraint is a time limit for the Mars 2020 Rover within which the rover's ejectable belly panel must be deployed.

2. *State of the sample acquisition tool during use*—Operational constraints may be placed on a sample if anomalies occur during the acquisition process. Faults or off-nominal events may trigger a pause in operations until it is determined if the integrity of the sample was affected. The collection of metadata from this process may allow the acquisition process to continue with the associated caveats that will then be linked to the sample.

3. *State of the sample acquisition tool between uses*—Upon completion of acquiring the sample, constraints may be put in place to protect the acquisition tool until its subsequent use to prevent contamination by direct transfer (particulate or gaseous). An example of an operational constraint is to maintain covers on the tool when not in use to keep surfaces clean between each use. This can extend the duration of a typical operation by adding time to remove or replace covers prior to starting the sample acquisition sequence.

4.3.6 OPERATIONAL CONSTRAINTS FOR SAMPLE EXTRACTION

The extraction of a sample from its native environment marks the start of a clock, of which each elapsed second poses a threat to the integrity of the sample. For some missions, the precise timing of the end-to-end acquisition process may be dictated by sample cleanliness limits set by the mission. In this case, an integral approach with the mission's systems engineering and surface operations team ensure proper planning and implementation of operational constraints. Limits on sample handling and exposure greatly benefit the fidelity of downstream science, as well as protect the integrity of the planetary protection approach. As an example, the Mars 2020 sampling and caching system was designed to minimize sample exposure and handling, with 25 times less surface area contacted throughout the sample acquisition chain than those on Mars Science Laboratory (MSL). Samples are collected directly into the storage tube in the center of the drill bit. Upon visual and volumetric inspection, the tube is tightly sealed, and would remain sealed through all operations needed to deliver the sample to Earth as part of the Sample Return Campaign planning.

Operational constraints for sample extraction can come in the form of:

1. *Time limits*—A maximum time duration may be imposed for operations between extraction and encapsulation/analysis of the sample.

2. *Consultation with the mission's science team*—Off-nominal extractions could pose a risk to the integrity of the sample, and the temporary halting of the sampling and encapsulation process and consultation with the science team is a possible operational constraint. It also may be possible as part of the mission operations planning that the encapsulation process is completed first to prevent further contamination to the sample prior to later ceasing further operations or discarding the sample.

3. *Documentation*—Off-nominal events which result in an exceedance of the required time limit should be documented and linked with the acquired sample should the project science team decide it is still of value. Independent of whether the execution of the extraction process was a success, the operational constraint of documenting the metadata associated with the extraction process should still be conducted.

4. *Collection of process blanks or calibration samples to characterize the contaminants that have deposited on the sample during the acquisition process*—Collection of process blanks or known calibration samples has been essential to parsing Earth-sourced contaminants from sample material in sample analysis. Projects such as Viking (Biemann and Lavoie, 1979), Phoenix (Ming et al., 2008), MSL (Conrad et al., 2012), and Mars 2020 all utilized blanks or known materials to characterize or calibrate against contaminants introduced from manufacturing, assembly, test, launch operations until sample collection.

4.4 BIOLOGICAL CONTAMINATION CONTROL CASE STUDIES

Mars-based spacecraft have been designed, assembled, tested, launched, and operated to include biological contamination controls to meet planetary protection requirements. Salient biological control program case studies are presented for the Mars Science Laboratory, InSight, Mars 2020, and a potential Mars Sample Return Campaign.

4.4.1 MARS SCIENCE LABORATORY

The Mars Science Laboratory (MSL) was categorized as a PP Category IVa mission to Mars. The mission was to send a small car-sized rover powered by a radioisotope thermoelectric generator to the surface of Mars to a non-special region. The concern is that this is a region conducive to replication of terrestrial organisms, should they be introduced here. For MSL, the risk meant that the rover could not horizontally or vertically traverse into a special region. Thus, the IVa driving biological cleanliness requirements were $<5 \times 10^5$ spores on the launched vehicle, $<3 \times 10^5$ spores on the landed hardware, and an average of <300 spores/m^2 on the planned landed and planned impacted hardware surfaces. The rover was built in an ISO 8 cleanroom using full coverall and surgical face mask stringent gowning protocols. To meet the biological requirements, most of the surface area underwent heat microbial reduction (88.7% of the surfaces). Other areas underwent alcohol cleaning (11%), precision cleaning on primary and secondary structures (0.1%), and destructive studies to determine actual bioburden (0.2%) as an alternative to cleaning. Notably, 0.01% of the surfaces did not undergo any microbial reduction measures due to material compatibility issues with solvents, high-temperature sensitivity, difficult-to-clean surfaces, commercial-off-the-shelf parts (COTS), and coefficient of thermal expansion constraints between multiple surfaces. For surface cleanliness verification, microbial assays were collected to verify cleanliness requirements throughout the hardware build and test. To meet the biological cleanliness requirements, the mission's non-metallic volumes underwent heat microbial reduction (60.3% of the volume) and destructive studies to determine the actual bioburden (38.3%). No microbial reduction was conducted for 1.4% of the non-metallic volumes on MSL due to COTS parts utilization and high-temperature sensitivities, with the unsterilized microbial cleanliness being accounted. After microbial reduction processes had taken place, recontamination prevention/recleaning measures were employed to maintain hardware cleanliness. These measures included hardware cleaning with isopropyl alcohol, covering the hardware when not in use, and the strategic positioning of hardware in the cleanroom as close to the HEPA filtered air as possible. The total bioburden at launch was 2.78×10^5 spores accounted on the launched vehicle, 5.64×10^4 spores on the landed hardware, and an average of 22 spores/m^2 accounted on all hardware surfaces (Benardini et al., 2014).

MSL's Mars material sampling hardware included the wheels as well as the drilling bits, and these components underwent special planetary protection measures. To enable the rover to "trench" the Martian regolith as part of nominal surface science operations (e.g., conduct science from the bottom of the trench) the rover wheels underwent a more detailed biological cleanliness implementation protocol. The rover had two sets of wheels: one was a ground testing set and the other was a flight-only set intended to only to touch the surface of Mars. The flight-only wheels were covered with bagging material after the heat microbial bakeout to protect the surfaces from recontamination. Installation of the flight wheels occurred at the last possible point in the mission integration operations to also prevent contamination. Given the drill bits would also touch the surface of Mars, the drill bits were precision cleaned, heat microbially reduced, and protected in a bit box to limit recontamination prior to launch (with the exception of one that was flown in the drill) (Benardini et al., 2014).

4.4.2 INSIGHT

Similar to MSL, the InSight mission launched a lander to a non-special region of Mars and was a PP Category IVa mission. The InSight mission implemented isopropyl alcohol cleaning, precision cleaning, destructive hardware studies, and dry heat microbial reduction for hardware surfaces and non-metallic volumes. The main difference in biological requirements management between MSL and InSight was the hardware build process. For MSL, the main flight system was built at JPL while the payloads were sourced from a distribution of suppliers around the world (including JPL). Whereas with InSight, the main lander flight system was built at Lockheed Martin Space Systems and the two major payloads were built by international partners. This presented a challenge for implementation of bioburden requirements, and required additional plans and interface documentation to ensure clear hardware handoffs between each hardware delivery. Each hardware handoff required surface verification microbial assays to be collected to verify cleanliness requirements. Additionally, this was the first interplanetary mission launched from the west coast of the United States, with the launch from Vandenberg Airforce Base. Thus, the launch support personnel had to be extensively trained, and the needed stringent bioburden protocols (e.g. isopropyl alcohol cleaning for biological removal, cleanroom tents on launch vehicle surfaces) had to be implemented. The total accountable bioburden at launch was 1.5×10^5 spores on the launched vehicle, 1.35×10^5 spores on the landed hardware, and an average of 129 spores/m^2 on all hardware surfaces (Hendrickson et al., 2020). The systems engineering approach for managing microbial requirements, clear definitions of biological cleanliness interfaces, and training and modifications of existing techniques contributed to the success of the planetary protection implementation for this mission.

The heat flow and physical properties probe (HP^3) on the InSight mission is a hammering probe with an internal hammer that will burrow up to 5 m below the Martian surface. Given the involvement in direct soil contact, the probe was heat microbially reduced and assembled and tested in a cleanroom to prevent recontamination. The probe was also directly cleaned with isopropyl alcohol, and surface verification microbial assays were collected to verify cleanliness requirements for final flight stow in launch configuration.

4.4.3 MARS 2020

The Mars 2020 mission enables scientists to expand our understanding of past habitable environments that may have sustained ancient microbial life on Mars. Furthermore, through technology demonstration and atmospheric analysis, this mission establishes major steps toward the future human exploration of Mars through the generation of a small amount of pure oxygen from Martian atmospheric carbon dioxide. The mission will deliver a rover to the surface of Mars that is designed to take scientific measurements of the environment, terrain, and material. The rover has been fitted with a set of hardware capable of acquiring, encapsulating, and caching samples, to include witness blank materials (Figure 4.3). NASA is planning to return the acquired sample to Earth by a later mission for subsequent analysis in terrestrial laboratories as part of a potential Mars Sample Return Campaign.

The Mars 2020 mission has been designated as a Category V Restricted Earth Return mission due to the possible future return of collected samples. As indicated in NPR 8020.12D, Section 5.3.3.2, the outbound leg of a Category V sample return mission would be expected to meet the requirements of a Category IVb mission. The project has taken a sample-centric system-level approach by starting with the cleanliness requirements for the return sample and working backwards to develop an implementation strategy. This approach represents a driving science requirement that meets a planetary protection requirement through the science verification approach. A detailed understanding of the contamination transport vectors from fabrication through operations drive the level of cleanliness to even more stringent levels than the forward Planetary Protection requirements dictate. Samples acquired at Mars must meet the following requirements:

- Less than 1 viable Earth-sourced organism per sample.
- Each sample in the returned sample set has more than a 99.9% probability of being free of any viable Earth-sourced organisms.

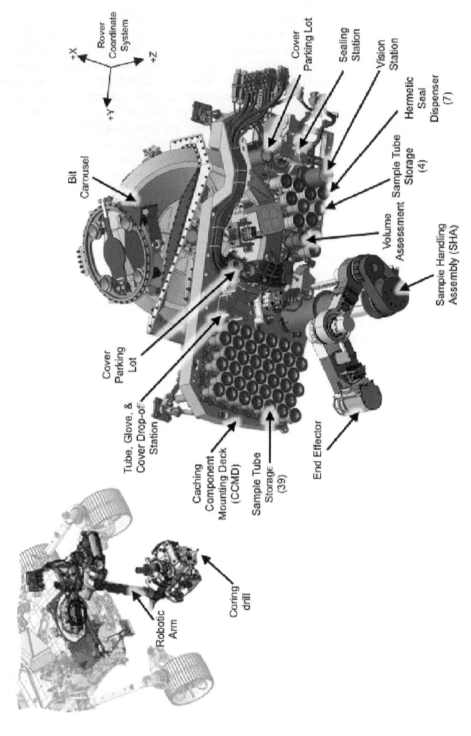

FIGURE 4.3 Sample Caching System (SCS) subsystem overview. The SCS acquires, encapsulates, and caches rock and regolith samples.

- The organic contamination levels in each sample in the returned sample set are less than:
  - Any Tier 1 compound (organic compounds deemed as essential analytes for mission success): 1 ppb
  - Any Tier 2 compound (organic compounds not categorized as Tier 1): 10 ppb
  - Total organic carbon: 10 ppb baseline, 40 ppb threshold
- Inorganic contamination limits are set based on the composition of three Martian meteorites: Shergottite, Nakhlite, and Chassignite (SNC).
  - Less than 1% of the average concentration in SNC meteorite of following elements: Zr, Nb, Ta, La, Ce, Eu, Gd, Li, B, Cs, Sc, Mn, Y, Mg, Zn, Ni, Co, Cl, Br, P, S
  - Less than 0.1% of the average concentration in SNC meteorite of following elements: K, Rb, Sr, Sm, Nd, U, Th, Re, Os, Lu, Hf, W, Pb

Engineering design solutions were also used to maintain the cleanliness levels to be achieved by the method of microbial and organic reduction/control. The fluid mechanical particle barrier (FMPB) was intended to keep all hardware surfaces that contact the acquired sample materials protected from particulate and biological contamination ingress. This is achieved through two mechanisms: reduction of the aerodynamic drag on particles and by a geometrical design that ensures that the filling time is short compared to the time it takes a particle to transit deeper into the orifice. Details on the FMPB principles, design, and validation can be found in Mikellides et al. (2017). Table 4.4 describes the approach taken on key hardware involved in the sampling and caching process. Elements of the hardware that contact the Martian soil are considered to be "sample intimate" hardware. The most contamination-sensitive hardware undergoes "late integration," where it is stored and transported in a nitrogen-purged environment until the latest point as possible in the rover assembly flow to minimize exposure of the hardware. The volume where the critical hardware is installed is subsequently purged with nitrogen until launch to maintain the organic cleanliness of the hardware. Figure 4.4 shows engineering test models of the tube, glove, and storage sheath. This tube assembly is a subset of the sample intimate hardware, which include the sample tubes, seal, coring/regolith bit, and the volume assessment probe.

The remainder of the Mars 2020 spacecraft takes a more standard approach to cleaning and recontamination prevention. Standard cleaning approaches for Planetary Protection Iva missions (e.g. Mars Exploration Rover, Phoenix, Mars Science Laboratory (MSL), and InSight) include cleaning the hardware to maintain and average of 300 spores/m^2 by leveraging heritage cleaning approaches, ISO 8 or better assembly environments, and covering the hardware while not in use to prevent recontamination. For MSL, heat microbial reduction was implemented on over 89% of the spacecraft surfaces yielding a cleanliness range of 0.001–30 spores/m^2, alcohol wiping was utilized 10.9% with hardware surfaces ranging from 30 to 300 spores/m^2, and precision cleaning constituted for the remaining 0.01% with hardware surfaces ranging from 30 to 50 spores/m^2 (Benardini et al., 2014). When analyzing internal project data and planned project data the following enhanced cleaning approaches will be covered: 4-log microbial reduction to achieved 0.001–0.003 spores/m^2, 6-log microbial reduction to achieved 0.00001–0.00003 spores/m^2, and the utilization of VHP to achieve sterility. Post cleaning and microbial reduction, the hardware will be in at least an ISO 8 but typically the hardware is in an ISO 7 with stringent full coveralls and cleanroom protocols of that of an ISO 5 environment. During hardware periods where electrical or mechanical integration and testing is not occurring, the hardware will be bagged and packaged for storage for shipment as not to contaminate the hardware surfaces. The Mars 2020 project has gone through great lengths to meet a set of unprecedentedly stringent contamination control requirements. Ultimately, the hardware was engineered to protect Mars from Earth-sourced contamination, enabling sensitive in situ measurements and ensuring a set of returnable samples.

TABLE 4.4

Cleaning and Recontamination Prevention Strategies for Contamination Critical Hardware

Hardware	Sample Intimate?	Microbial Reduction Technique	Organic Contamination Control Method	Recontamination Control Technique
Sample Tube Interior	Yes	350°C in oxidizing environment	Titanium Nitride (TiN) coating on all interior surfaces, 350°C in oxidizing environment	Fluid Mechanical Particle Barrier (FMPB), Isolated storage, nitrogen purge environment through launch, late integration
Seal	Yes	350°C in oxidizing environment	TiN and Gold coating, 350°C in oxidizing environment	FMPB, Isolated storage, nitrogen purge environment through launch, late integration
Seal Dispenser	No	350°C in oxidizing environment	Solvent cleaning, 350°C in oxidizing environment	FMPB, Isolated storage nitrogen purge environment through launch, late integration
Coring Bit, Regolith Bit, and Abrasion Tool	Yes	4-order-of-magnitude microbial reduction	TiN coating on bit teeth and bit surfaces, high-temperature bakeout	HEPA filter/tortuous path, late integration
Sample Tube Exterior	No	350°C in oxidizing environment	350°C in oxidizing environment, thermal Al_2O_3 exterior also serves as a getter	FMPB, Isolated storage nitrogen purge environment through launch, late integration
Tube Storage Sleeve	No	350°C in oxidizing environment	TiN coating on interior surfaces, 350°C in oxidizing environment	FMPB, Isolated storage nitrogen purge environment through launch, late integration
Bit Storage	No	4-order-of-magnitude microbial reduction	Solvent cleaning, high-temperature bakeout	HEPA filter/tortuous path, late integration
Volume Assessment Probe	Yes	350°C in oxidizing environment	TiN coating on probe, 350°C in oxidizing environment	FMPB, Isolated storage nitrogen purge environment through launch, late integration
Adaptive Caching Assembly Volume	No	Solvent cleaning, 3-order-of-magnitude microbial reduction post-assembly	Low-outgassing material selection, solvent cleaning, high temperature bakeout post-assembly	Tenax molecular getter, deployable bottom panel post-landing to limit organic buildup, Nitrogen purge from final access to launch, closed volume, with sample intimate hardware protected by FMPB, HEPA filter/tortuous path
Corer	No	4-order-of-magnitude microbial reduction	Solvent cleaning, high-temperature bakeout	Corer launch cover installed from final cleaning through the first sample acquisition on Mars

FIGURE 4.4 Test models of the sample tube assembly includes the sheath (left), tube (middle), and Fluid Mechanical Particle Barrier (FMPB) glove (right), which the sample handling arm utilizes as a protective barrier for tube manipulation while maintaining its level of cleanliness.

4.4.4 Mars Sample Return Technology Studies

Mars 2020 is the first step of Mars exploration that sets the stage for a potential sample return mission. The samples tubes built and launched for Mars 2020 must then be able to be located on the surface of Mars and interface with subsequent missions, thus highlighting the importance of the interface control documents and knowledge capture. Samples coming back from Mars would have the categorization of Category V Restricted Earth Return based on planetary protection policy. With this designation come requirements from the outbound mission all the way through the receipt of samples in the handling facility. A major motivation behind the requirements of a sample return mission is to ensure the safety of Earth life by preventing the inadvertent release of unsterilized Martian material. This can be achieved in two ways: sterilization of exposed surfaces of the returned vehicle, and complete isolation (containment) of the unsterilized Martian particles.

A probabilistic risk analysis was conducted in the 1990s in support of developing sample return capabilities (Gershman et al., 2004), which helped tailor further exploration of technologies that filled technical gaps or mitigated the identified risks. These efforts have been revived through the development of concepts that focus on both sterilization and encapsulation, with some technologies achieving both goals (Gershman et al., 2018). The final configuration has yet to be decided, but the following are leading technologies in consideration: brazing, bagging, and plasma sterilization.

4.4.4.1 Brazing

Brazing technology has been selected as a leading candidate based on the process maturity and capability to subsequently sterilize the seam in the process of the high temperatures generated during the sealing process. A novel brazing method has been developed that simultaneously seals, separates the clean surface, and sterilizes the seam where possible Martian particles may possibly be accumulated (Figure 4.5). Upon melting, surface tension/capillary forces allow the braze material to flow while the spring force between the outer and inner walls of the container drives the separation. Using a prototype double wall container, it has been successfully demonstrated in a vacuum environment.

Brazing materials are able to melt at a range of temperatures at and far above 500°C. This temperature exposure, even at the shortest (>0.5 seconds) durations may qualify the exposed surface to be considered sterile. Future work involves a discussion with the scientific community to gain a consensus on the range of sterilization temperatures acceptable for the returned sample exterior in

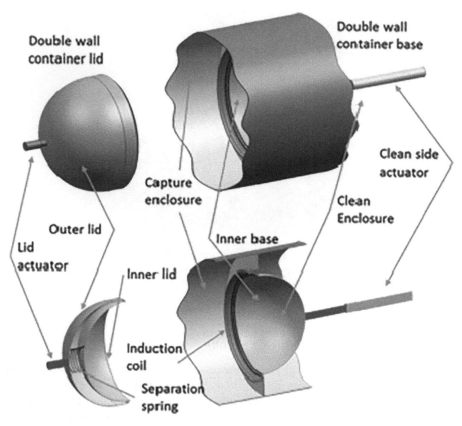

FIGURE 4.5 The components of the brazing mechanism with the cone-within-cone configuration concept. *Source:* Courtesy of Gershman et al. (2018).

order to revise down the upper design performance requirement in brazing temperature down and open up option spaces for the bagging concept.

4.4.4.2 Bagging

Bagging technology has been selected as an alternate approach to encapsulate the acquired samples with the capability to also reduce the microbial contamination at the seam based on the sealing temperature. Polymeric bagging has been used extensively within the space industry as well as commercially, where thousands of storage bags are sealed per hour for consumer use. This concept is implemented in a similar way to the brazing technology, which involves separating the contaminated "Mars dirty" environment from the "Earth clean" zone. After the container is placed into the bag, the bag is sealed by the application of heat, which also microbially reduces the microorganisms present at that interface. The bag is then cut at the seam line in a manner to preserve the seal, and packed into a secondary container for return to Earth (Figure 4.6).

4.4.4.3 Plasma

The plasma technology is an alternative approach that may be utilized as an in-flight active sterilization technology. The plasma technology would be conducted on the exterior of the orbiting sample and the Mars-exposed sample containment chamber to ensure that any potential Mars dust and associated biological particles are microbially reduced. Conceptually this could be generated multiple times in the engineering flow but ideally right before the sealing operation.

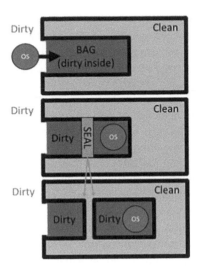

FIGURE 4.6 Bagging concept schematic showing the sealing of the Orbiting Sample (OS) container. *Source:* Courtesy of Gershman et al. (2018).

4.5 CONCLUSIONS

Biological contamination control and planetary protection will continue to be a necessary engineering discipline to address Earth-return and interplanetary missions as they continue to increase their scientific mission goals for in situ life detection and in seeking to bring back samples to terrestrial labs for analysis. This discipline, like any other, should be handled in the system engineering trade space to enable the mission and ensure that the project requirements and goals are characterized and have verifiable results. At a minimum, this trade space discussion should include the participation of the science team, biological contamination control/planetary protection, contamination control, materials and process engineering, and mechanical engineering. The implementation tradespace should consider the following:

1. Influencing the design where there is still possibility for major subsystem changes like terminal sterilization or adding a biobarrier as a mechanical system to prevent recontamination.
2. The utilization/exclusion of commercial off-the-shelf equipment in the case that these components compromise this design upfront. Or acceptance of alternative hardware flows or interface designs required may need to be considered. For example, if a part is heat resistant, this part may have to undergo an alternative treatment to the rest of the instrument before it is cleanly integrated.
3. Assessing credible contamination vectors and applying appropriate contamination models to drive contamination thresholds or establish cleanliness zones.
4. The effect at each mission phases (i.e. prior to microbial reduction, during microbial reduction, recontamination prevention during assembly and integration, recontamination prevention during testing, re-work or contingency scenarios, transportation, deployment, and operation) of planetary protection implementation.

Given the sensitivity increase in instrument detection, biological methods for cleaning and verification are proving far more (i.e. 2–3-orders-of-magnitude) sensitivity than the traditional chemical contamination control analysis methodologies (e.g. Fourier-transform infrared spectroscopy, mass spectroscopy, etc.). Hence, microbiologists and molecular biologists are becoming necessary members of the interdisciplinary engineering team for mission design and implementation, analogous to what occurs in pharmaceutical or medical device manufacturing. Nucleic acid, fatty acid, protein,

ATP, and LAL are among the types of biotic compounds where modern microbiology, molecular biology, and biochemical techniques should be considered to be employed for cleanliness verification. These techniques may prove advantageous to enable the mission if complete sterilization may not be achieved or even possible, as they may be able to define these limits and characterize the baseline biological contamination in a system.

Component-level cleaning and microbial reduction, in general, is a relatively straight-forward and simple process, compared to the challenge of preventing recontamination of the component post-cleaning. From the NASA spacecraft mission case studies presented herein, a majority (>50%) of the resources for ensuring planetary protection cleanliness went into recontamination prevention (e.g., bagging, alcohol cleaning, cleanroom assembly, and testing preparations and planning) and verifying cleanliness throughout the hardware development process. Although this may take additional staffing and scheduling to prevent hardware recontamination, it has enabled these missions to be successful in meeting requirements without need of terminal sterilization or full-system biobarriers.

To continue to enable future missions for life detection, sample return and human exploration policy updates, agency standards for microbial reduction standards, and verification techniques will need to be validated and updated, and continued investments in understanding and adapting industry-based technologies for biological contamination will be required. These missions will likely require a multi-pronged approach to assess the biological contamination in terms of the abundance, the types of organisms present, viability of organisms, potential functionality or metabolic capability, and risk potential to survive mission phases without compromising sample integrity. Annual COSPAR workshops from 2015 to 2019 have been conducted to develop the international planetary protection policy requirements for future human missions at Mars. These workshops have addressed knowledge gaps and technology needs for microbial detection, transport, modeling, human health, and environmental monitoring. Significant technology development has continued in nucleic acid-based metagenomic techniques for the International Space Station microbial monitoring and in robotic spacecraft assessments (e.g. InSight, Mars 2020, Europa Clipper). Subject matter experts are also starting to recognize the need for updated policy through establishing agency-level committees to evaluate planetary protection requirements and implementation, conducting mission-specific workshops to address gaps in standards for microbial reduction/control/monitoring, and increasing knowledge transfer through establishing a Microbiology of Human Spaceflight Conference. In conclusion, the recognition to evaluate planetary protection policy, increase knowledge transfer amongst experts, conduct workshops to identify and fill knowledge gaps, and invest in industry technologies for biological contamination control have been critical in enabling the current mission set and will need to continue to advance at an accelerated pace to inform mission design and enable future life-detection, sample return, and human exploration missions.

ACKNOWLEDGMENTS

Some of the research reported in this chapter was conducted at the Jet Propulsion Laboratory (JPL), California Institute of Technology, under a contract with the National Aeronautics and Space Administration (NASA). The authors would like to thank Andy Spry, NASA Planetary Protection Officer, and Elaine Seasley, deputy Planetary Protection Officer, NASA Headquarters, Washington DC, for reviewing this chapter and providing valuable technical comments and suggestions.

Any information about future missions presented in this Chapter, including Mars Sample Return, is predecisional, and is provided for planning and discussion only.

REFERENCES

Abraham, N.S. and Jallice, D.E. (2018) Using NASA's molecular adsorber coating technology during thermal vacuum testing to protect critical laser flight optics on the ATLAS instrument. *SPIE Optics + Photonics*, August 19–23, San Diego, CA.

An, J.J. (2008) *Study on Surface Kinetics in PECVD Chamber Cleaning Using Remote Plasma Source.* Cambridge, MA: Massachusetts Institute of Technology.

Anastasi, A., Varese, G.C., and Filipello Marchisio, V. (2005) Isolation and identification of fungal communities in compost and vermicompost. *Mycologia* 97:33–44.

Appelbaum, J. and Flood, D.J. (1990) Solar-radiation on mars. *Sol Energy* 45:353–363.

Bakermans, C. and Nealson, K.H. (2004) Relationship of critical temperature to macromolecular synthesis and growth yield in *Psychrobacter cryopegella*. *J Bacteriol* 186:2340–2345.

Bakermans, C., Tsapin, A.I., Souza-Egipsy, V., Gilichinsky, D.A., and Nealson, K.H. (2003) Reproduction and metabolism at −10 degrees C of bacteria isolated from Siberian permafrost. *Environ Microbiol* 5:321–326.

Barengoltz, J. and Edgars, D. (1975) The relocation of particulate contamination during space flight. NASA-CR-143503, JPL-TM-33-737. September 1.

Barton, H.A. and Northup, D.E. (2007) Cave geomicrobiology: Past, present and future perspectives. *J Cave Karst Stud* 69:163–178.

Barton, H.A., Taylor, M.R., and Pace, N.R. (2004) Molecular phylogenetic analysis of a bacterial community in an oligotrophic cave environment. *J Geomicrobiology* 21:11–20.

Beaudet, R. (2000) Simple mathematical models for estimating the bio-contamination transported from a lander or a rover to the Martian soil, SAE Technical Paper 2000-01-2422. doi:10.4271/2000-01-2422.

Benardini, J.N., La Duc, M., Ballou, D., and Koukol, R. (2014) Implementing planetary protection on the Atlas V fairing and ground systems used to launch the Mars science laboratory. *Astrobiology* 14(1):33–41.

Benardini, J.N., La Duc, M., Beaudet, R., and Koukol, R. (2014) Implementing planetary protection measures on the Mars science laboratory. *Astrobiology* 14(1):27–32.

Benardini, J.N., Vaishampayan, P.A., Schwendner, P., Swanner, E., Fukui, Y., Osman, S., Satomi, M., and Venkateswaran, K. (2011) *Paenibacillus phoenicis* sp. Nov., isolated from the Phoenix Lander assembly facility and a subsurface molybdenum mine. *Int J Syst Evol Microbiol* 61:1338–1343.

Biemann, K., and Lavoie, J.M. (1979) Some final conclusions and supporting experiments related to the search for organic compounds on the surface of Mars, *J. Geophys. Res.*, 84(B14), 8385–8390, doi:10.1029/JB084iB14p08385.

Borson, E.N. (2002) Model for particle redistribution during spacecraft launch, *Proceedings of SPIE 4774, Optical System Contamination: Effects, Measurements, and Control VII*, September 11.

Breezee, J., Cady, N., and Staley, J.T. (2004) Subfreezing growth of the sea ice bacterium "Psychromonas ingrahamii". *Microbial Ecol* 47:300–304.

Bryant, M.P. (1972) Commentary on the Hungate technique for culture of anaerobic bacteria. *Am J Clin Nutr* 25:1324–1328.

Calvin, M. and Gazenko, O.G. (1975) *Foundations of Space Biology and Medicine.* Volume 1 Space as a Habitat. National Aeronautics and Space Administration. U.S. Government Printing Office.

Chang, C.W. (1999) Application of general sticking coefficient models to spacecraft contamination analysis, *Proceedings of SPIE 3784, Rough Surface Scattering and Contamination*, October 25. doi:10.1117/12.366714

Checinska, A., Probst, A.J., Vaishampayan, P., White, J.R., Kumar, D., Stepanov, V.G., Fox, G.E., Nilsson, H.R., Pierson, D.L., Perry, J., and Venkateswaran, K. (2015) Microbiomes of the dust particles collected from the International Space Station and Spacecraft Assembly Facilities. *Microbiome* 3(50): 1–18.

Chelius, M.K. and Moore, J.C. (2004) Molecular phylogenetic analysis of archaea and bacteria in Wind Cave, South Dakota. *Geomicrobiol J* 21:123–134.

Chung, S., Barengoltz, J., Kern, R., Koukol, R., and Cash, H. (2009) The validation of vapor phase hydrogen peroxide microbial reduction for planetary protection and a proposed vacuum process specification. JPL Publication 06-06 Rev. 1.

Conrad, P.G., Eigenbrode, J.L., Von der Heydt, M.O. *et al.* (2012) The Mars Science Laboratory Organic Check Material. *Space Sci Rev* 170, 479–501. https://doi.org/10.1007/s11214-012-9893-1

Cooper, M., La Duc, M.T., Probst, A., Vaishampayan, P., Stam, C., Benardini, J.N., Piceno, Y.M., Andersen, G.L., and Venkateswaran, K. (2011) Comparison of innovative molecular approaches and standard spore assays for assessment of surface cleanliness. *Appl Environ Microbiol* 77:5438–5444.

Cox, M.M. and Battista, J.R. (2005) Deinococcus radiodurans—The consummate survivor. *Nat Rev Microbiol* 3:882–892.

Daly, M.J., Gaidamakova, E.K., Matrosova, V.Y., Kiang, J.G., Fukumoto, R., Lee, D.Y., Wehr, N.B., Viteri, G.A., Berlett, B.S., and Levine, R.L. (2010) Small-molecule antioxidant proteome-shields in *Deinococcus radiodurans*. *PloS One* 5:e12570.

Demaere, M.Z., Williams, T.J., Allen, M.A., Brown, M.V., Gibson, J.A., Rich, J., Lauro, F.M., Dyall-Smith, M., Davenport, K.W., Woyke, T., Kyrpides, N.C., Tringe, S.G., and Cavicchioli, R. (2013) High level of intergenera gene exchange shapes the evolution of haloarchaea in an isolated Antarctic lake. *Proc Natl Acad Sci U S A* 110:16939–16944.

Deming, J.W. (2002) Psychrophiles and polar regions. *Curr Opin Microbiol* 5:301–309.

Eastburn, D.J., Sciambi, A., and Abate, A.R. (2013) Ultrahigh-throughput mammalian single-cell reverse-transcriptase polymerase chain reaction in microfluidic drops. *Anal Chem* 85:8016–8021.

ECSS-Q-ST-70-56C. (2013) Vapour phase bioburden reduction for flight hardware. European Cooperation for Space Standardization.

ECSS-Q-ST-70-57C. (2013) Dry Heat bioburden reduction for flight hardware. European Cooperation for Space Standardization.

ESSB-ST-U-001. (2013) ESA planetary protection requirements. European Cooperation for Space Standardization.

Fredrickson, J.K., Li, S.M., Gaidamakova, E.K., Matrosova, V.Y., Zhai, M., Sulloway, H.M., Scholten, J.C., Brown, M.G., Balkwill, D.L., and Daly, M.J. (2008) Protein oxidation: Key to bacterial desiccation resistance? *ISME J* 2:393–403.

Gershman, R., Adams, M., Mattingly, R., Rohatgi, N., Corliss, J., Dillman, R., Fragola, J., and Minarick, J. (2004) Planetary protection for Mars sample return, *Adv Space Res* 34(11), 2328–2337.

Gershman, R., Bar-Cohen, Y., Hendry, M., Stricker, M., Dobrynin, D., and Morrese, A. (2018) Break-the-chain technology for potential Mars sample return, *2018 IEEE Aerospace Conference*, Big Sky, MT, pp. 1–21. doi:10.1109/AERO.2018.8396744

Ghosh, S., Osman, S., Vaishampayan, P., and Venkateswaran, K. (2010) Recurrent isolation of extremotolerant bacteria from the clean room where Phoenix spacecraft components were assembled. *Astrobiology* 10:325–335.

Giorgi, T.A., Ferrario, B., and Storey, B. (1985) An updated review of getters and gettering. *J Vac Sci Technol* 3:2, 417–423.

Gogate, P.R., Tatake, P.A., Kanthale, P.M., and Pandit, A.B. (2002) Mapping of sonochemical reactors: Review, analysis, and experimental verification. *AIChE J* 48: 1542–1560.

Golombek, M. et al. (2017) Selection of InSight landing site. *Space Sci Rev* 211:5–95.

Hall, R. G. (2010) *Lunar Impact: The NASA History of the Project Ranger*. Mineola, New York: Dover Publication, Inc.

Haugland, R.A., Brinkman, N., and Vesper, S.J. (2002) Evaluation of rapid DNA extraction methods for the quantitative detection of fungi using real-time PCR analysis. *J Microbiol Methods* 50:319–323.

Hendrickson, R., Kazarians, G., and Benardini, J.N. (2020) Planetary protection implementation on the insight mission. *Astrobiology*. http://doi.org/10.1089/ast.2019.2098.

Hendrickson, R., Kazarians, G., Crux, S., Guan, L., Seuylemezian, A., Matthias, L.L., Schrepel, T., and Benardini, J. (2020) Planetary protection implementation of the InSight Mission Launch Vehicle and Associated Ground Support Hardware. *Astrobiology*. https://doi.org/10.1089/ast.2019.2099.

Hespell, R.B., Wolf, R., and Bothast, R.J. (1987) Fermentation of xylans by *Butyrivibrio fibrisolvens* and other ruminal bacteria. *Appl Environ Microbiol* 53:2849–2853.

Hess, S.L.G. et al. (2015) "New SPIS capabilities to simulate dust electrostatic charging, transport, and contamination of lunar probes", *IEEE Trans Plasma Sci* 43(9):2799–2807.

International Organization for Standardization. (2009) Sterilization of health care products—General requirements for characterization of a sterilizing agent and the development, validation and routine control of a sterilization process for medical devices. ISO 14937:2009(en).

Jakosky, B.M., Nealson, K.H., Bakermans, C., Ley, R.E., and Mellon, M.T. (2003) Subfreezing activity of microorganisms and the potential habitability of Mars' polar regions. *Astrobiology* 3:343–350.

Junge, K., Imhoff, F., Staley, T., and Deming, J.W. (2002) Phylogenetic diversity of numerically important Arctic sea-ice bacteria cultured at subzero temperature. *Microb Ecol* 43:315–328.

Kempf, M.J., Chen, F., Kern, R., and Venkateswaran, K. (2005) Recurrent isolation of hydrogen peroxide-resistant spores of *Bacillus pumilus* from a spacecraft assembly facility. *Astrobiology* 5:391–405.

Kminek, G., Conley, C., Hipkin, V., Yano, H. (2017) COSPAR's Planetary Protection Policy. *Space Res. Today*, 200(December 2017):22.

Krisko, A. and Radman, M. (2010) Protein damage and death by radiation in *Escherichia coli* and *Deinococcus radiodurans*. *Proc Natl Acad Sci U S A* 107:14373–14377.

La Duc, M.T., Dekas, A.E., Osman, S., Moissl, C., Newcombe, D., and Venkateswaran, K. (2007) Isolation and characterization of bacteria capable of tolerating the extreme conditions of clean-room environments. *Appl Environ Microbiol* 73:2600–2611.

La Duc, M.T., Kern, R., and Venkateswaran, K. (2004) Microbial monitoring of spacecraft and associated environments. *Microbial Ecol* 47:150–158.

La Duc, M.T., Nicholson, W., Kern, R., and Venkateswaran, K. (2003) Microbial characterization of the Mars Odyssey spacecraft and its encapsulation facility. *Environ Microbiol* 5:977–985.

La Duc, M.T., Vaishampayan, P., Nilsson, H.R., Torok, T., and Venkateswaran, K. (2012) Pyrosequencing-derived bacterial, archaeal, and fungal diversity of spacecraft hardware destined for Mars. *Appl Environ Microbiol* 78:5912–5922.

Lalime, E. and Canham, J. et al. (2018) Development and implementation of aseptic operations to meet planetary protection requirements on the MOMA-mass spectrometer. *SPIE Conference Proceeding*, September 18, San Diego, CA.

Lim, S.W. and Abate, A.R. (2013) Ultrahigh-throughput sorting of microfluidic drops with flow cytometry. *Lab Chip* 13:4563–4572.

Mahnert A., Vaishampayan P., Probst A.J. et al. (2015) Cleanroom maintenance significantly reduces abundance but not diversity of indoor microbiomes. *PLoS One* 10(8):e0134848. doi:10.1371/journal.pone.0134848

Martin, A., Anderson, M.J., Thorn, C., Davy, S.K., and Ryan, K.G. (2011) Response of sea-ice microbial communities to environmental disturbance: An in situ transplant experiment in the Antarctic. *Mar Ecol Prog Ser* 424:25–37.

Marx, J.G., Carpenter, S.D., and Deming, J.W. (2009) Production of cryoprotectant extracellular polysaccharide substances (EPS) by the marine psychrophilic bacterium Colwellia psychrerythraea strain 34H under extreme conditions. *Can J Microbiol* 55:63–72.

Mattimore, V. and Battista, J.R. (1996) Radioresistance of *Deinococcus radiodurans*: Functions necessary to survive ionizing radiation are also necessary to survive prolonged desiccation. *J Bacteriol* 178:633–637.

Melnik, A.V., da Silva, R.R., Hyde, E.R., Aksenov, A.A., Vargas, F., Bouslimani, A., Protsyuk, I., Jarmusch, A.K., Tripathi, A., Alexandrov, T., Knight, R., and Dorrestein, P.C. (2017) Coupling targeted and untargeted mass spectrometry for metabolome-microbiome-wide association studies of human fecal samples. *Anal Chem* 89(14):7549–7559.

Meltzer, M. (2012) *When Biospheres Collide: A History of NASA's Planetary Protection Programs*. National Aeronautics and Space Administration. U.S. Government Printing Office.

Mendes, G., Brandao, T., and Silva, C. (2007) Ethylene oxide sterilization of medical devices: A review. *Am J Infect Contr* 35(9):574–581.

Ming, D. et al. (2008) Mars 2007 Phoenix scout mission organic free blank (OFB): Method to distinguish Mars organics from terrestrial organics. *J. Geophys. Res.* doi:10.1029/2007JE003061

Mikellides, I.G. et al. (2017) The viscous fluid mechanical particle barrier for the prevention of sample contamination on the Mars 2020 mission. *Planet Space Sci* 142:53–68.

Minich, J.J., Zhu, Q., Janssen, S., Hendrickson, R., Amir, A., Vetter, R., Hyde, J., Doty, M.M., Stillwell, K., Benardini, J., Kim, J.H., Allen, E.E., Venkateswaran, K., and Knight, R. (2018) KatharoSeq enables high-throughput microbiome analysis from low-biomass samples. *mSystems* 3(3):e00218-17, doi: 10.1128/mSystems.00218-17.

Moissl, C., Rudolph, C., and Huber, R. (2002) Natural communities of novel archaea and bacteria with a string-of-pearls-like morphology: Molecular analysis of the bacterial partners. *Appl Environ Microbiol* 68:933–937.

NASA. (2005) Planetary Protection provisions for robotic extraterrestrial missions. NPR 8020.12C, April 2005. National Aeronautics and Space Administration, Washington, DC.

National Academies of Sciences, Engineering, and Medicine. 2018. *Review and Assessment of Planetary Protection Policy Development Processes*. Washington, DC: The National Academies Press. http://doi.org10.17226/25172.

National Research Council 2000. *Preventing the Forward Contamination of Europa*. Washington, DC: The National Academies Press.

Newcombe, D.A., Schuerger, A.C., Benardini, J.N., Dickinson, D., Tanner, R., and Venkateswaran, K. (2005) Survival of spacecraft-associated microorganisms under simulated Martian UV irradiation. *Appl Environ Microbiol* 71:8147–8156.

Niemczewski, B. (2007) Observations of water cavitation intensity under practical ultrasonic cleaning conditions. *Ultrason Sonochem* 14(1):13–18.

Nocker, A., Mazza, A., Masson, L., Camper, A.K., and Brousseau, R. (2009) Selective detection of live bacteria combining propidium monoazide sample treatment with microarray technology. *J Microbiol Methods* 76:253–261.

Otterbein, I.N. and Pflug, I. (2010) *Microbiology and Engineering of Sterilization Processes*, 14th Edition. Otterbein, IN: Environmental Sterilization Laboratory.

Pasumarthi, P. (2016) Investigation of geometric effect on the ultrasonic processing of liquids. All Theses. 2447.

Petrosino, J.F., Highlander, S., Luna, R.A., Gibbs, R.A., and Versalovic, J. (2009) Metagenomic pyrosequencing and microbial identification. *Clin Chem* 55:856–866.

Pillai, S.D., Venkateswaran, K., Cepeda, M., Soni, K., Mittasch, S., Maxim, J., and Osman, S. (2006) Electron beam (10 MeV) irradiation to decontaminate spacecraft components for planetary protection. *Aerospace Conference, 2006 IEEE*, Big Sky, Montana.

Planetary Protection Provisions for Robotic Extraterrestrial Missions, NPR 8020.12D, April 20, 2011.

Probst, A.J., Holman, H.Y., Desantis, T.Z., Andersen, G.L., Birarda, G., Bechtel, H.A., Piceno, Y.M., Sonnleitner, M., Venkateswaran, K., and Moissl-Eichinger, C. (2013) Tackling the minority: Sulfate-reducing bacteria in an archaea-dominated subsurface biofilm. *ISME J* 7:635–651.

Probst, A., Vaishampayan, P., Osman, S., Moissl-Eichinger, C., Andersen, G.L., and Venkateswaran, K. (2010) Diversity of anaerobic microbes in spacecraft assembly clean rooms. *Appl Environ Microbiol* 76:2837–2845.

Puleo, J.R., Bergstrom, S.L., Peeler, J.T., and Oxborrow, G.S. (1978) Thermal resistance of naturally occurring airborne bacterial spores. *Appl Environ Microbiol* 36(3):473–479.

Rabbow, E. et al. (2012) EXPOSE-E: An ESA astrobiology mission 1.5 years in space. *Astrobiology* 12(5):374–386. doi: 10.1089/ast.2011.0760.

Rainey, F.A., Ray, K., Ferreira, M., Gatz, B.Z., Nobre, M.F., Bagaley, D., Rash, B.A., Park, M.J., Earl, A.M., Shank, N.C., Small, A.M., Henk, M.C., Battista, J.R., Kampfer, P., and da Costa, M.S. (2005) Extensive diversity of ionizing-radiation-resistant bacteria recovered from Sonoran Desert soil and description of nine new species of the genus Deinococcus obtained from a single soil sample. *Appl Environ Microbiol* 71:5225–5235.

Rummel, J.D., Beaty D.W., Jones, M.A. et al. (2014) A new analysis of Mars "Special Regions": Findings of the second MEPAG Special Regions Science Analysis Group (SR-SAG2). *Astrobiology* 14(11):887–968. doi:10.1089/ast.2014.1227

Rutala, W.A., Webster, D.J., and the Infection Control Practices Advisory Committee (HICPAC). (2008) Guideline for Disinfection and Sterilization in Healthcare Facilities. https://www.cdc.gov/infectioncontrol/guidelines/disinfection/, last accessed November 1, 2019.

Sagan, C. and Coleman, S. (1966) Decontamination standards for Martian exploration programs, pp. 470–481 in NRC, *Biology and the Exploration of Mars*, National Academy Sciences, Washington, DC.

Salinas, Y., et al. (2007) Bio-barriers: Preventing forward contamination and protecting planetary astrobiology instruments, *IEEE Aerospace Conference Proceedings, 2007 IEEE Aerospace Conference Digest*, Big Sky, Montana, p. 4161294.

Schubert, W. and Beaudet, R.A. (2001) Determination of lethality rate constants and D-values for heat-resistant Bacillus spores ATCC 29669 exposed to dry heat from 125C to 200C. *Astrobiology* 11(3):213–223. doi: 10.1089/ast.2010.0502.

Schuerger, A.C., Moores, J.E., Smith, D.J., and G. Reitz. (2019) A lunar microbial survival model for predicting the forward contamination of the Moon. *Astrobiology* 19(6):1–27.

Schuerger, A.C., Richards, J.T., Newcombe, D.A., and Venkateswaran, K. (2006) Rapid inactivation of seven Bacillus spp. under simulated Mars UV irradiation. *Icarus* 181(1):52–62.

Shabala, L., Bowman, J., Brown, J., Ross, T., McMeekin, T., and Shabala, S. (2009) Ion transport and osmotic adjustment in Escherichia coli in response to ionic and non-ionic osmotica. *Environ Microbiol* 11:137–148.

Singh, N.K., Wood, J.M., Karouia, F., and Venkateswaran, K. (2018) Correction to: Succession and persistence of microbial communities and antimicrobial resistance genes associated with International Space Station environmental surfaces. *Microbiome* 6:214.

Slade, D. and Radman, M. (2011) Oxidative stress resistance in *Deinococcus radiodurans*. *Microbiol Mol Biol Rev* 75:133–191.

Suzuki, M.T., Beja, O., Taylor, L.T., and Delong, E.F. (2001) Phylogenetic analysis of ribosomal RNA operons from uncultivated coastal marine bacterioplankton. *Environ Microbiol* 3:323–331.

Suzuki, M.T., Taylor, L.T., and DeLong, E.F. (2000) Quantitative analysis of small-subunit rRNA genes in mixed microbial populations via 5'-nuclease assays. *Appl Environ Microbiol* 66:4605–4614.

ten Kate, I.L., Canham, J.S., Conrad, P.G., Errigo, T., Katz, I., and Mahaffy, P.R. (2008) Mitigation of the impact of terrestrial contamination on organic measurements from the Mars Science Laboratory. *Astrobiology* 8(3):571–582.

Vaishampayan, P., Osman, S., Andersen, G., and Venkateswaran, K. (2010) High-density 16S microarray and clone library-based microbial community composition of the Phoenix spacecraft assembly clean room. *Astrobiology* 10:499–508.

Vaishampayan, P., Probst, A.J., La Duc, M.T., Bargoma, E., Benardini, J.N., Andersen, G.L., and Venkateswaran, K. (2013) New perspectives on viable microbial communities in low-biomass cleanroom environments. *ISME J* 7:312–324.

Vaishampayan, P., Rabbow, E., Horneck, G., and Venkateswaran, K. (2012) Survival of *Bacillus pumilus* spores for a prolonged period of time in real space conditions. *Astrobiology* 12:487–497.

Venkateswaran, K., La Duc, M.T., and Vaishampayan, P. (2012) Genetic inventory task: Final report, JPL Publication 12-12. Jet Propulsion Laboratory, California Institute of Technology, Pasadena, CA, pp. 1–117.

Venkateswaran, K., Hattori, N., La Duc, M.T., and Kern, R. (2003) ATP as a biomarker of viable microorganisms in clean-room facilities. *J Microbiol Methods* 52:367–377.

Venkateswaran, A., McFarlan, S.C., Ghosal, D., Minton, K.W., Vasilenko, A., Makarova, K., Wackett, L.P., and Daly, M.J. (2000) Physiologic determinants of radiation resistance in *Deinococcus radiodurans*. *Appl Environ Microbiol* 66:2620–2626.

Venkateswaran, K., Vaishampayan, P., Cisneros, J., Pierson, D.L., Rogers, S.O., and Perry, J. (2014) International Space Station environmental microbiome—Microbial inventories of ISS filter debris. *Appl Microbiol Biotechnol* 98(14), 6453–6466.

Wang, J., He, X.M., and Cao, Y. (2007) Modeling spacecraft charging and charged dust particle interactions on lunar surface. *Proceedings of the 10th Spacecraft Charging Technology Conference*, Biarritz, France.

Weinmaier, T., Probst, A.J., Duc, M.T., Ciobanu, D., Cheng, J.F., Ivanova, N., Rattei, T., and Vaishampayan, P. (2015) A viability-linked metagenomic analysis of cleanroom environments: Eukarya, prokaryotes, and viruses. *Microbiome* 3(62). https://doi.org/10.1186/s40168-015-0129-y

Index